Nuclear Electronics

Nuclear Electronics

P. W. NICHOLSON

Lecturer in Physics,
Middlesex Hospital Medical School,
University of London

A Wiley-Interscience Publication

JOHN WILEY & SONS

LONDON · NEW YORK · SYDNEY · TORONTO

Library of Congress Cataloging in Publication Data:

Nicholson, Paul W.
Nuclear electronics.

"A Wiley–Interscience publication."
1. Nuclear physics—Instruments. 2. Electronic measurements. 3. Nuclear counters. I. Title.

QC786.N47 539.7'7 73-8196
ISBN 0 471 63697 5

Set in 'Monophoto' 569 Mathematical Times and printed offset Litho in Great Britain by Page Bros (Norwich) Ltd, Norwich.

Preface

Electronic techniques have always been closely associated with nuclear radiation measurements and in recent years such techniques have become extensively used in many disciplines. At the same time the scope and complexity of the techniques has greatly increased. This has been occasioned partly by the need to exploit the characteristics of semiconductor detectors, but also by the ready availability of circuit elements such as field effect transistors, tunnel diodes and integrated circuits. The user is now faced with a formidable array of techniques; in this book I have attempted to provide the basis for a systematic appraisal of electronic techniques associated with nuclear particle detection.

In writing the book it has proved impossible to cite all of the vast number of original publications relevant to the subject. The approach therefore has been necessarily selective. For each topic precedence has generally been given to any theoretical analyses, where such exist. I have then attempted to isolate the major trends in experimental techniques and these have been illustrated with what appeared to be the best examples of circuits published in recent years.

By adopting this scheme the treatment should appeal to a spectrum of readership. Commercial electronic equipment is readily available, but users will need to know the scope of the techniques available in order to assemble equipment appropriate to any given application. Such users require a knowledge of the possibilities and limitations of the techniques; this often involves the familiar exchange of performance in one respect with performance in other respects. For the more specialist applications the reader should find that the topics are dealt with in sufficient depth to enable him to design new circuits.

This book arose out of a course of lectures given for a number of years to M.Sc. students. Some of the fundamentals of the subject nominally fall within the fields of electronic engineering and statistics and, for that reason, may be unfamiliar or inaccessible to many readers. In order to provide a reasonably self-contained treatment I have therefore included introductory outlines of several of the relevant topics.

London 1973

P. W. NICHOLSON

Acknowledgements

I am indebted to the following authors for permission to reproduce or adapt figures from previously published material: T. K. Alexander (Figure B.3); N. Balaux (Figure 7.31); A. Barna (Figures 6.9, 6.15(d)); Z. Bay (Figure 7.23); M. Bertolaccini (Figures 3.39, 3.40, 4.15, 4.33, 7.10); G. C. Best (Figure 9.5); T. V. Blalock (Figures 5.2, 5.3, 5.14, 5.32); S. L. Blatt (Figure 7.40); A. M. Boyarski (Figures 9.6, 9.7, 9.8); F. D. Brooks (Figures 1.18, 1.19, 6.26); R. L. Chase (Figures 3.36, 7.19, B.2); J. E. Cline (Figures 5.11, 5.15); B. Collinge (Figure 5.31); C. Dardini (Figure 7.30); C. M. Davisson (Figure 1.4); G. Dearnaley (Table 1.1, Figure 1.8); E. DeBlust (Figure 6.17(c)); M. O. Deighton (Figure 4.32); G. Dimmler (Figure 9.2); E. Fairstein (Figure 3.21); W. Franzen (Figure 4.16); E. Gatti (Figures 4.15, 4.33, 6.21); E. A. Gere (Figure 3.37); F. S. Goulding (Figures 5.16, 5.17, 5.18, 5.19, 5.22); V. Goursky (Figures 8.8, 8.9); M. Goyot (Figure 5.34); R. L. Graham (Figure 7.8); P. K. F. Grieder (Figure 6.16(a)); K. Hatch (Figures 4.21, 4.22); L. Heistek (Figure 6.24); L. J. Herbst (Figure 6.15(c)); W. A. Higinbotham (Figure 6.17(b)); T. Ichimori (Figure 6.29); K. Kandiah (Figures 5.20, 8.10, 8.11); S. Kinbara (Figure 6.28); H. Koeman (Figure 8.16); M. Konrad (Figure 4.20); F. T. Kuchnir (Figures 6.19, 6.20); F. J. Lynch (Figures 6.19, 6.20); G. W. McBeth (Figure 6.25); W. J. McDonald (Figures 7.11, 7.12); P. F. Manfredi (Figures 5.33, 7.36); P. Maranesi (Figures 5.33, 7.36); R. Meunier (Figure 7.24); J. K. Millard (Figure 5.35); G. L. Miller (Figure 3.37); A. P. Mills (Figure 7.33); M. Moszyński (Figure 7.38); R. B. Murray (Figure 1.17); A. Muser (Figure 6.15(a)); H. U. Nachbar (Figure 6.13); J. H. Neiler (Figure 1.6); C. H. Nowlin (Figures 3.26, 3.27, 3.28); M. Obu (Figure 6.29); R. Patzelt (Figures 8.12, 8.13, 8.14); V. Radeka (Figures 4.31, 5.21); L. B. Robinson (Figure 3.32); S. J. Rudnick (Figures 6.12, 6.18); C. J. Rush (Figure 5.36); B. Sabbah (Figure 7.39); H.-J. Schuster (Figure 7.35); K. Shirakata (Figure 6.29); P. C. Simms (Figure 7.22(b)); B. Smith (Figure 7.34); B. Souček (Figure 9.3); U. Tamm (Figure 6.27); I. J. Taylor (Figure 6.22); R. B. Tomlinson (Figure 7.22(a)); R. Van Zurk (Figures 6.15(b), 6.16(b), 6.17(a)); W. H. Venable (Figure 7.26); C. M. York (Figure 6.8).

I am also grateful to the following publishers for permission to reproduce or adapt material that has appeared in their publications: Academic Press Inc. (Figure 4.16); American Institute of Physics, Review of Scientific

Instruments (Figures 3.26, 3.27, 3.28, 3.32, 4.31, 5.35, 5.36, 6.12, 6.18, 7.26, 7.28, 7.36); American Physical Society, Review of Modern Physics (Figure 1.4); Associated Book Publishers Ltd. (Table 1.1); Atomic Energy of Canada Ltd. (Figure B.3); E.M.I. Ltd. (Figure 1.16); Gesellschaft für Kernforschung m.b.H. (Figure 9.2); Institute of Electrical and Electronics Engineers Inc., IEEE Transactions on Nuclear Science (Figures 2.11, 3.36, 3.37, 4.20, 4.21, 4.22, 4.32, 5.2, 5.3, 5.11, 5.14, 5.15, 5.21, 5.22, 5.29, 5.32, 6.19, 6.20, 7.22(a), 7.23, 8.10, 8.11, 9.5); International Atomic Energy Agency (Figures 6.17(b), 6.21, 7.24); McGraw-Hill Book Company (Figures 1.8, 2.9, B.2); Morgan–Grampian (Publishers) Ltd. (Figure 7.25); Motorola Semiconductor Products Inc. (Figure 2.20); Mullard Ltd. (Figure 2.21); North-Holland Publishing Company, Nuclear Instruments and Methods (Figures 1.6, 1.17, 3.39, 3.40, 4.15, 4.33, 5.16, 5.17, 5.18, 5.19, 5.31, 5.33, 5.34, 6.8, 6.9, 6.13, 6.15(a), 6.15(b), 6.15(c), 6.15(d), 6.16(a), 6.16(b), 6.16(c), 6.17(a), 6.22, 6.24, 6.25, 6.26, 6.27, 6.28, 6.29, 7.10, 7.11, 7.12, 7.19, 7.22(b), 7.30, 7.31, 7.33, 7.34, 7.35, 7.38, 7.39, 7.40, 8.12, 8.13, 8.14, 8.16, 9.3); Nuclear Enterprises Ltd. (Figures 1.1, 1.2, Table 1.2); Pergamon Press Ltd. (Figures 1.18, 1.19); The Radiochemical Centre Ltd. (Figure 1.3); Texas Instruments Ltd. (Figure 2.22); U.S. National Academy of Sciences (Figure 5.20).

Contents

I

An Outline of Detection Methods

1.1 Interaction of nuclear radiation with matter

1.1.1 Introduction

Nearly all detection methods, the Cerenkov detector being a notable exception, make use of the ionization or excitation produced in a detection medium as a result of the absorption of all or part of the energy of the nuclear particle. In the case of charged particles, ionization and excitation is produced directly by the interaction of the electromagnetic field of the particle with the electrons of the detection medium and the resultant ionization and excitation is distributed as a track centred on the track of the particle. In the case of uncharged particles such as X- and γ-ray photons, the particle must first undergo some process, such as a photoelectric or Compton interaction, which transfers all or part of its energy to an electron which in turn produces the track of ionization or excitation. Similarly neutrons must undergo an interaction, such as a collision with a nucleus, the charged product of which then produces the ionization or excitation. In consideration of their interaction with the detection medium the commoner nuclear particles may be divided into four groups:

(i) heavy charged particles (protons, alpha particles, heavy ions, mesons)
(ii) electrons and positrons
(iii) X-rays and γ-rays
(iv) neutrons

1.1.2 Heavy charged particles

A heavy charged particle entering the detection medium loses its energy by a succession of interactions, mainly between its electromagnetic field and that of electrons in the medium, resulting in electronic excitation and ionization. According to classical mechanics the maximum energy E_{max} that an electron of mass m_0 can acquire in a collision with a particle of mass M and energy E is given by

$$E_{max} = \frac{4m_0ME}{(m_0 + M)^2} \simeq 4E\frac{m_0}{M} \tag{1.1}$$

The quantum mechanical treatment shows that there is a small probability of the electron acquiring an energy slightly higher than the classical maximum E_{max}. The more energetic of the electrons, often termed 'delta' rays, can

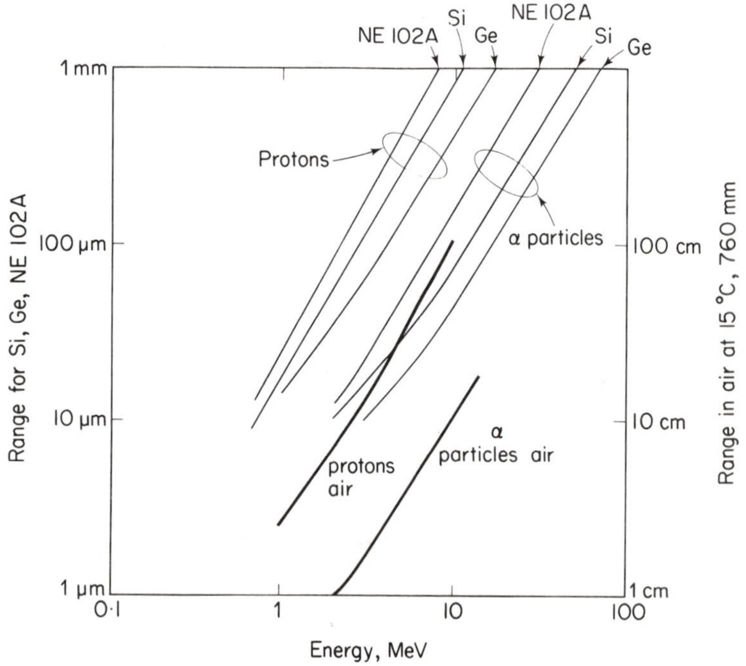

Figure 1.1 Range of α particles and protons in various detector media. Data taken from Goulding[4] for Si and Ge, Evans[2] for air, and Nuclear Enterprises Ltd[3] for plastic scintillator type NE 102A

themselves produce further ionization; it is this factor that governs the width of the track of ionization. Since the fraction of energy lost per collision of the particle, given by (1.1), is small, the resulting deflections of the particle path are small and the track is comparatively straight. Owing to the large number of collisions necessary to bring the particle to rest, the range is comparatively well defined (typically with a standard deviation of 1–5%).

For a given particle type and energy, if the range R_1 is known for a medium of density ρ_1 and atomic weight A_1, then the range R_2 in a medium of density ρ_2 and atomic weight A_2 may be estimated by a simple empirical relation known as the Bragg–Kleeman rule:

$$\frac{R_1}{R_2} = \frac{\rho_2\sqrt{A_1}}{\rho_1\sqrt{A_2}} \tag{1.2}$$

For mixtures and compounds the same rule can be used if an effective atomic weight A is used where

$$\sqrt{A} = \frac{n_1A_1 + n_2A_2 + \ldots}{n_1\sqrt{A_1} + n_2\sqrt{A_2} + \ldots} \tag{1.3}$$

where $n_1, n_2, \ldots$ are the atomic fractions of the constituent elements of atomic weights $A_1, A_2, \ldots$. Figure 1.1 shows the ranges of protons and alpha particles calculated for various detection media.

1.1.3 Electrons and positrons

For energies up to 10 MeV, electrons lose their energy to the detection medium mainly by excitation and ionization of the electrons of the medium, as in the case of heavy charged particles. For higher energy electrons the loss of energy as bremsstrahlung becomes increasingly important and the intensity of this varies as Z^2 where Z is the atomic number of the medium. Thus, for example, 9 MeV electrons in lead lose as much energy due to bremsstrahlung as due to ionization.

In absorption due to ionization, owing to the lower mass of the electron compared to that of a heavy charged particle, a much greater fraction of the particle energy is transferred to the absorber electron in each collision so that the delta rays are correspondingly more energetic and are capable of causing considerable secondary ionization at a greater distance from the original particle track. For a 0·1 MeV electron in air about two-thirds of the total ionization is due to secondary ionization.

Owing to the large energy loss per collision, the path of the particle shows considerable deflections and the range is not well defined. Since an electron travels at a much higher velocity than a heavy charged particle of the same energy, it spends less time in the vicinity of the absorber atoms and so the rate of energy loss and the density of ionization are correspondingly less. For this reason the range of electrons is much greater than that of heavy particles of the same energy. Figure 1.2 shows the range (essentially an approximate figure) for several detector media. For various medium atomic weight absorbers the range, for a given energy, expressed in terms of mass per unit area of the absorber, is approximately constant provided bremsstrahlung does not accout for significant energy loss. Finally for beta rays having a maximum energy of E_{max}, the nomogram in Figure 1.3 gives the approximate percentage transmission for absorbers having different mass per unit area when the absorber is interposed between source and detector (external absorber), or where the source itself is thick and causing self-absorption.

Positrons have ranges differing only slightly from those of electrons of the same energy. When brought to rest a positron will annihilate with a neighbouring electron and emits the characteristic annihilation radiation of two γ-rays, each of energy m_0c^2 (511 keV), in mutually opposite directions.

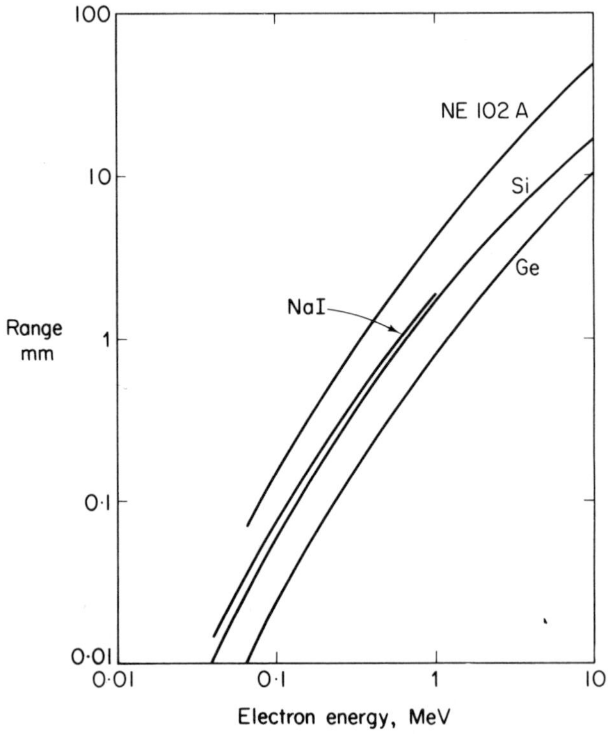

Figure 1.2 Range of electrons in various detector media. Data taken from Goulding[4] for Ge and Si, Nelms[5] for NaI, and Nuclear Enterprises Ltd[3] for NE 102A plastic scintillator

1.1.4 X- and γ-rays

X- and γ-rays lose energy in matter by three main processes:

(i) photoelectric effect
(ii) Compton effect
(iii) pair production

In the photoelectric effect, all the energy of the photon is transferred to an electron, the original photon disappearing in the process. The process is

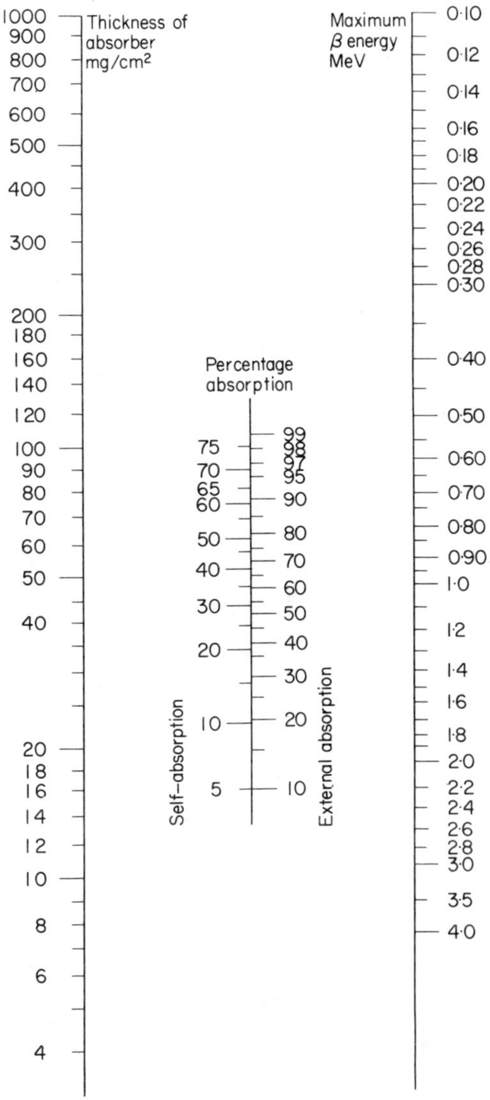

Figure 1.3 Nomogram for the absorption of continuous β rays according to Radiochemical Centre, Amersham[6], reproduced with permission. A straight edge placed between the absorber thickness and the maximum β energy enables the percentage absorption to be read off on one of the centre scales. The external absorption scale refers to an absorber interposed between source and detector, and the self-absorption scale refers to the case of a thick source with uniformly distributed radioactivity

most likely to occur with a K electron, in which case the energy imparted to the photoelectron by a photon of energy $h\nu$ would be

$$h\nu - E_b^K \tag{1.4}$$

where E_b^K is the binding energy of the K electron. This is immediately followed by de-excitation of the atom with the emission of one or more X-rays of total energy E_b^K. These X-rays may escape from the detection medium but, due to their low energy, are likely to be reabsorbed in other photoelectric processes. Thus the K X-ray may interact with an L electron of another atom which in turn would de-excite and give an L X-ray. This in turn would probably be reabsorbed with the eventual emission of an X-ray of still lower energy. Thus after this rapid sequence of events a whole family of photoelectrons are produced whose combined energy is very nearly equal to the original photon energy $h\nu$. Of course any of the family of X-rays may escape the detector medium or may undergo a Compton process rather than a photoelectric process. However, owing to the decreasing energies of the sequence of X-rays, the probability rapidly decreases and even the escape of the K X-ray is an event rare enough to be usually unimportant.

In the Compton effect only part of the photon energy is transferred to an electron and the photon is deflected and decreased in energy. The angle at which the photon is scattered can take on any value and the energy imparted to the electron correspondingly has a continuous distribution from zero to a maximum of E_{max}. The value of E_{max} (which corresponds to the case of the photon being back-scattered) is given by

$$E_{max} = \frac{2(h\nu)^2}{m_0c^2 + 2h\nu} \tag{1.5}$$

Figure 1.4 shows the distribution of energy imparted to the electron by gamma rays of three different energies.

Pair production can occur if the gamma ray has an energy exceeding $2m_0c^2$ (1·02 MeV). In this case the original photon disappears and the electron–positron pair carries away the excess energy $h\nu - 2m_0c^2$. The electron and positron lose energy through collisions and eventually come to rest. The positron then annihilates with a neighbouring electron and emits the characteristic annihilation radiation, a pair of γ-rays each of energy m_0c^2 (=511 keV) that are emitted in opposite directions. One or both of these may be further absorbed by photoelectric or Compton processes so that the total energy imparted to detector medium can vary widely. It is usually possible, however, to see three major components in the energy imparted to the medium of $h\nu$, $h\nu - m_0c^2$ and $h\nu - 2m_0c^2$ corresponding to the total absorption of both, one, or neither of the annihilation γ-rays.

The relative magnitude of the Compton, photoelectric and pair production processes will depend on the energy of the photon and the nature of the detection medium. If the photon traverses a thickness dx of the medium we will define $\xi_T\,dx$ as the total probability of any of the three absorption processes occurring, where ξ_T is known as the total coefficient of absorption or macroscopic cross section. In this case a beam of photons of intensity, I_0 after traversing an absorber thickness of x will have a primary beam intensity of $I_0 e^{-\xi_T x}$. Since the total probability of absorption $\xi_T\,dx$ in a thickness dx is

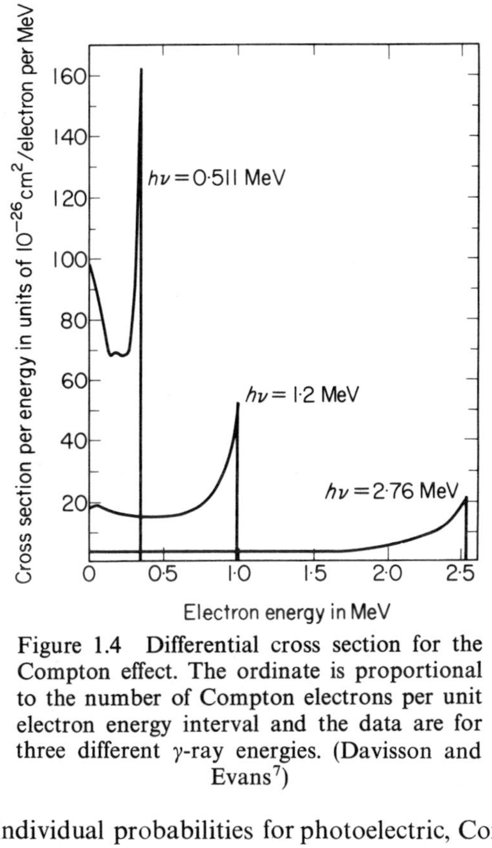

Figure 1.4 Differential cross section for the Compton effect. The ordinate is proportional to the number of Compton electrons per unit electron energy interval and the data are for three different γ-ray energies. (Davisson and Evans[7])

the sum of the individual probabilities for photoelectric, Compton and pair production processes we may write

$$\xi_T = \xi_{Photo} + \xi_{Compt} + \xi_{Pair} \tag{1.6}$$

For the photoelectric process with γ-rays of energy above the K shell binding energy the magnitude of ξ_{Photo} is given approximately by

$$\xi_{Photo} \propto NZ^5 (h\nu)^{-3\cdot 5} \tag{1.7}$$

where the absorber has N atoms per unit volume of atomic number Z. Thus the use of a high atomic number absorber and of low energy photons greatly increases the probability of photoelectric interaction.

For the Compton process the probability is mainly governed by the electron density of the absorber and ξ_{Compt} is given approximately by

$$\xi_{\text{Compt}} \propto \frac{NZ}{h\nu} \tag{1.8}$$

For pair production the magnitude of ξ_{Pair} for gamma rays of energy near the threshold is given approximately by

$$\xi_{\text{Pair}} \propto NZ^2 (h\nu - 2m_0c^2) \tag{1.9}$$

Figures 1.5 and 1.6 show plots of values of ξ_{Photo}, ξ_{Compt} and ξ_{Pair} against $h\nu$ for various detector media. Since the ideal process from the detection point

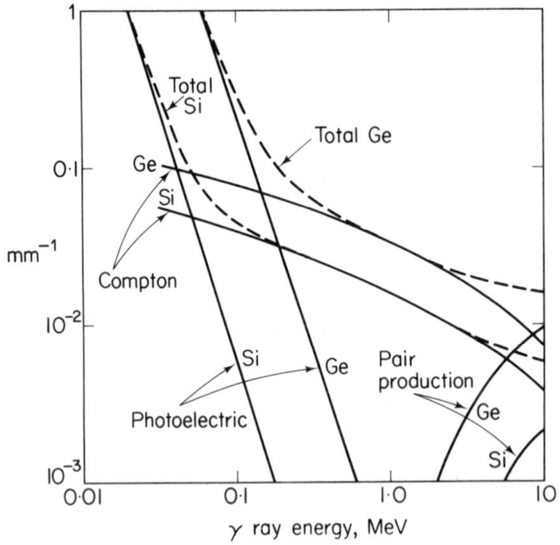

Figure 1.5 Partial and total γ-ray absorption coefficients for Si and Ge. Data taken from Storm, Gilbert and Israel[18]

of view is one where all the photon energy is transferred to the absorber electrons, it will be apparent that the photoelectric effect provides this and, from Figures 1.5 and 1.6, the importance of using a medium of high atomic number to achieve this may be appreciated.

1.1.5 Neutrons

Neutrons differ from the nuclear particles already discussed in that they interact negligibly with electrons. However various nuclear reactions can occur which produce a charged particle or a gamma photon and there are a wide range of such reactions that can be utilized for neutron detectors.

One of the commonest methods for the detection of fast and intermediate neutrons is to detect proton recoils from a hydrogenous absorber. The reaction has a reasonably large cross section for these neutron energies (see Figure 1.8) and, according to the angle of scatter, the fraction of neutron

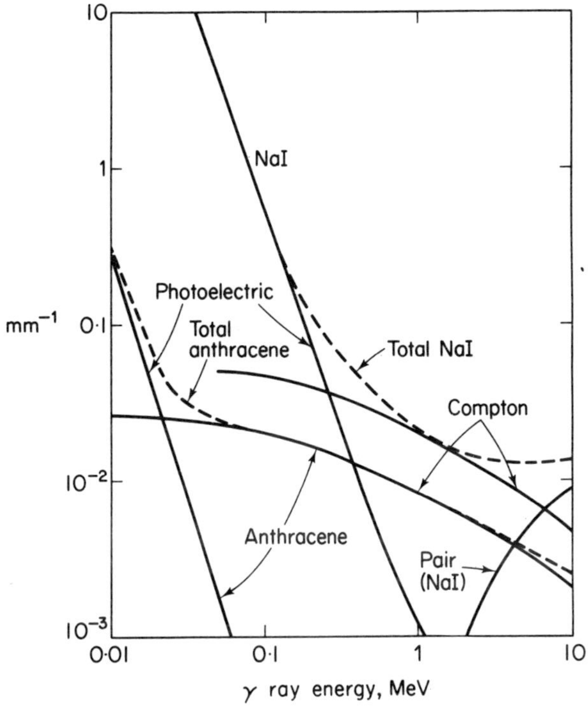

Figure 1.6 Partial and total absorption coefficients for γ-rays in NaI and anthracene. Data taken from Neiler and Bell[8]

energy transferred to the proton varies between 0 % and 100 % with an almost uniform distribution (see Figure 1.7) for energies up to 10 MeV. It is therefore possible to infer the neutron energy distribution if the proton distribution is found, although this involves a certain amount of computation.

Apart from the proton recoil method a number of other methods utilizing charged particle reactions are often used. Usually the reaction is chosen for its high cross section to the neutrons of interest. For thermal neutrons, spectroscopy is impracticable (at least by electronic means) and so the reaction is chosen to have a positive Q value (i.e. exothermic) so that the charged

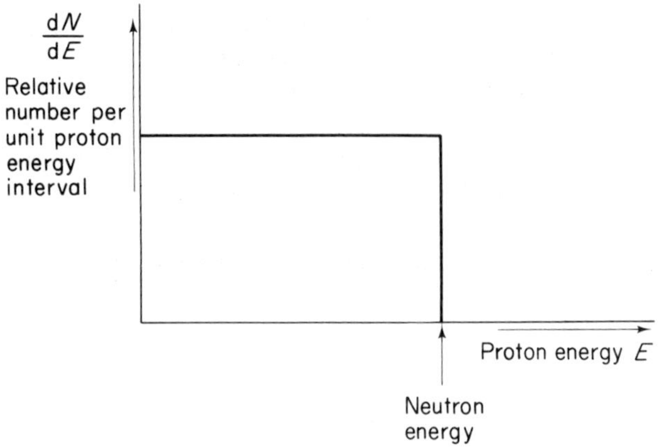

Figure 1.7 Distribution of protons produced by recoil from mono-energetic neutrons

particle product has an energy high enough to be easily discriminated from any γ or other background radiation that may be about. The following are reactions which are commonly used for thermal neutron detection

$$^{3}\text{He} + \text{n} \rightarrow {}^{3}\text{H} + \text{p} + 765\ \text{keV}$$
$$^{10}\text{B} + \text{n} \rightarrow {}^{7}\text{Li} + \alpha + 2{\cdot}78\ \text{MeV}$$
$$^{6}\text{Li} + \text{n} \rightarrow {}^{3}\text{H} + \alpha + 4{\cdot}78\ \text{MeV}$$
$$^{235}\text{U} + \text{n} \rightarrow \text{fission products} + \sim 160\ \text{MeV}$$

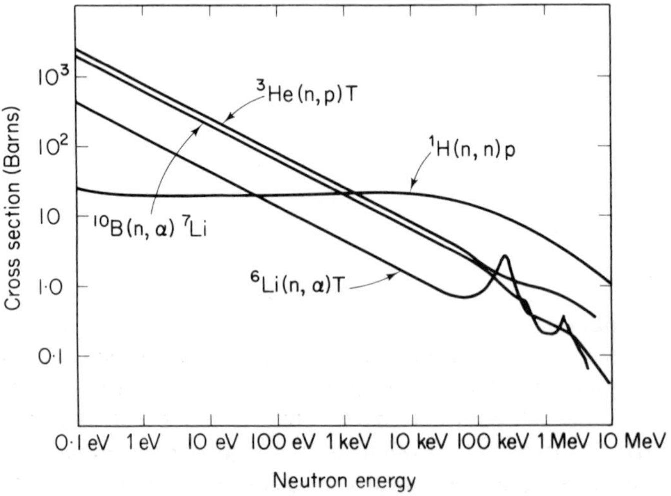

Figure 1.8 Cross sections for some reactions used for the detection of neutrons (from Dearnaley and Ferguson[9])

For fast neutrons the above reactions may also be used and, provided the combined energies of the products of the reaction can be measured, there is the possibility of neutron spectroscopy being performed. For fast neutrons there is also the possibility of using a threshold reaction so that neutrons below a certain energy do not undergo the reaction. Thus the fission of ^{237}Np has a reaction threshold of 0·75 MeV and this is used as the basis of a fast neutron detector which is not sensitive to lower energy neutrons.

Figure 1.8 shows some atomic cross sections for the more commonly used reactions in the detection of neutrons.

1.2 Statistics of the production of ionization

In the case of gas filled and semiconductor detectors the ionization produced by the nuclear particle is the quantity which is measured and it is from this that the energy of the nuclear particle is inferred. An important parameter for such detection media is the 'w value', that is, the mean particle energy per ion pair produced. It is fortuitous that the w value is fairly constant over a wide range of particle energies and is nearly the same for different particle types in the same medium. It is this constancy of w, and the resulting proportionality between particle energy and ionization, that is the basis of spectroscopy.

In a spectroscopy system which measures the amount of ionization produced in the medium, the accuracy of the measurement will depend on the statistical fluctuations of the number n of ion pairs produced and so the energy resolution is directly related to the standard deviation of n. Table 1.1 gives the w values for several gases, together with their ionization potentials, and for silicon and germanium, together with their energy gaps (i.e. the energy necessary to elevate an electron from the valence band to the conduction band). It will be apparent from this table that in all cases a fraction of only a half to a quarter of the particle energy is utilized in the production of ionization, the rest being expended in excitation and other processes.

It is possible to view the production of ionization in two extreme cases. In the first we may assume the creation of each ion pair has a constant probability, independent of the creation of other ion pairs. For such a random phenomenon the total number of ion pairs would be governed by a Poisson distribution, as in the case of radioactive emission, and if the mean number of ion pairs is $\bar{n}$ the standard deviation would be $\sqrt{\bar{n}}$ (see Appendix B.3). At the second extreme, that part of the particle energy producing ionization would be exactly apportioned in the production of ion pairs and the standard deviation would be zero. The true state of affairs lies somewhere between the two extremes and the standard deviation may be written as $F\sqrt{\bar{n}}$ where F is the Fano factor.

The difficulty of accurate determination of the Fano factor for detector media has led to widely varying estimates whose magnitude, in general, has reduced as better techniques have become available. For many gases the resolution obtained with ionization chambers and proportional counters generally indicate a Fano factor of about 1/3 is applicable. However, a much lower Fano factor, 0·09, has been indicated by recent work[10] on a mixture

Table 1.1 w values for various detection media, together with their ionization potential or energy gap (from Dearnaley and Northrop[1])

Gases

Medium	w for electrons	w for 5 MeV alphas	Mean ionization potential
	eV	eV	eV
Hydrogen	36·9	36·6	15·4
Argon	26·3	26·3	15·8
Air	34·2	35·2	—
Methane	28·1	29·1	13·1

Solids

Medium	w	Type of radiation	Energy gap
	eV		eV
Silicon	3·79	365 keV electrons	1·09
	3·61	5MeV alphas	
Germanium	2·94	several	0·75
Cadmium sulphide	7·3	5 MeV alphas	2·5

of argon with a small amount of acetylene. For both Si and Ge, values of F in the region of 0·08 have been reported.[11, 12]

From the above figures it will be seen that the larger value of $\bar{n}$ obtained with Si and Ge is the controlling factor that determines the better resolution attainable with semiconductor detectors over that attainable in gas filled detectors. Furthermore, since the noise from the amplifier causes some additional loss of resolution, this contribution would be expected to be less for a semiconductor detector as the signal is larger.

The factors determining the energy resolution in a scintillation counter are not readily compared with the above values of F and $\bar{n}$. The light emitted from the scintillator is due to the de-excitation of molecular levels which in turn causes the emission of photoelectrons from the photocathode of the photomultiplier tube. Thus a rough equivalent to the w values discussed earlier might be obtained by considering the particle energy per cathode photoelectron produced. For a NaI(Tl) crystal this may be about 0·5–1 keV per photoelectron and, for organic scintillators excited by electrons, several keV per photoelectron and considerably higher for densely ionizing particles. For a scintillation counter several additional factors contribute to give the relatively poor energy resolution. Birks[13] discusses the contributions to this and indicates that variation in light output from multiple absorption events and non-proportional response, variation in light collection efficiency and electron multiplier variation all contribute additional components. For NaI(Tl) the energy resolution for 661 keV γ-rays is at best about 7% (full width of peak at half height), or at 1850 keV is about 5·5%.

1.3 Gas filled detectors

1.3.1 Ionization chambers

The presence of an ionizing particle in an ionization chamber creates a track of electron positive ion pairs. Under the influence of the electric field between the two plates of the chamber, the electron and ion are attracted in opposite directions and we will assume that the probability of recombination is negligible. We will consider the case of an ion chamber, having a capacity C between the plates, whose upper plate is charged up and isolated and has a positive potential V with respect to the grounded lower plate (Figure 1.9(a)). Consider now the creation of a single electron positive ion pair at a point in the chamber volume having a potential fV, where of course $f < 1$. The resultant motions of the electron to the anode and the positive ion to the cathode will both induce charge on the plates and will cause the anode potential to be changed slightly. Considering only the electron, in going from its point of generation to the anode an amount of work $(1 - f)Ve$ will be done on it. This work would appear first as kinetic energy of the electron and then would be dissipated in collisions with gas molecules and, ultimately, with the anode. The energy $(1 - f)Ve$ can only be obtained at the expense of the electrostatic energy of the system which originally was $\frac{1}{2}CV^2$. If, after collection of the electron by the anode, the anode voltage is reduced to $V - \Delta V$ the change in electrostatic energy is

$$\tfrac{1}{2}CV^2 - \tfrac{1}{2}C(V - \Delta V)^2 \simeq CV\Delta V \tag{1.10}$$

$$\therefore (1 - f)Ve = CV\Delta V \;\therefore \Delta V = \frac{(1 - f)e}{C} \tag{1.11}$$

So the voltage induced on the anode by the electron is proportional to $(1 - f)$, the fraction of voltage through which the electron falls before collection.

The same argument applies to the voltage change induced by the motion of the positive ion. In this case the voltage induced is given by fe/C. Thus the total voltage charge induced due to electron and ion is

$$\frac{(1-f)e}{C}+\frac{fe}{C}=\frac{e}{C} \tag{1.12}$$

and is therefore independent of the original position of the ion pair.

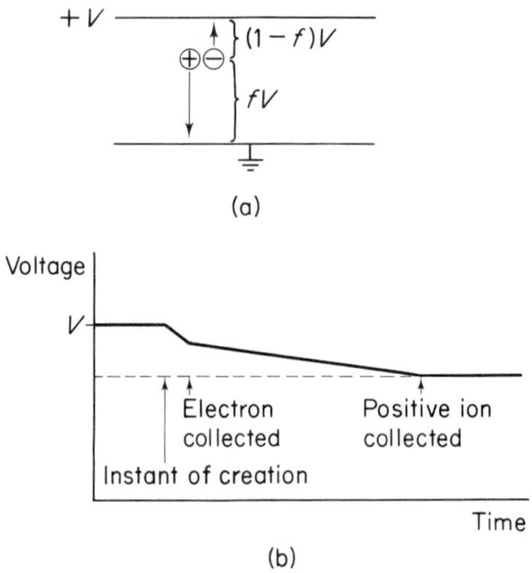

Figure 1.9 (a) Creation of ion pair in an electrically isolated ion chamber; (b) the voltage change induced on the upper electrode

Owing to the different mobilities of the electron and the ion the two contributions to the total induced voltage will not have the same time course. The mobility of electrons depends on the probability of attachment to neutral gas molecules or atoms, and this occurs predominantly with electronegative gases such as water vapour, oxygen and halogens. Assuming that these gases are excluded from the chamber, the mobility of the electrons is about two or three orders of magnitude larger than that of the positive ions and Figure 1.9 (b) shows the time course of the anode voltage for a single ion pair, assuming constant drift velocities. Figure 1.10 shows the voltage change alone, where the ion pair is initially at different locations. In general the ion pairs produced in the detection of a nuclear particle will be located over a finite region and the observed pulse shape will be compounded of many such superimposed curves. Thus, particularly when the drift velocities are not constant, as in a chamber with non-uniform electric field, the observed pulse shape will have a more rounded shape than those of Figure 1.10 (b) but will nevertheless still show a systematic variation of shape with the location and orientation of the track of ionization.

It is not possible to use an isolated ionization chamber in a practical system as the chamber would eventually become completely discharged, so

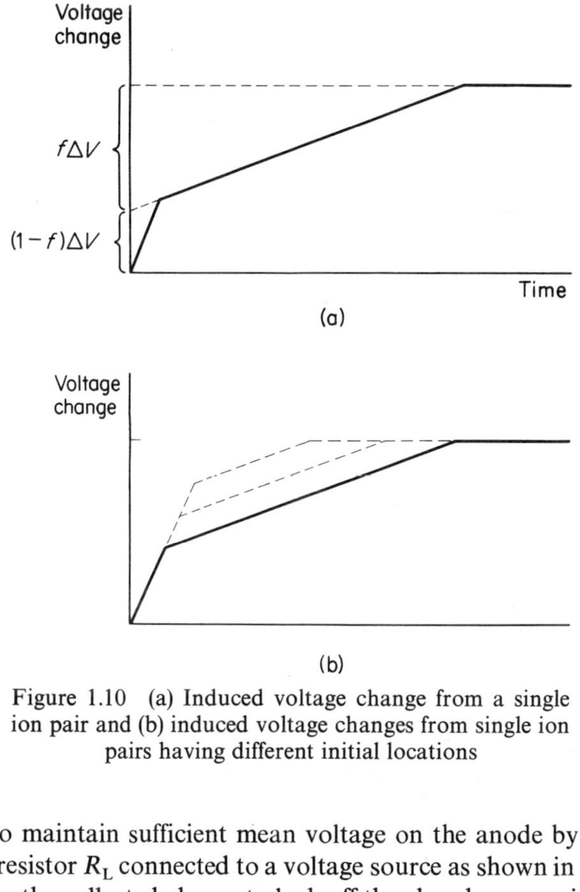

Figure 1.10 (a) Induced voltage change from a single ion pair and (b) induced voltage changes from single ion pairs having different initial locations

it is usual to maintain sufficient mean voltage on the anode by means of a large value resistor R_L connected to a voltage source as shown in Figure 1.11, which allows the collected charge to leak off the chamber capacitance C at a rate controlled by the time constant R_LC. However, from the point of view of individual signals, R_LC is usually large enough to justify the approximation of the anode being isolated.

If, after the production of ionization by a nuclear particle, both the electrons and the ions are completely collected, the final induced voltage signal will be proportional to the total ionization, and hence to the particle energy, and will be independent of the distribution of the ionization in the chamber volume. Usually, however, this requires too long a time ($\sim$ ms) if the detector is to be capable of handling reasonably high counting rates; so it is more common to derive the ultimate signal from that component of induced voltage due to the electron collection only, which, as we have seen, takes 10^{-2} to 10^{-3} of the time for ion collection. This can be achieved by using pulse shaping circuits of suitably short time constant. The proportionality between ionization and induced voltage is therefore lost, since the voltage

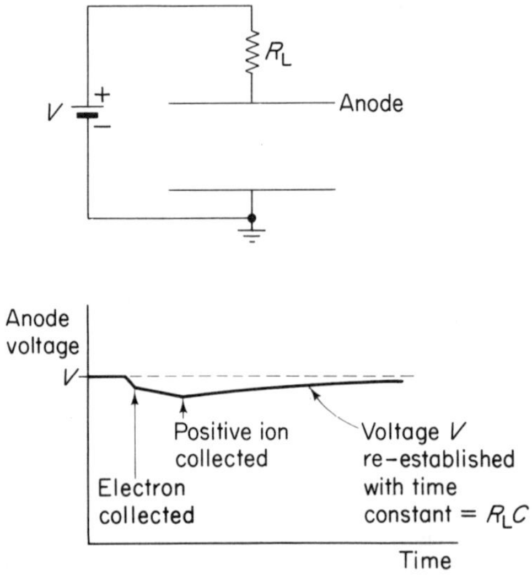

Figure 1.11 Use of resistive load and source of e.m.f. to allow re-establishment of anode voltage following collection of an ion pair

through which the electrons (Equation (1.11)) fall prior to collection is variable. The design of chambers to be operated in the electron collection mode therefore aims to keep this voltage nearly constant for the majority of electrons. The commonest configuration is a cylindrical cathode and a thin wire anode arranged coaxially. The voltage at a radius r in the chamber is then given by $V(\log b/r)/(\log b/a)$, where a and b are the radii of the anode and cathode, and so most of the voltage drop is confined to the central area near the anode. Electrons originating in the main body of the volume fall through a large fraction of the total voltage drop and this fraction is fairly insensitive to their initial position. For this purpose the centre wire should have a small diameter, but this means the field in the main body of the chamber, and hence the collection speed, is low if multiplication and breakdown in the intense field near the anode are to be avoided.

A better arrangement is to use a gridded chamber (Figure 1.12). If it can be arranged that the ionizing particles are confined to the lower section, then the electrons are attracted through the grid to the anode. Owing to the screening action of the grid the signal induced on the anode is due only to the electrons falling through the grid–anode potential difference, this being constant for any initial position of the ionization. A further advantage of the gridded chamber is the shorter collection time which can be of the order of 1 μs.

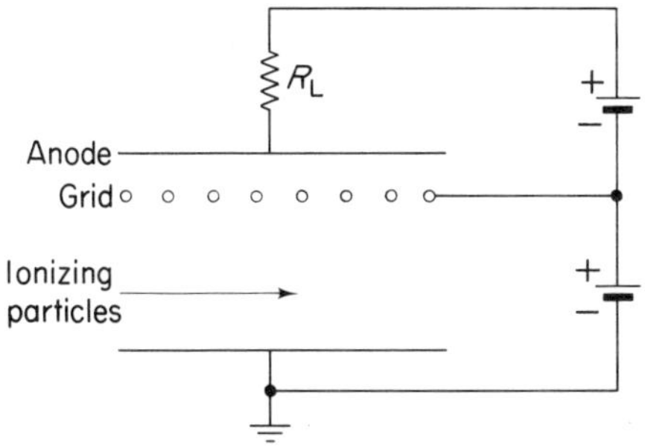

Figure 1.12 Principle of gridded ionization chamber

The gas filling an ionization chamber should be free from electronegative impurities and may be chosen to give a reaction with the particles, as in the case of BF_3 gas for the detection of neutrons. The gas pressure may be chosen to give a convenient range for the charged particles and is typically between 1/10 atm and 10 atm. Gas flow may also be employed to give a windowless counter. The walls of the chamber may be coated internally with a substance designed to react with the particles, with lead, for instance, for gamma detection or with hydrogenous or fissile material for neutron detection.

The essential disadvantages of the ion chamber are (a) the poor resolution compared to semiconductor detectors, (b) the small output signal compared to semiconductor detectors and proportional counters. Using a gridded chamber a resolution of better than 50 keV is rarely seen although, using a gas mixture of 0·5 % acetylene in argon, Alkhazov *et al.*[10] obtained a resolution of 13 keV fwhm (full width of peak at half maximum) for 5·68 MeV alphas.

1.3.2 Proportional counters

If the electric field in an ionization chamber is increased sufficiently, the electrons migrating to the anode can acquire sufficient kinetic energy between collisions to cause the further ionization of gas molecules. The electrons produced in this secondary ionization are in turn accelerated towards the anode and themselves can produce further ionization, so that an avalanche of electrons and positive ions results. Provided the electric field is not too high the charge collected by the plates is proportional to the initial ionization, and may be up to 10^4 times that of the initial ionization, this factor being known as the gas multiplication factor. The gas multiplication factor increases as the electric field is increased, but at high values the relationship between collected charge and initial ionization becomes increasingly non-proportional, due to the onset of space charge effects. Figure 1.13 shows the magnitude of the collected charge for two particles of different energy as the electric field is increased. The regions corresponding to the ionization chamber region, the proportional region and the Geiger region may be seen.

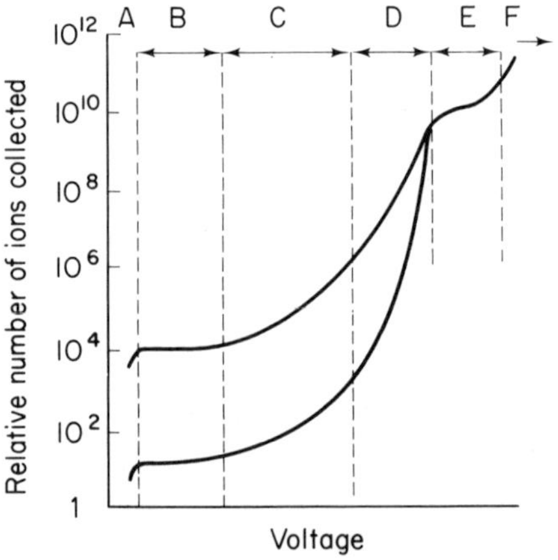

Figure 1.13 Relative number of ions collected as a function of applied voltage for charged particles of two different energies. A, number reduced by recombination; B, ionization chamber region; C, proportional counter region; D, non-proportional region; E, Geiger counter region; F, continuous discharge region

In order that the gas multiplication factor be as independent as possible of the initial position of the ionization, a thin wire anode is often used, usually coaxial to a cylindrical cathode. In the intense field around the anode the multiplication takes place. Since about half of the total number of ion pairs are formed within one mean free path of the surface of the anode, and since the potential in this region is little different from the anode itself, of these ion pairs the positive ions fall through a much greater potential before collection than the electrons. The signal induced on the anode is therefore predominantly due to the positive ions rather than the electrons.

In spite of the low mobility of the positive ions the rise time of the signal is quite rapid (~ μs), owing to the high field near the centre wire which constitutes the major part of the potential drop for the positive ions.

Proportional counters may be used with a wide variety of gas fillings and pressures. Because of the multiplication action the signal is much larger than in an ion chamber or a semiconductor detector. Hence the amplifier noise is not as important in limiting the resolution and the energy threshold as it is in the case of the ionization chamber. The proportional counter is therefore particularly useful for the detection of low energy radiations, down to about 250 eV. Although semiconductor counters can better this performance, it can only be done with considerable attention to the design of a low noise preamplifier. For use at higher energies a resolution of 6% fwhm at 70 keV is typical.

In general, proportional counters find application wherever an adequate efficiency can be obtained, i.e. for the detection of charged particles. For neutron detection the filling gas may be BF_3 or the counter may contain a hydrogenous proton radiator material. In this form they find application in reactor monitoring where the high temperature and neutron flux would make the use of other types of detector unsuitable.

1.3.3 Geiger counters

If the voltage on a proportional counter is increased sufficiently, the amount of charge collected becomes independent of the amount of the initial ionization and, in fact, a single initial ion pair can trigger the discharge. A Geiger counter (or Geiger–Müller counter) is designed to operate in this region and, like the proportional counter, has a wire anode. Compared to the proportional counter the operation of the Geiger counter is not so well understood. However, it appears that, when the electrons are accelerated in the intense field surrounding the anode, as well as producing an avalanche of electrons and ions through collision there is considerable excitation of atoms and molecules. These rapidly de-excite with the emission of light and UV photons, some of which produce photoelectrons from other parts of the gas or the counter walls. Thus, although the avalanche is initially in one part of the tube, it rapidly spreads so the gas in most of the tube is involved in the discharge. During this phase the electrons are continually being collected by the anode but the positive ions, having a lower mobility, are left in the chamber volume as a positive sheath which, owing to the space charge effect, lowers the field. This positive sheath continues to build up until the field is reduced to a level where further multiplication cannot take place.

In all Geiger counters some method of 'quenching' must be used. This is because when, at the end of the discharge, the positive ions arrive at the cathode, they cause some photon emission, which in turn can cause the whole discharge to start again. Without quenching, the tube would undergo continual repetitive discharging. Quenching may be by electronic means or by adding a quenching agent to the gas in the tube. In the electronic method a reduction in the voltage on the tube is brought about immediately after a pulse arrives so that the tube has insufficient voltage to sustain a second discharge. After allowing time for all the ions to be cleared from the tube volume ($\sim$ 300 μs) the voltage is restored to its full working value.

The other method of quenching involves the addition of a compound which, when ionized, loses its energy by dissociation rather than by photo-emission. In addition it should have an ionization potential below that of the main gas so that at the end of a discharge, when the positive ions are migrating to the cathode, the ionization is transferred in collisions to molecules of the quenching gas. Thus only ions of the quenching gas reach the cathode where they dissociate rather than photoemit.

A typical filling for a Geiger counter is 90% argon and 10% alcohol as a quenching gas. The alcohol gas eventually becomes totally dissociated and the tube will operate for about 10^8 to 10^9 counts. Another common filling uses a small quantity ($\sim$ 0·1%) of halogen gas, such as chlorine or bromine, as a quenching agent. In this case it appears that diatomic molecules of the halogen gas are dissociated in the quenching process but, as there is some regeneration process, the gas does not become used up. The life of halogen quenched tubes is much longer than organic quenched tubes; they are more robust with respect to high values of anode voltage but they require special construction techniques to obviate the highly corrosive action of the halogen gas.

In the period immediately following the discharge, the Geiger counter is unable to trigger a second discharge from another incoming particle. After this there is a progressive recovery over 100–200 μs when the counter can be triggered again, but the resulting discharge does not attain its full magnitude. Whether such a partial discharge is recognized by the counting circuits depends on the sensitivity of the following circuits. Hence it is usual to arrange the following circuit to have a defined dead time so that the loss of counts due to dead time may be accurately assessed.

Most Geiger counters use gas pressures of the order of 10 cm of mercury and may have window thicknesses down to 1·5 mg cm^{-2}. Windowless types with gas flow have also been constructed but these require a higher operating voltage. A large number of geometrical configurations are available for various applications.

1.4 Semiconductor detectors

1.4.1 Introduction

Nearly all the semiconductor detectors in current use are, in principle, solid state ionization chambers. The ionization takes the form of electron-hole pairs in the solid state and, as in the case of the gas ionization chamber, the object is to collect both positive and negative charge carriers on their respective electrodes by the application of a biasing voltage. Complete collection of holes and electrons is desirable so that the voltage induced on the electrodes is proportional to the number of electron-hole pairs and is independent of their initial locations as already considered (Equation (1.12)).

Although it may appear at first sight that many insulating or semiconducting materials may be used to form the basis of a solid state ionization chamber this is not the case for a number of reasons. In most solids the electrons and holes would very quickly be trapped or would recombine at defect centres so that they could only migrate a negligible distance under the influence of the biasing field and so would induce a negligible voltage change on the electrodes. Even if this were not the case, the conductivity of many materials is too high so that the leakage current due to the biasing voltage would swamp any signals present. These two considerations alone limit the choice of materials to those available in high purity i.e. with an impurity concentration of, at most, 1 part in 10^8.

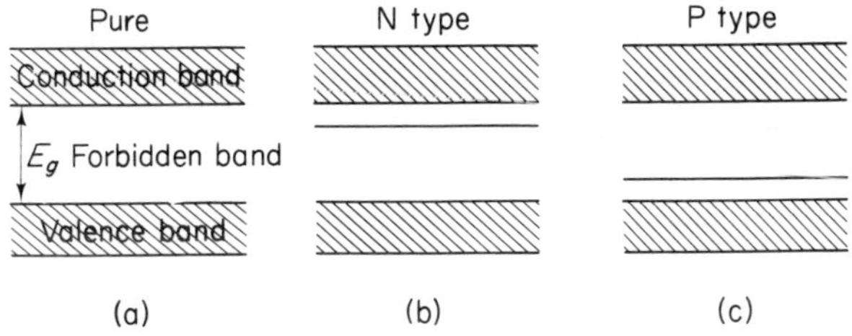

Figure 1.14 Energy level diagram for (a) a pure semiconductor material, (b) a material with donor impurities, (c) a material with acceptor impurities

Figure 1.14 (a) shows a simplified energy level diagram for a perfectly pure semiconductor material. In such a pure material an electron can only exist in the valence band, where it is immobile, or in the conduction band, where it is free to migrate under the influence of an applied field. At absolute zero temperature all the electrons are in the valence band so no conduction can take place. As the temperature is raised increasing numbers of electrons can acquire enough energy from thermal excitation to occupy the conduction band and the quiescent conductivity shows a gradual increase. The action of a nuclear particle is to raise electrons from the valence band to the conduction band by supplying an energy to each of E_g, equal to the energy gap across the forbidden band. In such a pure material the elevation of an electron to the conduction band by any means leads to the concurrent creation of a positive 'hole' which, under the influence of an applied field, can also move by a mechanism that may be imagined as arising from the field-induced hopping of valence electrons between neighbouring atoms. The mobility of the holes is usually comparable to that of the electrons so that they contribute a significant part of the overall conductivity. The production of conduction electrons and holes by thermal excitation of valence electrons is known as the intrinsic mechanism.

In practice a semiconductor material contains impurity centres in the crystal lattice. At these centres electrons can take on energy values which fall within the forbidden band of the pure material, as shown in Figure 1.14. Impurities having energy levels that are initially filled near the bottom of the conduction band (Figure 1.14 (b)) are known as donor impurities, the resulting material being known as N type, and may be produced by impurity atoms having an outer electronic structure with one more electron than the host material. An electron occupying such an impurity level is in a very favourable position for thermal excitation into the conduction band compared to a valence electron, and the quiescent conductivity (i.e. in the absence of nuclear radiation) may therefore be much greater than would be given by the intrinsic mechanism alone. Unlike intrinsic conductivity the positive sites left behind are not free to move, since the hopping of a neighbouring valence electron is energetically unfavourable, so that the only contribution by holes to conduction is that arising from the intrinsic mechanism.

Impurities which introduce energy levels which are initially vacant just above the top of the valence band (Figure 1.14 (c)) are termed acceptors, the resultant material being known as P type, and may be produced by the addition of atoms having an outer electronic structure with one electron less than the host material. In this case a vacant impurity level may be easily filled by thermal excitation of a valence electron from a neighbouring atom so creating a hole on that atom. The numbers of such holes may greatly exceed the number available from the intrinsic mechanism and so increase the conductivity.

In practice a material will always contain both donor and acceptor impurities and the effects of these will partly cancel one another out, since the holes produced by the acceptor impurities will combine with the electrons produced by the donor impurities and so are unavailable for conduction. The material can therefore be regarded as nett P type or nett N type depending on whether the number of holes produced by the acceptor exceeds the number of electrons produced by donors or vice versa. The degree of purity attainable by refining techniques often does not give material of low enough conductivity. If this material corresponds to, say, P type it is possible to 'compensate' it by the addition of the correct amount of a donor impurity so that the material then approximates more closely to the intrinsic material.

The action of nuclear radiation on a semiconductor may be considered as due to interaction with host atoms only, since the fraction of impurity atoms is always negligible in this context. Thus equal numbers of holes and electrons are produced. These may recombine with one another, be trapped at 'trapping centres', or be collected by the electrodes of the detector. Direct recombination is rare; the mean carrier lifetime due to this cause alone would be as high as ~ 1 s in silicon[14]. In fact the purest silicon available has a carrier lifetime of 10^{-2} s and this reduction is due to indirect recombination after trapping. It is therefore the concentration of these recombination traps which critically determines the lifetimes of the carriers.

Other types of trapping can occur where the carrier is trapped for a long period, compared to the collection time, but is eventually released. One effect of such trapping is to give a long tail to the voltage signal induced on the electrodes. This is often made worse by an additional effect since the trapped carriers constitute a space charge and, to maintain electroneutrality, an associated number of carriers of the opposite sign are retained in the material. Since these are free to move they give an increased conductivity during the trapping period. This effect is therefore equivalent to amplification of the slow component and is responsible for the large currents observed in, for instance, cadmium sulphide when exposed to light or nuclear radiation.

In summary, the properties required for a semiconductor detector material are listed below:

(a) Small energy gap E_g to give a large yield of electron-hole pairs from the nuclear particle.
(b) Low quiescent carrier concentration to give low leakage current.
(c) High mobilities of holes and electrons and long carrier lifetime to give efficient collection and a rapid rise time of the signal.
(d) Freedom from delayed trapping to give a rapid rise time and freedom from space charge effects.
(e) High atomic number to give a good photoelectric cross section for γ-rays.

Some of these requirement are mutually incompatible, as in the case of (a) and (b) for example. The materials that are almost exclusively used in current practice are silicon ($Z = 14$) and germanium ($Z = 32$), since it is only with these materials that an adequate technology for purification is available.

1.4.2 Homogeneous (bulk) detectors

The most obvious configuration for a semiconductor detector is that of a homogeneous block of material between suitable electrodes. The essential problems with this arrangement are that it requires the use of a material with a very low carrier density in order to give an acceptably low leakage current, and the construction of the electrode–semiconductor interface presents problems. Junction counters (see below) are not as critical in these respects and the performance of homogeneous counters has never approached that obtainable with junction counters. Nevertheless the homogeneous counter configuration is inherently attractive since large active volumes can be accommodated. All the work to date has been with compensated material usually operated at low temperatures.

1.4.3 PN junction detectors

The configuration that is currently almost universally used is that of a reverse biased PN junction. In this case a depletion layer is created extending across the junction so that nearly all the biasing voltage appears across this depletion layer. The free carriers in both the P and the N type material are cleared from the layer due to the action of the field and the only contribution to conduction in the quiescent state can come from carriers generated by, for example, recombination centres and surface effects[1].

Since the biasing field is effectively all located across the depletion layer, it is only this layer that is active for nuclear particle detection. The depletion layer width is determined by the biasing voltage and the carrier densities. If ρ_1 and ρ_2 are the resistivities of the regions on either side of the junction, then the depletion layer extends into sides 1 and 2 distances proportional to ρ_1 and ρ_2. If $\rho_1 \ll \rho_2$, as in the case of the surface junction detector, the depletion layer extends almost entirely into the high resistance side and, with a biasing voltage V, is proportional to $\rho_2\sqrt{V}$. Any reverse biased PN junction behaves as a capacitance to small signals and for the case $\rho_1 \ll \rho_2$ the capacitance may be shown[1] to be approximately twice the value for the geometrically equivalent parallel plate capacitance filled with dielectric having the same dielectric constant as the semiconductor material.

It is impossible to consider here the many different considerations that have led to the methods of fabrication of the types of semiconductor detectors in current use. Even with the types to be described the performance with respect to resolution and noise is dependent on many factors such as the method of forming the electrodes and surface current leakage. Silicon is chosen for charged particle detection, where a high atomic number of detector medium is unimportant, because the energy gap of silicon detectors is large enough to ensure that the leakage current at room temperature is sufficiently low. For gamma ray detection, germanium is to be preferred on account of its higher atomic number but the detector usually requires cooling, often to liquid nitrogen temperature (77° K), to obtain an acceptably low leakage current.

1.4.4 Surface junction detectors

Since the regions outside the depletion layer have negligible field they act as a dead layer for the purposes of particle detection. For the detection of charged particles, a thin dead layer or window is desirable and this has led to the surface junction form of construction which gives a window thickness of 0·1–1 μm.

The method of fabrication may take several forms. For example, one method is to start with P type silicon and allow a donor impurity such as phosphorus to diffuse at a high temperature (~800 °C) through the front face. The resulting superficial layer is then N type and a PN junction results close to the front surface. Electrical contact to the front face is made by a metallic film evaporated on. Since the N type region is much more heavily doped then the original P type region, the depletion layer extends mainly into the P type. In a similar manner it is possible to start with N type material and diffuse an acceptor impurity, boron for example, to give a superficial P type layer.

To avoid the high temperatures involved in the diffusion process another method is to start with N type silicon and allow the surface to oxidize, which creates a highly doped P type region. Another technique involves the creation of a superficial doping by ion implantation where ions from an accelerator impinge on the surface and penetrate a short distance.

Surface junction detectors can be made with depletion layers of up to 1 mm and are therefore satisfactory for the detection of many charged particles.

1.4.5 Lithium drifted detectors

The width of the depletion layer in the surface junction construction is limited by the resistivity of the starting material and by the bias voltage applicable without breakdown. The lithium drift method effectively compensates the material in the junction region so that its resistivity, and consequently the depletion layer thickness, is larger. The process is usually carried out on germanium, because of its higher photoelectric cross section for γ-rays, but may also be used on silicon. The method makes use of the fact that lithium is a fast interstitial diffusant in germanium and silicon and acts as a donor impurity. The coefficient of diffusion is some 10^7 greater than the diffusant employed in surface barrier construction (e.g. boron and phosphorus).

In the case of germanium the purest material available is, in fact, P type, mainly because of the presence of boron impurity. Lithium is diffused into the front surface of a block of this germanium at a temperature of up to 500 °C. So far the process has been similar to that used in the fabrication of a surface barrier detector, Figure 1.15(a). The resulting PN junction is now back-biased at approximately room temperature when the lithium ions, having a high mobility, are free to move under the influence of the internal fields and so redistribute themselves to give a concentration profile as shown in Figure 1.15(b). This is essentially due to the fact that the lithium ions move at a drift velocity proportional to the electric field and so will tend to drift away faster from local regions of positive charge. Thus the tendency is to produce a uniform electric field. In the central region this will only be attained if there is no nett space charge, i.e. the number of lithium ions compensates nearly exactly the number of acceptor impurities.

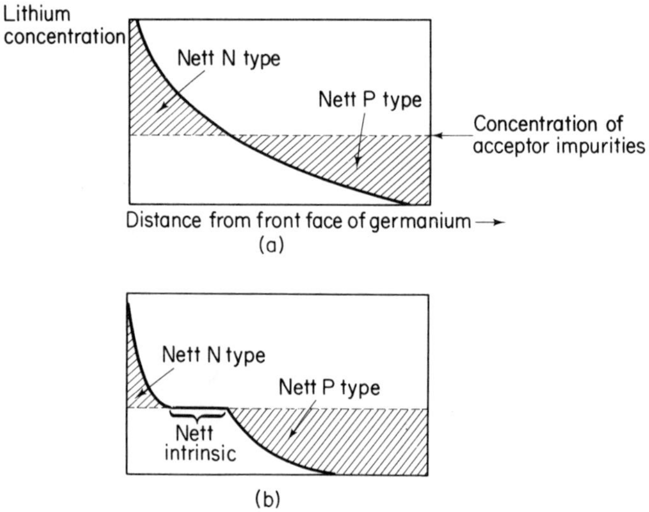

Figure 1.15 Ge(Li) detector fabrication. Distribution of lithium (a) after diffusion and (b) after drifting

Lithium drifted detectors, often denoted by Ge(Li) or by PIN where the I signifies the intrinsic region, may be made with active volumes of up to 120 cm^3. The window of such detectors may be quite thick (up to 100 μm) but this is not usually very important for gamma detection. Ge(Li) detectors must be operated at low temperature, usually that of liquid nitrogen (77° K), to reduce the noise. In addition in order to prevent redistribution of the lithium ions it is necessary to store the detector at low temperature.

1.4.6 Thin, totally depleted d*E*/d*x* *detectors*

In this form the detector offers a way of identifying particles on the basis of their rate of energy loss, dE/dx. The dE/dx detector is arranged to be thin so that the particle emerges with small energy loss and can be further detected by another detector; the energy of the particle can then be deduced by adding the signals from the two detectors. The dE/dx detector should ideally have thin windows on both the entrance side and the exit side. These detectors may be fabricated by any of the three usual techniques, with the difference that the starting material is a thin wafer and the process of forming the PN junction is controlled so that, when biased, the depletion layer extends almost completely through the wafer. Currently this type of detector can be fabricated in thicknesses down to 10 μm. The rate of energy loss of ions in crystalline materials depends on the orientation of the ion beam relative to the crystal planes, the so-called channelling effect. In order to avoid this sensitivity to orientation a wafer used in a dE/dx detector is usually cut so that its axis does not lie in any crystal plane.

1.5 Scintillation counters

1.5.1 Introduction

The scintillation counter is probably the most widely used detector of nuclear particles. This partly arises from the fact that the detection medium, the scintillator, can be made to be large and of any geometry, and is dense so that a good detection efficiency can be obtained, particularly for gamma rays. The photomultiplier gives an output signal pulse of several volts so that the use of a sensitive amplifier is unnecessary. When used with fast scintillators the scintillation counter can be made to respond faster than

Table 1.2 Properties of scintillators[3]

Scintillator	Specific gravity	Wavelength of maximum emission	Decay time constant light	Relative light output for γ-rays	α/β ratio*
		Å	s		%
NaI(Tl) 0·1% crystal	3·67	4130	$0{\cdot}23 \times 10^{-6}$	230	50
NaI pure at 77°K crystal		3030	$0{\cdot}06 \times 10^{-6}$	440	
CsI(Tl) crystal	4·51	5800	$1{\cdot}1 \times 10^{-6}$	95	80
ZnS(Ag) multi-crystalline	4·09	4500	$0{\cdot}2 \times 10^{-6}$	300	100
Anthracene crystal	1·25	4470	30×10^{-9}	100	10
Trans-stilbene crystal	1·16	4100	$4{\cdot}5 \times 10^{-9}$	50	8
Liquid scintillators	~0·9	3800–4200	$2–8 \times 10^{-9}$	50–80	~10
Plastic scintillators	~1·0	3800–4200	$2–4 \times 10^{-9}$	50–70	~10

* α/β ratio is the ratio of light output per unit energy for α particles to that for β particles. The data in this column was compiled from various sources containing widely different estimates and must be regarded as approximate.

any other type of counter, with the exception of a Cerenkov counter, and is therefore chosen for applications involving high counting rates or precise timing.

Many materials luminesce under the influence of nuclear or ultraviolet radiation. For use as a scintillator for nuclear particles the decay time of the luminescence must be short, the scintillator must be transparent to its own emitted light and the light should have a spectral distribution that nearly matches the photocathode response of the most sensitive photomultiplier tubes. A large variety of materials have been used as scintillators in gaseous, liquid and solid form and the properties of some commonly used scintillators are given in Table 1.2. These materials may be conveniently divided into organic and inorganic materials, organic materials are not as suitable for gamma detection due to their lower photoelectric cross section and do not have as high a light output; but they do have a faster response and shorter light decay time and in plastic or liquid form may be fabricated in large and irregular geometries. The size of inorganic scintallators is restricted by the maximum size that a single fabricated crystal may reach. They show less 'ionization quenching' with heavily ionizing particles and may be preferable on this account. Since the mechanisms of luminescence are rather different in organic and inorganic scintillators they will be considered separately.

1.5.2 Organic scintillators

Many conjugated and aromatic molecules exhibit the phenomenon of luminescence. Whether in the gaseous, crystalline, liquid, plastic or glassy state the molecules are loosely bound and retain their individual identity, electronic structure and luminescence. In general there are always several competing processes for the de-excitation, which give luminescence components of different decay times. The most rapid process, fluorescence, arises in organic materials when the excited molecule loses some energy by a radiationless transition to an intermediate state, which in turn de-excites with the accompaniment of photoemission. This takes place in a time of the order of 10^{-8} to 10^{-9} s, which is the same order as the lifetime of an atomic state for an allowed transition. Slower processes, usually termed phosphorescence although the terminology is confused (see Birks[13]), arise from a similar pair of de-excitations but where the intermediate level is metastable. This component may have a decay time of 10^{-4} s or longer and emits at a longer wavelength than the fluorescence. In order that high counting rates may be accepted the slow components of the decay should ideally be negligible. Finally the molecule may de-excite entirely by radiationless transitions. The fluorescent component is the one that is utilized in the scintillation counter and ideally the incident particle energy should be efficiently converted into fluorescently emitted light.

Of the pure materials used as organic scintillators aromatic hydrocarbons with linked benzene ring structure, particularly anthracene and trans-stilbene, give the required fluorescent efficiency with very little phosphorescence. Both these materials may be fabricated as single crystals having a side of several centimetres.

If a small amount of impurity is present there may be a high probability that the excitation energy be transferred to these molecules. If these molecules de-excite by radiationless transition the process reduces the overall fluorescent efficiency and the effect is known as 'quenching'. On the other hand if the impurity de-excites by photoemission it may increase the fluorescent efficiency; this mechanism is utilized in a wide variety of plastic and liquid scintillators, both of which may be regarded as solutions. The primary interaction and excitation occurs mainly in the host material or solvent and is transferred to solute molecules having a lower energy level by a mechanism that is poorly understood. These solute molecules then de-excite by photoemission so that the resulting light emission in such solutions has a spectral distribution characteristic mainly of the solute and is of longer wavelength than that of the pure solvent. Commonly used solutes are p-terphenyl, PPO (2,5-diphenyloxazole) and PBD (2-phenyl-5-(4-biphenylyl)-1,3,4-oxadiazole), and commonly used solvents include toluene, xylene, benzene and 1,4-dioxan. For plastic materials the host 'solvent' may be polyvinyltoluene or polystyrene. Often a second solute, sometimes termed a 'wavelength shifter', is added and the excitation is further transferred to this from the primary solute. The spectral distribution of the resulting photoemission from the wavelength shifter then more nearly matches the spectral response of the photomultiplier tube and gives an increase in the overall conversion efficiency. Probably the most commonly used compound for this purpose is POPOP (1,4-di-(2-(5-phenyloxazolyl))-benzene).

In the case of plastic or liquid organic scintillators it is possible to incorporate additional compounds in the material. Thus, for instance, a boron or lithium compound may be incorporated to give a thermal neutron detector. Probably the commonest application of liquid scintillators is in the assay of radioactivity by so called 'internal counting'. The radioactive sample is dissolved or dispersed in the scintillator which then acts as a very efficient detector for the emitted beta particles. The windowless nature of the method makes it one of the few available for the assay of activity of low energy beta rays emitted by isotopes such as tritium and ^{14}C. For use in chemical assay work, water miscible scintallators are available, usually based on dioxan solvent, in which small quantities of aqueous solution containing the labelled compound may be dissolved.

Whenever such additional compounds, and also oxygen, are added to the scintillator a certain amount of quenching is introduced. This may be offset by the inclusion of compounds such as naphthalene which act as a primary solute to which the excitation is rapidly transferred before quenching can take place. When using the internal counting method it is usually necessary, nevertheless, to make a correction for the quenching effect of the dissolved sample.

1.5.3 Inorganic scintillators

Although it is possible to use pure inorganic crystals as scintillators, in distinction to pure organic crystals their luminescent efficiency is improved by the addition of a small amount of a suitable impurity or 'activator'. The excitation in the host material is transferred to these impurity centres in the crystal lattice, which de-excite with the accompaniment of photoemission characteristic of the impurity.

The primary interest in using inorganic scintillators is in the detection of gamma rays where the higher density and higher atomic number gives a much better efficiency of detection by the photoelectric effect than in organic materials. By far the most common material for this purpose is sodium iodide activated with 0·1 % mole fraction of thallium. Such NaI(Tl) crystals have a high luminescent efficiency (see Table 1.2) and may be fabricated in large sizes (having up to 20 cm side) of single crystals. The decay time of the fluorescence is long ~230 ns compared to organic materials however. Sodium iodide is deliquescent and must be encapsulated in an airtight enclosure.

Pure sodium iodide has a poor luminescent efficiency at room temperature but has a value higher than NaI(Tl) if it is cooled to liquid nitrogen temperature. Since pure NaI has a shorter luminescence decay time, 60 ns, at this temperature than has NaI(Tl) at room temperature it is sometimes used where high counting rates are encountered or where precise time information about the nuclear events is required.

More recently CsI(Tl) and CsI(Na) have become commercially available for use as gamma detectors and have the advantage of a higher average atomic number than NaI(Tl). Caesium iodide crystals are physically more robust than sodium iodide and are not as deliquescent.

The other main application for inorganic crystals is in the detection of heavy charged particles since their response to these particles is considerably greater than in organic crystals. For these purposes it is usually desirable to use a scintillator thickness only large enough to stop the particles, so that the efficiency of detection of stray gamma rays and electrons is minimized. The mechanical properties of caesium iodidc lend themselves to the production of such thin wafers and, as noted above, they are not as deliquescent as NaI(Tl) and so do not require encapsulation.

Zinc sulphide, although having a high luminescent efficiency, can only be obtained as a multicrystalline powder which is semi-opaque to its emitted light. It finds application in the detection of heavy charged particles where the powder is dispersed as a thin layer. For the purposes of fast neutron detection an efficient detector of the proton recoil variety may be made out of a dispersion of zinc sulphide in a hydrogenous medium. For thermal neutrons a mixture of a compound of ^{6}Li or of boron with zinc sulphide may be used. The zinc sulphide then detects the charged particle products of the reaction of neutrons with the lithium or boron. Alternatively the reaction with ^{6}Li may be exploited by the use of a $^{6}LiI(Eu)$ scintillator crystal. When cooled to liquid nitrogen temperature this scintillator may be used for fast neutron spectroscopy.

Quite apart from inorganic crystal scintillators a number of inorganic glass scintillators have been developed. Although they have luminescent efficiencies rather less than even organic scintillators they find application in corrosive or high temperature environments. Types are available containing ^{6}Li for the purpose of neutron detection.

1.5.4 Photomultiplier tubes

The photomultiplier tube consists of a photocathode, a series of dynodes having successively higher potentials and, finally, a collector electrode. Each of the photoelectrons emitted from the photocathode by the action of light is accelerated through 100–200 V, or higher in special purpose tubes, onto the surface of the first dynode where it produces a number of secondary electrons, typically ~5 for caesium–antimony dynodes, which are in turn accelerated onto the next dynode, where further multiplication takes place. The number of dynodes may be between 9 and 14 and give an overall multiplication of the initial photoelectrons by a factor of between 10^5 to 10^7 by the time they arrive at the collector. This gain is sufficient to produce a signal pulse of several volts on the collector when the collector is connected to its voltage source via a resistor. If this resistor is high valued the collector signal will approximate to the charge arriving on the collector capacitance. If the collector load is made small the signal approximates to the collector current, and this is normally preferable for high counting rates or for fast timing of the pulses. In the latter application it is possible to use a collector load of 50 to 100 ohms so that it may match the characteristic impedance of a coaxial cable connected to it.

Various forms of photomultiplier tube are in use. The photocathode material is usually deposited as a semi-transparent layer on the inside of the glass envelope as this allows good optical coupling between the scintillator and photocathode to be made. Until recently the photocathode materials having the highest sensitivity were based on Cs–Sb alloys which gave a peak sensitivity to light in the wavelength region of 3000–4500 Å and a peak quantum efficiency (i.e. photoelectrons per incident light photon) of up to 23%. More recently bi-alkali photocathodes (K–Cs–Sb for example) have given slightly higher quantum efficiencies and have the advantage that the thermionic emission, and hence the noise, is considerably less than that of the Cs–Sb photocathodes.

For the dynode surface, materials such as Cs–Sb, Cs–Ag–Mg–O, or more recently GaA and GaP have been used. The geometry of the tube must be arranged so that the electric field at the dynode surface is such as to accelerate the secondary electrons away from the surface and onto the next dynode.

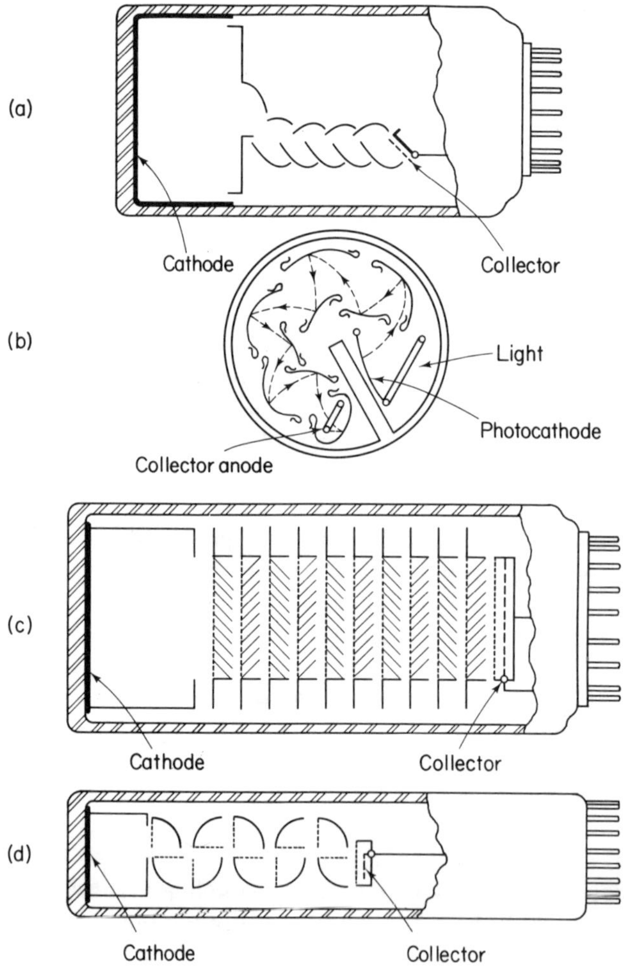

Figure 1.16 Different photomultiplier tube geometries. (a) and (b) show two types of focused dynode arrangements, (c) the venetian blind geometry and (d) the box and grid arrangement (Reproduced by permission of E.M.I Ltd)

This may be attained by the 'focused' arrangement of Figure 1.16(a) and 1.16(b), the venetian blind, Figure 1.16(c) or box and grid, Figure 1.16(d), geometries. The spread in electron transit times, and hence the rise time of the output pulse, is greater in the unfocused type, typically giving a rise time of ~10 ns. For high speed applications the focused type is to be preferred and can give rise times down to a fraction of a nanosecond.

It is important to efficiently collect the photoelectrons onto the first dynode since the statistical fluctuations in this small number directly affect the energy resolution. For this reason the potential difference between the photocathode and first dynode is often made larger than that between pairs of other dynodes.

There is always a certain amount of noise produced in a photomultiplier tube. Such a 'dark current' is mainly the product of thermionic emission from the photocathode and its magnitude varies widely with the photocathode type and also from tube to tube. The advent of bi-alkali photocathodes has considerably reduced this component of noise and it may be reduced still further by cooling to around −20 °C. The existence of this noise is mainly troublesome where it is necessary to detect very low energy nuclear particles such as the soft beta radiation of tritium. For this application two cooled photomultiplier tubes are often used which view the same scintillator and whose outputs are summed and are arranged so that a count is only registered when coincidence occurs between pulses in the two outputs. In this way the contribution of noise can be greatly reduced.

1.5.5 Response to different particle energies and types

The variation of the light output of a scintillator with particle energy is slightly non-linear especially for low energies. Figure 1.17 shows the light output per unit energy for various energies of electrons in NaI(Tl). For gamma rays of higher energy, where the gamma energy is transmitted to the electrons mainly by multiple events, this non-linearity of response is responsible for some loss of energy resolution. Figure 1.18 shows the light output of anthracene for electrons, protons and alpha particles of various energies and it will be immediately apparent from this that heavily ionizing particles are less efficient in producing light, the so-called ionization quenching effect. The controlling factor appears to be the specific energy loss of the particle dE/dx and Figure 1.19 shows the rate of light output dL/dx against the specific energy loss for various types of particles. Ionization quenching is less for the inorganic scintillators, and the possible mechanisms responsible in both organic and inorganic scintillators are fully discussed by Birks[13].

Quite apart from the difference in light output with different particle types there is a difference in the time course of the pulse of light. This is particularly noticeable in organic scintillators and may be made the basis of a particle discrimination system. Section 6.5 deals with such systems and

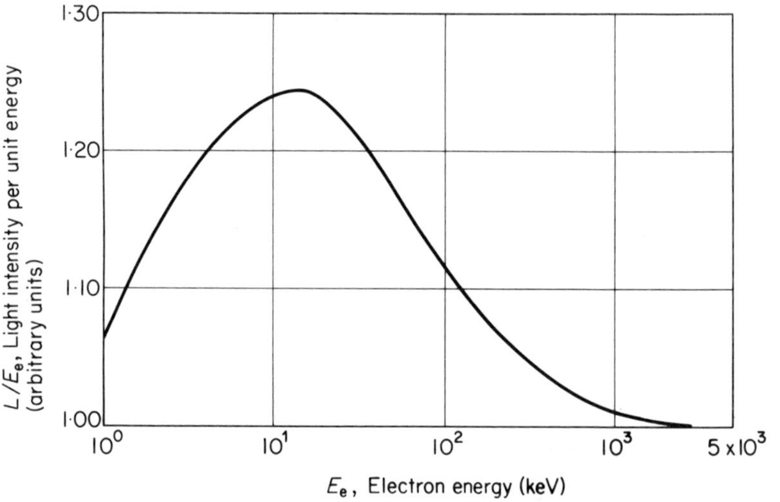

Figure 1.17 Calculated relative light output per unit electron energy of NaI(Tl) crystals plotted against electron energy E_e (Zerby, Meyer and Murray[15])

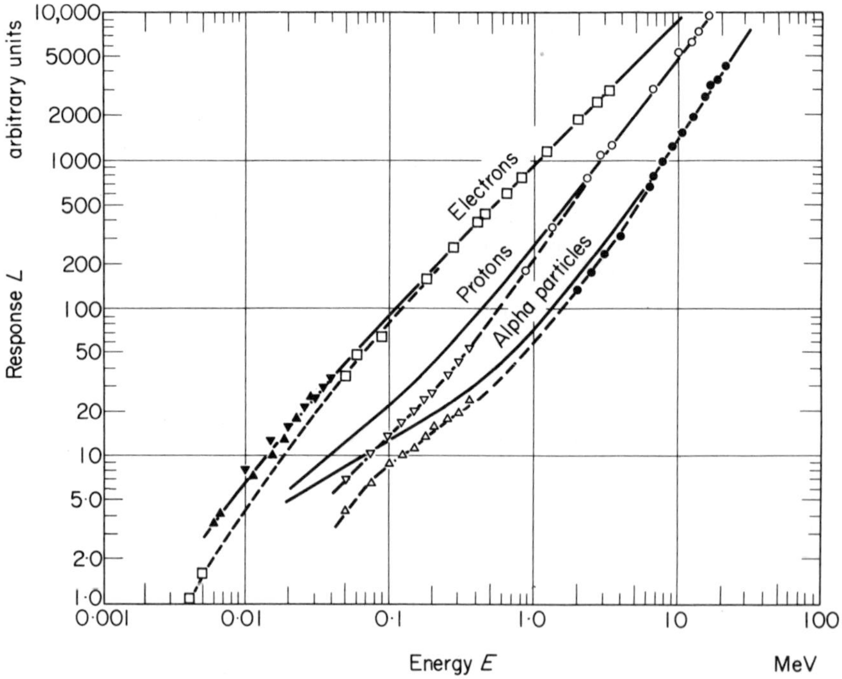

Figure 1.18 Light output of anthracene plotted against particle energy for electrons, protons and alpha particles (Brooks[16]). The continuous curves represent the theoretically calculated responses

Figure 6.19 shows the response of different organic scintillators to various particles. It will be observed that, unlike the fast component, the slower components do not decay in an exponential manner and the relative amounts of light in the fast and slow components depend on the particle type. It appears that it is primarily the fast component of the luminescence that is subject to the ionization quenching.

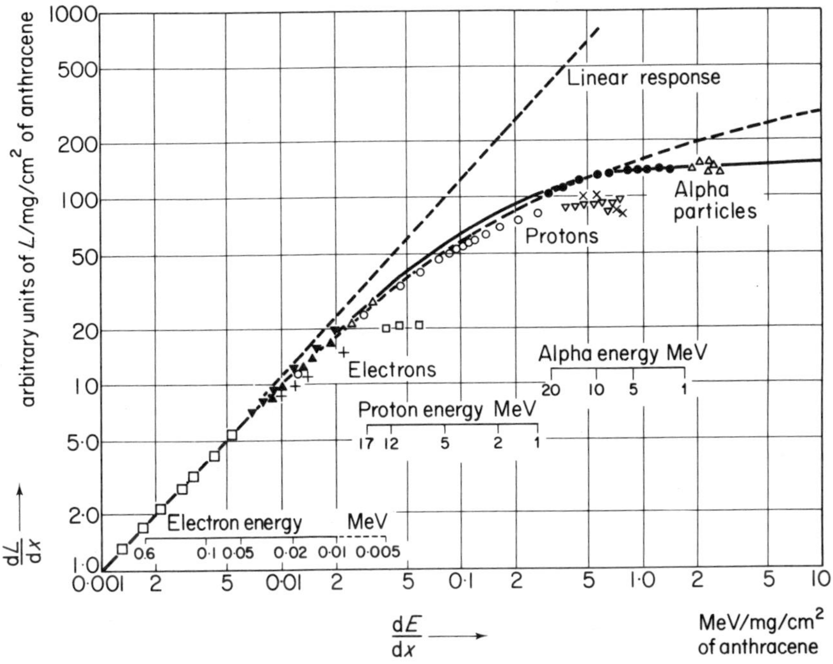

Figure 1.19 Variation of specific fluorescence dL/dx of anthracene plotted against specific energy loss dE/dx (Data collected from various sources by Brooks[16])

1.6 Other detection methods

1.6.1 Cerenkov detectors

Cerenkov radiation occurs whenever a charged particle travels through a transparent medium at a speed faster than light in that medium. The Cerenkov radiation may be regarded as an electromagnetic shock wave and, as in the case of an acoustic shock wave, the wavefront has a conical shape centred on the particle path (Figure 1.20). If the particle has a velocity which is a fraction β of the velocity of light in vacuo, then in a medium of refractive index μ the light cone has a semi-angle θ given by $\cos\theta = 1/\mu\beta$. The intensity of the light per unit particle path length in the wavefront increases with μ and the light is essentially white over the visible region.

Cerenkov detectors are probably most used in the identification of high energy particles, sometimes associated with magnetic analysis. Since they only detect particles having $\mu\beta > 1$ the medium may be chosen to have a value of μ giving a suitable threshold and may simply be a block of the material coupled to a photomultiplier tube in a similar manner to that used for scintillators. It may be noted that the light pulse accepted by the photomultiplier is of extremely short duration and the output pulse of the photomultiplier has a width determined essentially by the spread in transit time of the tube itself. Compared to scintillators, however, the light output from the Cerenkov effect is much smaller and, for example, for electrons travelling near the velocity of light, in a medium having $\mu = 1{\cdot}5$ about 270 photons per centimetre are produced[17].

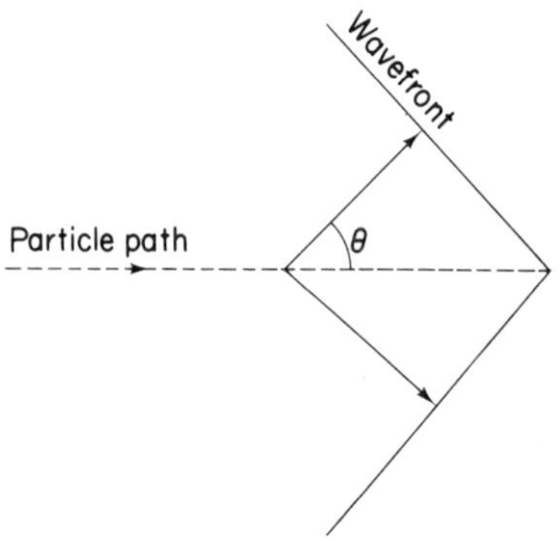

Figure 1.20 Cerenkov effect

Spectrometry of high energy gammas or charged particles may be approximately performed if a dense medium is employed having sufficient size to contain the entire shower development. However a more useful technique is to derive information about the particle speed by means of the cone angle of the radiated light. For this various 'focusing' types have been used which serve, by optical means, to select only a limited range of θ, so confining detection to particles having the corresponding range of β. For a discussion of these and other techniques the reader is referred to the review by Moyer[17].

1.6.2 Spark chambers

In essence the spark chamber consists of a pair of parallel plates, to which a high potential difference may be applied, and the volume between the plates is filled with gas. An ionizing particle traversing the gap between the plates will leave a trail of ionization which, if the electric field is high enough, will initiate a spark discharge in the region of the track. Spark chambers are normally used in conjunction with another detector which, on detecting a particle of interest, causes the high voltage to be applied to the spark chamber plate. Slightly later, the voltage is reduced to a lower level in order to terminate the discharge, the residual electric field being necessary to clear ionization from the chamber. Spark chambers, therefore, offer a means of spatial localization of a nuclear particle and the position of the spark may either be photographed or inferred by acoustic means. In order that the particle track may be more accurately defined it is common to use a stack of spark chambers each having a small electrode spacing.

References

1. G. Dearnaley and D. C. Northrop, *Semiconductor Detectors for Nuclear Radiations,* 2nd ed., Spon, London, 1966.
2. R. D. Evans, *The Atomic Nucleus,* McGraw–Hill, 1955.
3. Nuclear Enterprises Ltd, Edinburgh, *Scintillator Catalogue* 1973.
4. F. S. Goulding, *IEEE Trans. Nucl. Sci.,* NS **11**, No. 3, 177 (1964).
5. A. T. Nelms, U.S. Nat. Bur. Stds., Circular No. 577 (1956).
6. *The Radiochemical Manual,* 2nd ed., The Radiochemical Centre, Amersham, England, 1966.
7. C. M. Davisson and R. D. Evans, *Rev. Mod. Phys.,* **24**, 79 (1952).
8. J. H. Neiler and P. R. Bell, in α-, β- *and γ-ray Spectroscopy,* (Ed. K. Seigbahn), Vol. 1, North Holland Publishing Co., 1965.
9. G. Dearnaley and A. T. G. Ferguson, *Nucleonics,* **20**, No. 4, 84 (1962).
10. G. D. Alkhazov, A. P. Komar and A. A. Vorob'ev, *Nucl. Instr. and Meth.,* **48**, 1 (1967).
11. J. B. Eberhardt, *Nucl. Instr. and Meth.,* **80**, 291 (1970).
12. R. H. Pehl and F. S. Goulding, *Nucl. Instr. and Meth.,* **81**, 329 (1970).
13. J. B. Birks, *The Theory and Practice of Scintillation Counting,* Pergamon, 1964.
14. J. M. Taylor, *Semiconductor Particle Detectors,* Butterworths, London, 1963.
15. C. D. Zerby, A. Meyer and R. B. Murray, *Nucl. Instr. and Meth.,* **12**, 115 (1961).
16. F. D. Brooks, in *Progress in Nuclear Physics* (Ed. O. R. Frisch), Vol. 5, Pergamon 1956.
17. B. J. Moyer, in *Nuclear Instruments and their Uses* (Ed. A. H. Snell), Vol. 1, Wiley, 1962.
18. E. Storm, E. Gilbert and H. Israel, Report LA 2237, U.S. Atomic Energy Commission (1958).

Selected bibliography on detectors

F. Adams and R. Dams, *Applied γ-ray Spectroscopy,* Pergamon 1970.

J. B. Birks, *The Theory and Practice of Scintillation Counting,* Pergamon, 1964.

G. Dearnaley and D. C. Northrop, *Semiconductor Detectors for Nuclear Radiations,* Spon, London, 1966.

E. Fenyves and O. Haiman, *The Physical Principles of Nuclear Radiation Measurement,* Academic Press, 1969.

W. J. Price, *Nuclear Radiation Detection,* 2nd ed., McGraw–Hill, 1964.

J. Sharpe, *Nuclear Radiation Detectors,* 2nd ed., Methuen, 1964.

R. P. Shutt (Ed.), *Bubble and Spark Chambers: Principles and Use,* Vol. 1 (1968) and Vol. 2 (1967), Academic Press.

K. Siegbahn (Ed.), α-, β- *and γ-ray Spectroscopy,* Vol. 1, North Holland Publishing Co., 1965.

A. H. Snell (Ed.), *Nuclear Instruments and their Uses,* Vol. 1, Wiley, 1962.

J. M. Taylor, *Semiconductor Particle Detectors,* Butterworths, 1963.

2

Basic Pulse Circuits

2.1 Linear (analogue) circuits

2.1.1 RC low pass filter

If the circuit of Figure 2.1 has an alternating voltage applied at the input of complex voltage e_i, the complex output voltage e_o will be given by $e_o/e_i = 1/(j\omega RC + 1)$ and the circuit acts as a low pass filter with a 3 dB frequency or 'corner frequency' $f_c = \omega_c/2\pi = 1/2\pi RC$. The response $|e_o/e_i|$ is plotted in

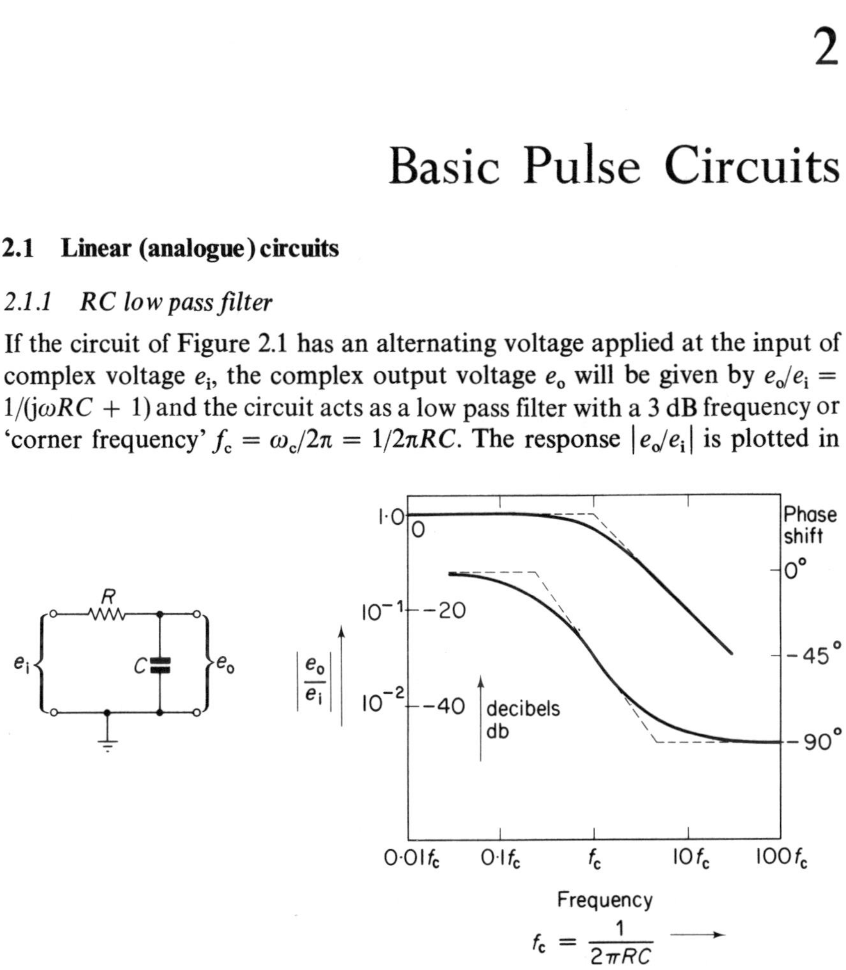

Figure 2.1 RC low pass filter; frequency response and phase shift

Figure 2.1 on a log–log scale and it may be noted that the response is asymptotic to two lines intersecting at f_c, the second of which has a slope of -1 which corresponds to -6 dB per octave. Figure 2.1 also shows the phase shift plotted on a linear–log scale.

If a pulse signal $v_i(t)$ is applied to the input, the differential equation governing the output voltage $v_o(t)$ is given by

$$\frac{dv_o}{dt} = \frac{i}{C} = \frac{v_i - v_o}{RC} \tag{2.1}$$

The case of a rectangular pulse input of height E volts and duration T may be considered by means of the arrangement in Figure 2.2. Initially in position A, the switch in Figure 2.2 moves to position B at $t = 0$ and back to position A at $t = T$. During the period $0 < t < T$ the solution of (2.1) for the appropriate initial conditions may easily be seen to be $v_o(t) = E(1 - e^{-t/RC})$ and during the period $t > T$ the solution is of the form $Ae^{-t/RC}$ where A is a constant. The complete output pulse is shown in Figure 2.2.

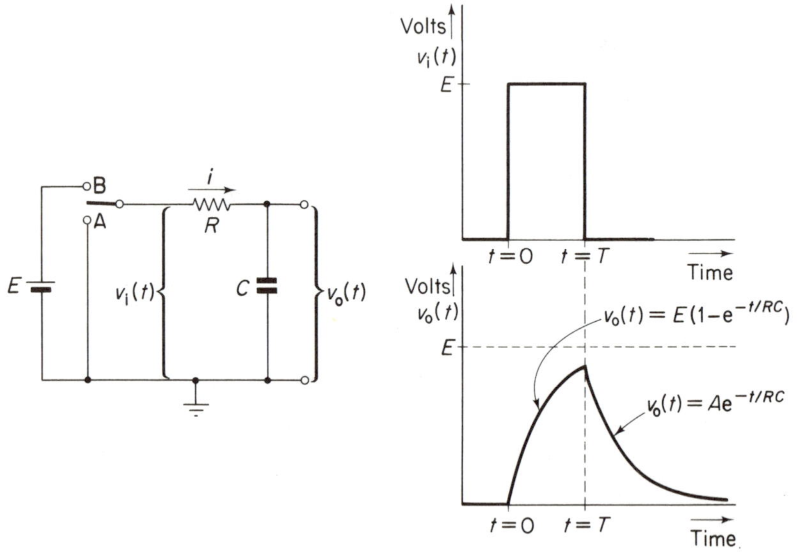

Figure 2.2 Application of a rectangular pulse to a RC low pass filter and the form of the output signal $v_0(t)$

The *rise time* t_r of a pulse is defined as the time taken for its leading edge to rise from 10% to 90% of the peak height and for the output pulse of Figure 2.2 may be shown to be $2{\cdot}2RC = 0{\cdot}35/f_c$. Thus one effect of the RC low pass filter has been to degrade the rise and fall times of the pulse. Later it will be shown how any such single pulse may be expressed as a continuous spectrum of component frequencies and that, as a general rule, the removal of the high frequencies by a filter degrades the pulse rise and fall times i.e. affects the rapidly changing parts of the signal.

Figure 2.3 shows the output pulse shape for various values of the ratio RC/T and in particular it may be noted that for the case $RC \gg T$ the output approximates to the integral of the input pulse. Provided RC is large enough the integrating property of the circuit is effective for any shape of input pulse since, if $RC \gg T$, then $v_o \ll v_i$ and (2.1) approximates to

$$\frac{dv_o}{dt} = \frac{v_i}{RC} \text{ i.e. } v_o = \frac{1}{RC}\int v_i\, dt$$

For this reason the low pass RC filter is often termed a RC integrator although the integrating action is only approximated when RC is much larger than the pulse length.

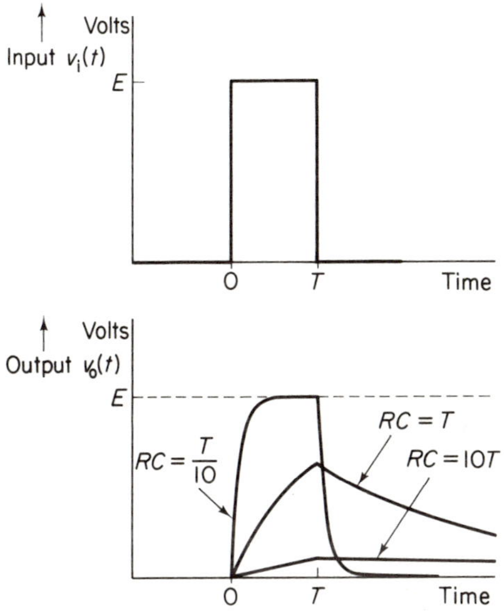

Figure 2.3 Effect of the value of RC of a low pass filter on the form of the output signal for a rectangular pulse input

2.1.2 *CR high pass filter*

The CR circuit of Figure 2.4 is a high pass filter having a response $e_o/e_i = j\omega RC/(j\omega RC + 1)$ and a 3 dB frequency of $f_c = 1/2\pi RC$. The amplitude and phase characteristics are plotted in Figure 2.4 together with their asymptotes. The differential equation relating the input voltage pulse $v_i(t)$ and the output voltage $v_o(t)$ is

$$\frac{dv_i}{dt} - \frac{dv_0}{dt} = \frac{v_0}{RC} \tag{2.2}$$

For the same rectangular pulse input as assumed in the previous section the solution may be found in a similar manner to be $v_o(t) = Ee^{-t/RC}$ for

$0 < t < T$ and $A^{-t/RC}$ for $t > T$ where A is a constant. Figure 2.5 will make the form of the solution clear and it will be seen that the effect of the circuit is to degrade the slowly changing parts of the signal.

The output pulse form is shown in Figure 2.5 for various values of the ratio RC/T and it will be seen that the output approximates to the differential

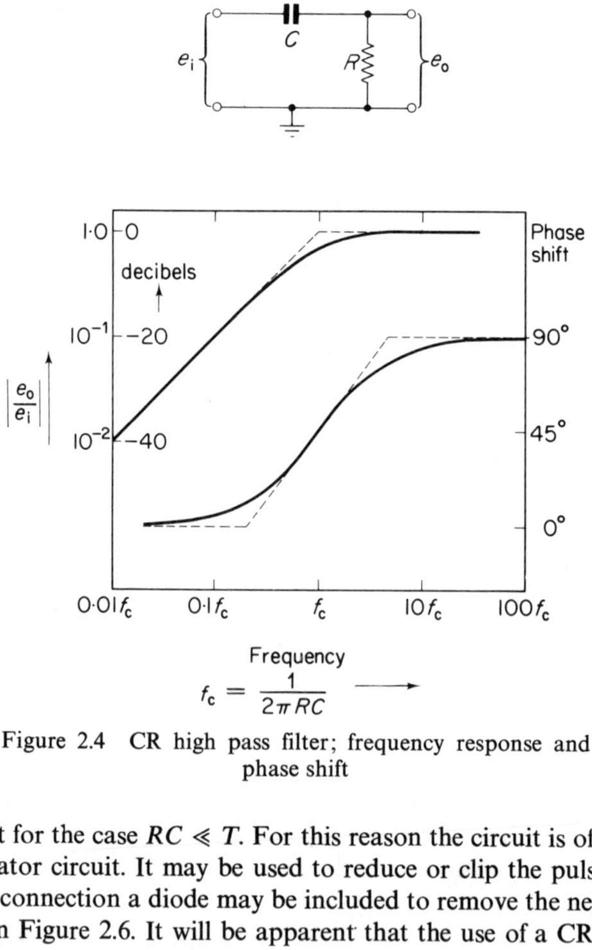

Figure 2.4 CR high pass filter; frequency response and phase shift

of the input for the case $RC \ll T$. For this reason the circuit is often termed a differentiator circuit. It may be used to reduce or clip the pulse duration and in this connection a diode may be included to remove the negative lobe as shown in Figure 2.6. It will be apparent that the use of a CR circuit for pulse clipping is only feasible if the input pulse has a short rise time compared to RC.

One important aspect of the simple CR circuit is that, since the capacitor can transmit no DC component of the signal, the area of the undershoot portion of the output pulse must be equal to the area under the main part of the pulse so the undershoot can never be entirely suppressed. Thus if RC is made very large, although the first part of the pulse is transmitted with

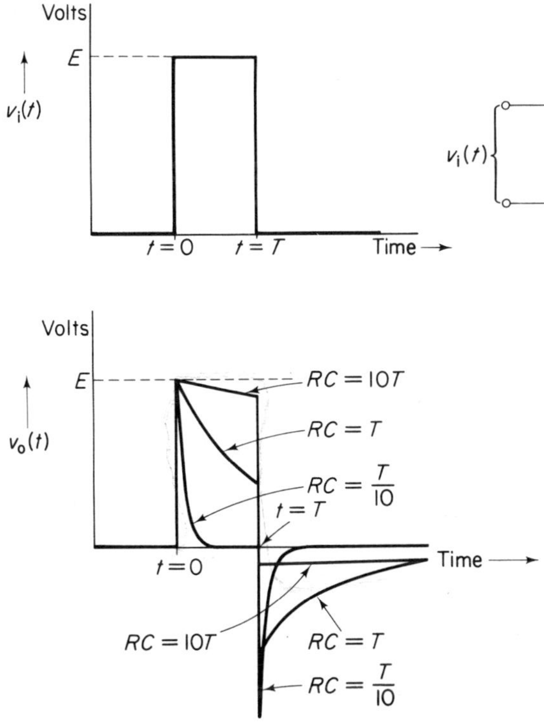

Figure 2.5 CR high pass filter and its effect on a rectangular pulse input

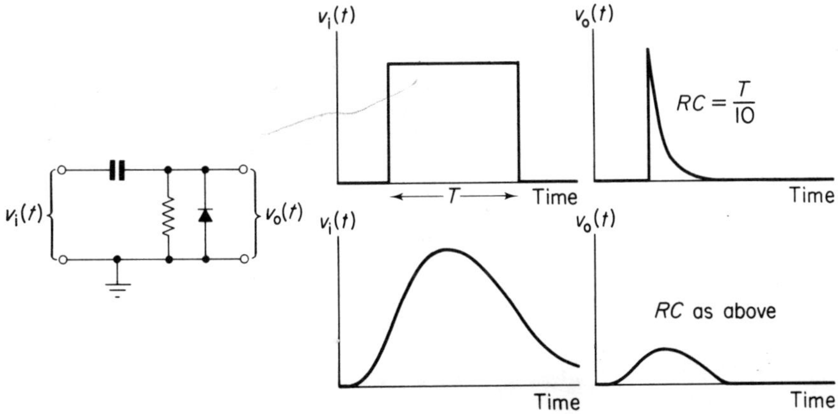

Figure 2.6 Use of CR circuit with a diode to give clipping of a pulse. The effect on an input pulse with a long rise time is shown in the lower diagram and illustrates the poorer clipping effect

negligible distortion, it will be followed by a long shallow undershoot of form $Ae^{-t/RC}$ having equal area. Since CR circuits cannot usually be avoided, when coupling amplifier stages and other circuits, the undershoot produced by a pulse will overlap with any subsequent pulses. Thus, for a regular train of rectangular pulses at the input as in Figure 2.7, since the capacitor can transmit no DC component, at equilibrium the mean output voltage must be zero. For the case of n rectangular pulses per unit time each of height E volts and duration T at the input, if RC has a large value the baseline of the pulses at the output is $-EnT$ volts. If, therefore, the height of the pulses is to be measured by means of a height sensing circuit, the measured height will be in error by the amount EnT and this is rate dependent.

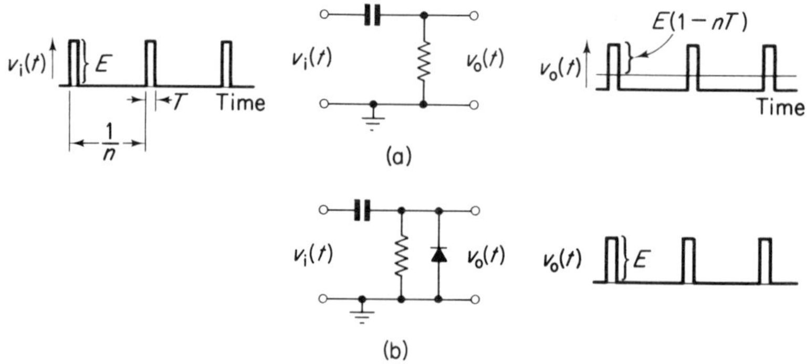

Figure 2.7 Effect of CR filter of large time constant on a regular train of rectangular pulses, showing (a) depression of baseline and (b) the baseline restoration effect of the diode

A simple way of avoiding this baseline shift is to place a diode across R as shown in Figure 2.7(b) which provides a current path for the rapid restoration of zero charge on the capacitor after each pulse. For the case of pulses from a nuclear detector, the non-rectangular shape, variable magnitude and random times of arrival make the analysis of the effect of a CR circuit more complicated and this is dealt with in Subsection 4.5.3. Although baseline restoration can still be effected by the use of the diode of Figure 2.7(b), the presence of any undershoot on the incoming pulses causes faulty operation and, for precision pulse height measurements, other arrangements, discussed in Section 3.3, are to be preferred.

2.1.3 Transmission of pulses in various systems

In an amplifier system the frequency response is normally limited at the upper end by capacities inherent in the devices used, which form low pass RC filters, and at the lower end by interstage coupling capacitors, which form high pass CR filters. It is usually possible to make these coupling capacitors large enough so that, apart from the addition of the slow undershoot and baseline shift discussed in the previous Subsection, negligible distortion of the pulse shape occurs. The high frequency limitation affects the rapidly changing parts of the signal. Figure 2.8(a) shows the frequency (Fourier) spectrum for a single rectangular pulse of duration T and for comparison Figure 2.8(b) shows the effect on the pulse shape of completely removing all the frequencies above $1/T$ and also above $5/T$. It should be noted however that the complete removal of such portions of the frequency spectrum cannot be accomplished by any physically realizable filter.

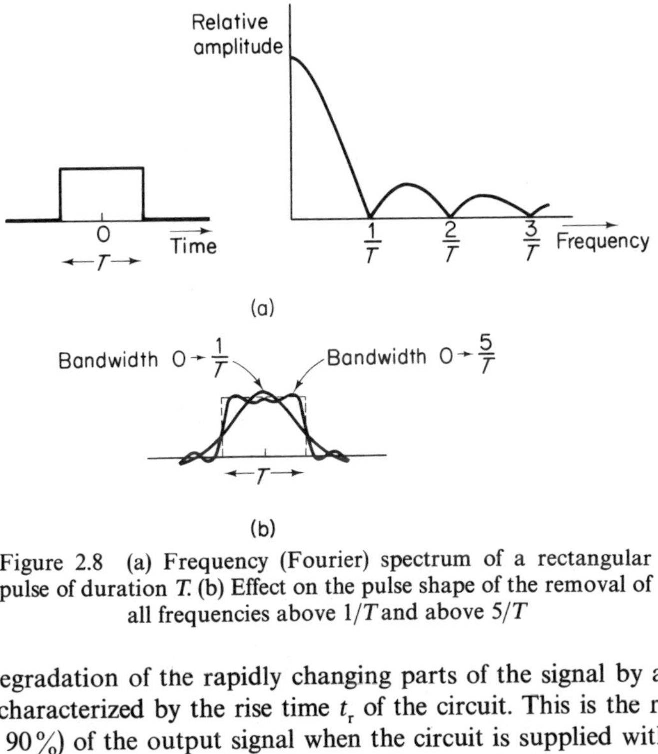

Figure 2.8 (a) Frequency (Fourier) spectrum of a rectangular pulse of duration T. (b) Effect on the pulse shape of the removal of all frequencies above $1/T$ and above $5/T$

The degradation of the rapidly changing parts of the signal by a circuit may be characterized by the rise time t_r of the circuit. This is the rise time (10% to 90%) of the output signal when the circuit is supplied with a step function input signal (Figure 2.9). A useful rule of thumb relates t_r to the upper 3 dB frequency f_h by

$$f_h t_r \simeq 0{\cdot}35 \tag{2.3}$$

a relation which was seen to be exact for the single RC filter. A second useful rule is for the rise time t_r of a series of cascaded stages each having rise times $t_1, t_2, \ldots t_n$ and states

$$t_r^2 = t_1^2 + t_2^2 \ldots + t_n^2 \tag{2.4}$$

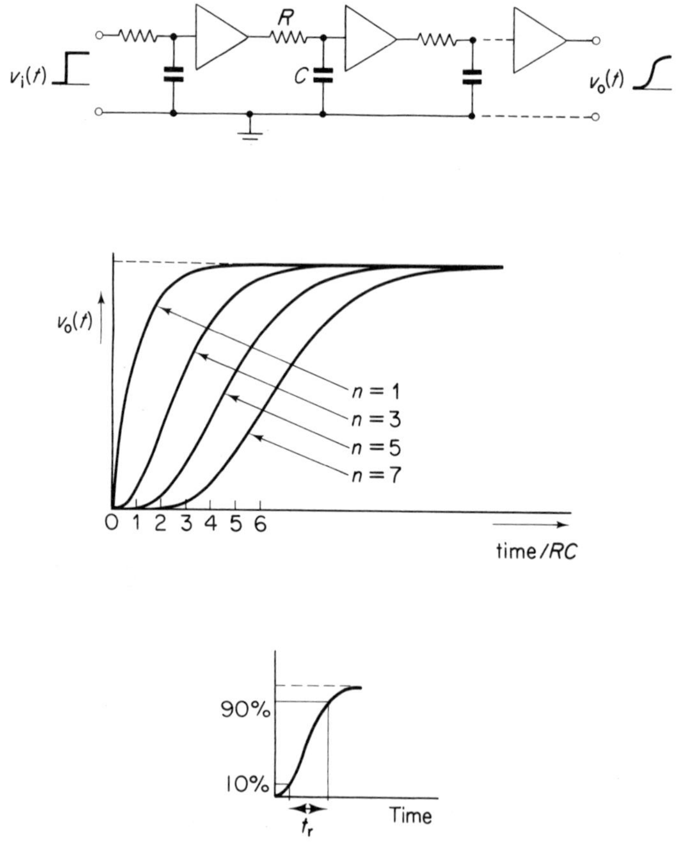

	Relative rise time, t_r			
	Number of stages n			
	1	3	5	7
Exact	1·0	1·9	2·5	3·0
$0{\cdot}35/f_h$	1·0	2·0	2·6	3·1
$\sqrt{n}$	1·0	1·7	2·2	2·6

Figure 2.9 Response of a cascade of n identical RC filters to an input step function from *Vacuum Tube Amplifiers* by G. E. Valley and H. Wallman[1]. Copyright 1948 by the McGraw–Hill Book Co. Inc. Reproduced by permission of the McGraw–Hill Book Company. The table shows the rise time of the output signal, computed by Millman and Taub[2], using different methods

The first rule is quite accurate for a surprising variety of circuits while the second rule is generally not too accurate for $n < 5$. By way of illustration Figure 2.9 gives the rise time for a cascade of n identical RC filters computed exactly and by the two rules of thumb and Figure 2.9 also shows the shape of the step response for these cases. These step responses show no overshoot and this is a characteristic of a cascade of any number of low pass RC stages, even if they are not identical.

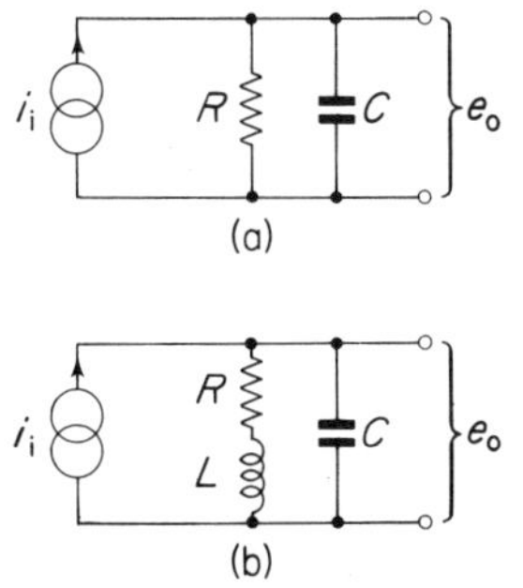

Figure 2.10 (a) Equivalent circuit for a transistor collector and the input of the following stage of an amplifier. (b) As above but with the inclusion of a compensating inductor

It is sometimes possible to extend the high frequency response, and hence shorten the rise time, of a stage by adding an extra component. This technique, known as peaking or compensation, may be applied in numerous ways and, since most amplifier stages have some negative feedback applied, the peaking component is often incorporated in the feedback loop. As a simple example we will take the circuit of Figure 2.10(a), a constant current generator, which might represent the output from a transistor collector, feeding into a parallel RC combination which might represent the load presented by the following stage. The complex voltage at the output is given by $e_o/i_i = R/j\omega RC + 1$ which, apart from a constant, is identical to that for the voltage transmittance of the low pass RC filter. The 3 dB frequency is $f_h = 1/2\pi RC$ and since C is fixed by the device characteristics the only way of increasing f_h in this circuit is by decreasing R i.e. at the expense of gain. If, however, an inductance is connected in series with the resistance as shown in Figure 2.10(b), then it is possible to extend the frequency response without loss of gain. Figure 2.11 shows the gain $|e_o/i_i|$ as a function of frequency for various values of the Q value of the circuit where $Q = (1/R)\sqrt{(L/C)}$. The corresponding step responses are shown in Figure 2.11 and it will be seen that increasing the Q of the circuit (by increasing L) increases f_h and reduces the rise time, but at the expense of some overshoot at the higher Q values. It may be noted that the value of Q giving the maximally flat gain/frequency response is $Q \simeq 0{\cdot}65$ and this has an overshoot of 3·8 % in the step response. For no overshoot Q must be no greater than 0·50 giving a rise time of 0·70 times the value without compensation (i.e. with $Q = 0$) (see Millman and Taub[2]).

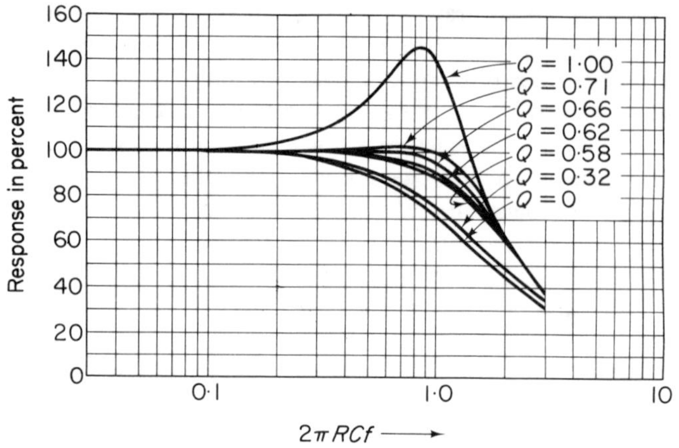

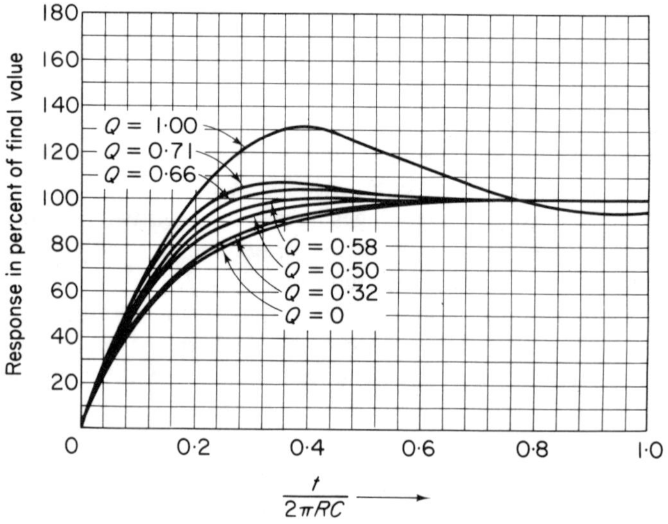

Figure 2.11 Effect of various degrees of compensation on the frequency response and step response. (Adapted from Bedford and Fredendall[3])

2.2 Transient response of linear circuits

2.2.1 Transform methods

The Laplace and Fourier transform methods provide systematic methods for finding the response of linear circuits to transient input signals. Only the briefest outline of the transform methods is given here and the reader is referred to Appendix A for a fuller description.

If we assume the input signal (voltage or current) is $f_1(t)$ and we require to find the output $f_2(t)$ of the circuit, the essence of the method is first to take the transform of $f_1(t)$ and then form the product of this and the 'transfer function' of the circuit. The inverse transform of this product then gives the output signal $f_2(t)$. The taking of the transform and of the inverse transform can often be done by recourse to standard tables. The transfer function of the circuit can then usually be written down so the transform method can often by used to give the output $f_2(t)$ very easily.

Considering first the Fourier transform method, if $\mathscr{F}_1(j\omega)$ is the Fourier transform of the input signal $f_1(t)$ then, by definition:

$$\mathscr{F}_1(j\omega) = \int_{-\infty}^{\infty} f_1(t)\,e^{-j\omega t}\,dt \tag{2.5}$$

and the inverse transform of $\mathscr{F}_1(j\omega)$ gives:

$$f_1(t) = \frac{1}{2\pi}\int_{-\infty}^{\infty} \mathscr{F}_1(j\omega)\,e^{j\omega t}\,d\omega = \text{Real part of}\,\frac{1}{\pi}\int_{0}^{\infty} \mathscr{F}_1(j\omega)\,e^{j\omega t}\,d\omega \tag{2.6}$$

Equation (2.6) expresses the fact that $f_1(t)$ can be represented as a continuous sum of sinusoidal terms with all values of ω with a weighting function $\mathscr{F}_1(j\omega)$. $\mathscr{F}_1(j\omega)$ therefore represents a spectral density function of the frequency components and as an example the values of $\mathscr{F}_1(j\omega)$ for a triangular shaped pulse are given in Figure 2.12. The spectral density of the components will be modified by the filtering action of the circuit and if $H(j\omega)$ is the complex transfer function of the circuit (as used in ordinary complex AC theory) the frequencies at the output of the circuit will have a spectral density $H(j\omega)\mathscr{F}_1(j\omega)$. These frequencies may now be recombined to give the output signal $f_2(t)$ by taking the inverse transform of $H(j\omega)\mathscr{F}_1(j\omega)$:

$$f_2(t) = \text{Real part of}\,\frac{1}{\pi}\int_{0}^{\infty} H(j\omega)\,\mathscr{F}_1(j\omega)\,e^{j\omega t}\,d\omega$$

The Laplace transform method is similar but the Laplace transform of $f(t)$ is defined as:

$$\mathscr{L}(s) = \int_{-\infty}^{\infty} f(t)\,e^{-st}\,dt, \qquad s = \sigma + j\omega \tag{2.7}$$

and the inverse transform as

$$f(t) = \frac{1}{2\pi j}\int_{\sigma - j\infty}^{\sigma + j\infty} \mathscr{L}(s)\,e^{st}\,ds \tag{2.8}$$

Thus the Laplace transform expresses the signal $f(t)$ as a continuous sum of complex exponentials with the exponent st where $s = \sigma + j\omega$. Unlike the Fourier representation s can range over the complex plane and the inverse transform (2.8) represents a line integral in the complex plane. The Fourier transform is thus a special case of the Laplace transform with $\sigma = 0$ and it may be noted that it is not possible to give a physical analogue to the taking of a Laplace transform nor strictly to plot 'spectral densities' since s ranges over the complex plane. The Laplace transform is usually restricted to time functions which are zero for $t < 0$; this covers most cases of interest and will always be assumed to be the case for the present applications. The Laplace method can be used for functions extending over all time but an ambiguity arising in taking inverse transforms must be first settled before the method yields the answer.

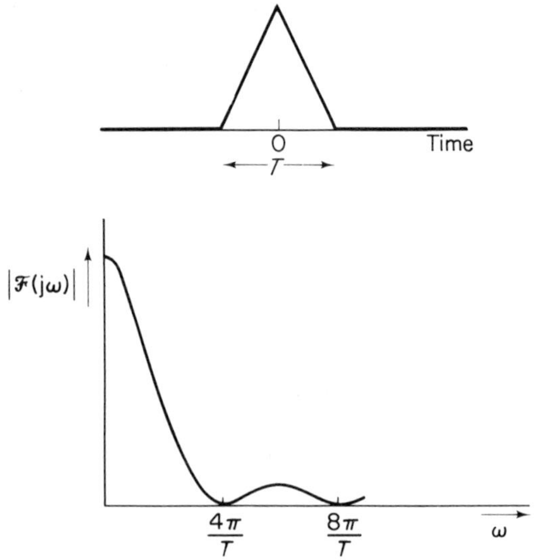

Figure 2.12 Triangular pulse and its frequency (Fourier) spectrum

It should be remarked right away that there is very little difference in practical terms between the use of the Fourier and Laplace methods and usually only entails the substitution of the symbol s for $j\omega$ or vice versa. Since Laplace transforms exist for a wider variety of functions than Fourier transforms it is normally preferable to use the Laplace formulation. For instance functions which increase indefinitely with time do not possess Fourier transforms and so such functions, and hence any unstable system, must be treated with Laplace methods.

The Laplace and Fourier transforms of various pulses is given in Figure 2.13. The transfer function of the circuit $H(j\omega)$, as mentioned earlier, is that used in ordinary complex AC theory, i.e. that obtained by replacing resistances R, capacitances C, and inductances L by impedances of R, $1/j\omega C$ and

$f(t)$	Fourier transform $\mathcal{F}(j\omega)$	Laplace transform $\mathcal{L}(s)$
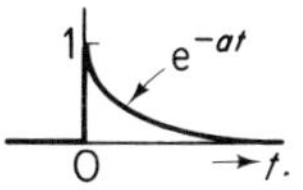	$\dfrac{1}{j\omega + a}$	$\dfrac{1}{s+a}$
Unit step 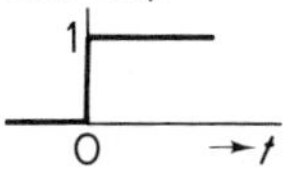	$\dfrac{1}{j\omega}$ *	$\dfrac{1}{s}$
Unit impulse 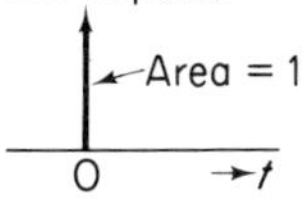	1	1
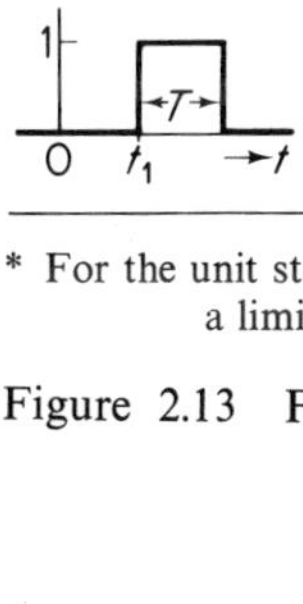	$\dfrac{(1-e^{-j\omega T})e^{-j\omega t_1}}{j\omega}$	$\dfrac{(1-e^{-Ts})e^{-t_1 s}}{s}$

* For the unit step the Fourier transform exists only as a limiting case, see Appendix A.3

Figure 2.13 Fourier and Laplace transforms of some pulses

$j\omega L$ respectively. If the Laplace transform is being used $H(s)$ is required; in other words impedances R, $1/sC$ and sL are used for resistances, capacitances and inductances. The transfer functions for some simple circuits are given in Figure 2.14.

The taking of the inverse transform can often be done by using standard tables. In particular it is often necessary to find the inverse transform of functions of the form $N(s)/D(s)$ where $N(s)$ and $D(s)$ are polynomials in s, and the degree of s in $N(s)$ is less than that in $D(s)$. If the denominator is factored we have:

$$\frac{N(s)}{D(s)} = \frac{N(s)}{(s - v_1)(s - v_2)\ldots(s - v_m)}$$

	Transfer function for $\frac{e_o}{e_i}$ or $\frac{e_o}{i_i}$	
	$H(j\omega)$	$H(s)$
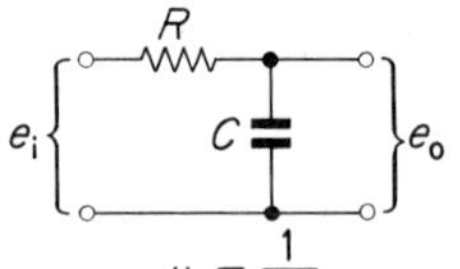 $u_a = \frac{1}{RC}$	$\frac{u_a}{j\omega + u_a}$	$\frac{u_a}{s + u_a}$
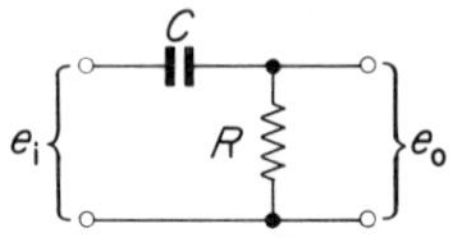 $u_b = \frac{1}{RC}$	$\frac{j\omega}{j\omega + u_b}$	$\frac{s}{s + u_b}$
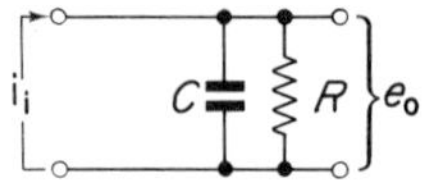 $u_c = \frac{1}{RC}$	$\frac{1}{C(j\omega + u_c)}$	$\frac{1}{C(s + u_c)}$

Figure 2.14 Transfer functions for some circuits

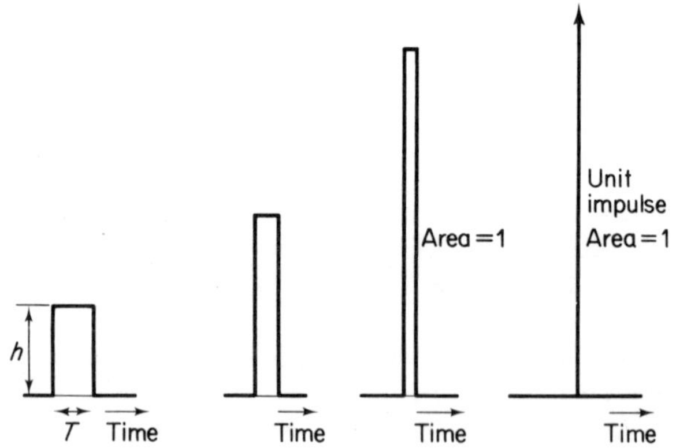

Figure 2.15 A unit impulse viewed as the limiting case of a rectangular pulse with $h \to \infty$ and $T \to 0$ so that $hT = 1$

and $\nu_1, \nu_2, \ldots, \nu_m$ are termed the poles of the function. Provided the poles are all different the inverse transform is given by:

$$f(t) = 0 \qquad \text{for } t < 0$$

$$f(t) = \sum_{i=1}^{m} \left[\frac{N(s)(s - \nu_i)\,\mathrm{e}^{st}}{D(s)} \right]_{s=\nu_i} \qquad \text{for } t > 0 \tag{2.9}$$

The inverse transform is therefore a sum of exponential terms, one for each pole, and the coefficient of $\mathrm{e}^{\nu_i t}$ is found by omitting the term $(s - \nu_i)$ in the denominator of $N(s)/D(s)$ and putting $s = \nu_i$. If the denominator contains any repeated terms such as $(s - \nu_k)^2$ then (2.9) still holds good for the coefficients of $\mathrm{e}^{\nu_i t}$ where $\nu_i \neq \nu_k$. The term containing $\mathrm{e}^{\nu_k t}$ however is now obtained by the following expression:

$$\left[\frac{N(s)(s - \nu_k)^2}{D(s)} \right]_{s=\nu_k} t\,\mathrm{e}^{\nu_k t} + \left[\frac{\mathrm{d}}{\mathrm{d}s} \left\{ \frac{N(s)(s - \nu_k)^2}{D(s)} \right\} \right]_{s=\nu_k} \mathrm{e}^{\nu_k t} \tag{2.10}$$

2.2.2 *The impulse response and the time convolution method*

For an arbitrary input function the pulse can only rarely be approximated by a pulse whose transform is known. Although it is possible to approximate an arbitrarily shaped pulse by a series of step functions and find the transform of this, for the present purposes a much more valuable procedure is to use the time convolution method[4]. With this method it is often possible to graphically determine the form of the output signal for many input pulse shapes and to see how the input pulse shape affects the output pulse shape. The time convolution method requires a knowledge of the impulse response of the circuit.

The idea of an impulse signal is analogous to that of an impulsive force in mechanics and consists of an infinitely large signal (current or voltage) which lasts for an infinitely short time. If we consider the rectangular pulse (current or voltage) of Figure 2.15 of height h and duration T such that $hT = 1$ the unit impulse may be imagined as the limiting case of $h \to \infty$ and $T \to 0$, while preserving the product hT equal to unity. It should be understood that the definition of a unit impulse is not dependent on the assumption of a rectangular shape before applying the limiting procedure. The Fourier or Laplace transform of a unit impulse is unity and so has a uniform spectral density over all frequencies. If a unit impulse is applied to the input of a circuit having a transfer function $H(s)$, since $\mathscr{L}_1(s)$ in this case is unity, it follows that the output, the so-called impulse response, is simply the inverse transform of $H(s)$.

Once the impulse response of the circuit is known, it is possible to apply the convolution method for the response to any other type of signal. If $I(t)$ is the impulse response of the circuit and an input $f_1(t)$ is applied, then the output $f_2(t)$ is given by

$$f_2(t) = \int_{-\infty}^{\infty} f_1(\xi)\, I(t - \xi)\, d\xi \tag{2.11a}$$

$$= \int_{-\infty}^{\infty} f_1(t - \xi)\, I(\xi)\, d\xi \tag{2.11b}$$

These integrals may often be interpreted graphically. If we take the case of the circuit having the impulse response shown in Figure 2.16 and we require to find the output when a rectangular input pulse is applied, suppose we are interested in the output $f_2(t_i)$ at some instant t_i. Figure 2.16 shows the

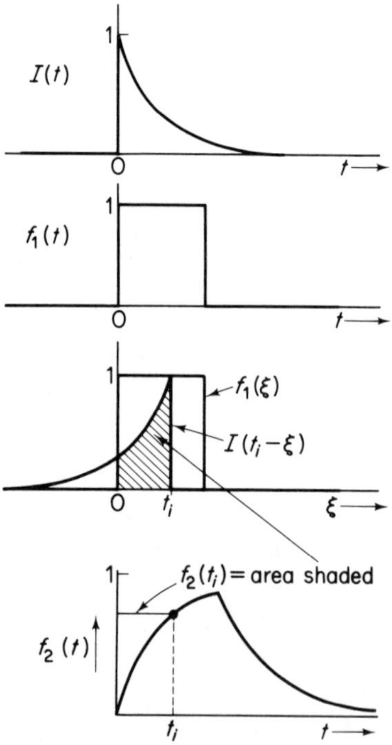

Figure 2.16 Time convolution method for finding output signal $f_2(t)$ for an input $f_1(t)$ applied to a circuit having an impulse response $I(t)$

functions $f_1(\xi)$ and $I(t_i - \xi)$ plotted against ξ for the chosen value of t_i, and of course the latter function is just the mirror image of the impulse response about the $t = 0$ axis and displaced by t_i, the time instant of interest. The integral, (2.11a), can now be evaluated numerically by taking products of the ordinates of the two functions in Figure 2.16 and for the present case of a rectangular pulse equals the shaded area in Figure 2.16. This procedure has yielded the value of $f_2(t)$ for a single value of t(i.e. t_i). The value of $f_2(t)$ for other values of t may now be evaluated by the same procedure i.e. displacing the two curves in Figure 2.16 by the appropriate new value of t_i.

For non-rectangular pulse inputs the value of the integral (2.11a) is no longer given by the area of overlap but must be computed as the sum of the products of the ordinates of the two curves. However several simplifications are possible. In particular if the input $f_1(t)$ is short compared to the impulse response $I(t)$ it should be obvious that the output pulse shape will differ little from the impulse response. Another observation of importance concerns the height of the output pulse being insensitive to the shape of the input pulse. This will be assured if the impulse response has a flat top at least as long as the total length of the input pulse.

2.3 Digital circuits

2.3.1 Digital logic

Pulses from a nuclear detector carry information in their existence, in their time position and in their pulse amplitude and shape, and are sometimes termed analogue or linear pulses. Since few satisfactory means exist of recording the values of such parameters by analogue means (the use of photographic film being one such example) it is usual to inspect the pulses according to whether their parameters of interest lie in various defined ranges. Thus a pulse height analyser accepts analogue input pulses and only delivers an output pulse if the input pulse height falls within prescribed limits, in which case the amplitude information is of no further interest and the output pulse is arranged to be of standardized height and duration. Such an output pulse, indicating the presence of a detector pulse satisfying the chosen criteria, is termed a logic pulse since its standardized form permits its use in performing logical operations. Such operations may involve the use of gating arrangements to test for time coincidence with logic pulses from another source. The final output pulses may then be counted up and the number stored in a scaler or register. Since all electronic counting arrangements use storage elements that can exist in two stable states the operations of digital logic are of relevance here also.

In addition to logic signals derived from detector pulses many nuclear electronic instruments utilize internally generated logic signals. Such is the case where internal processes have to be programmed so as to be executed

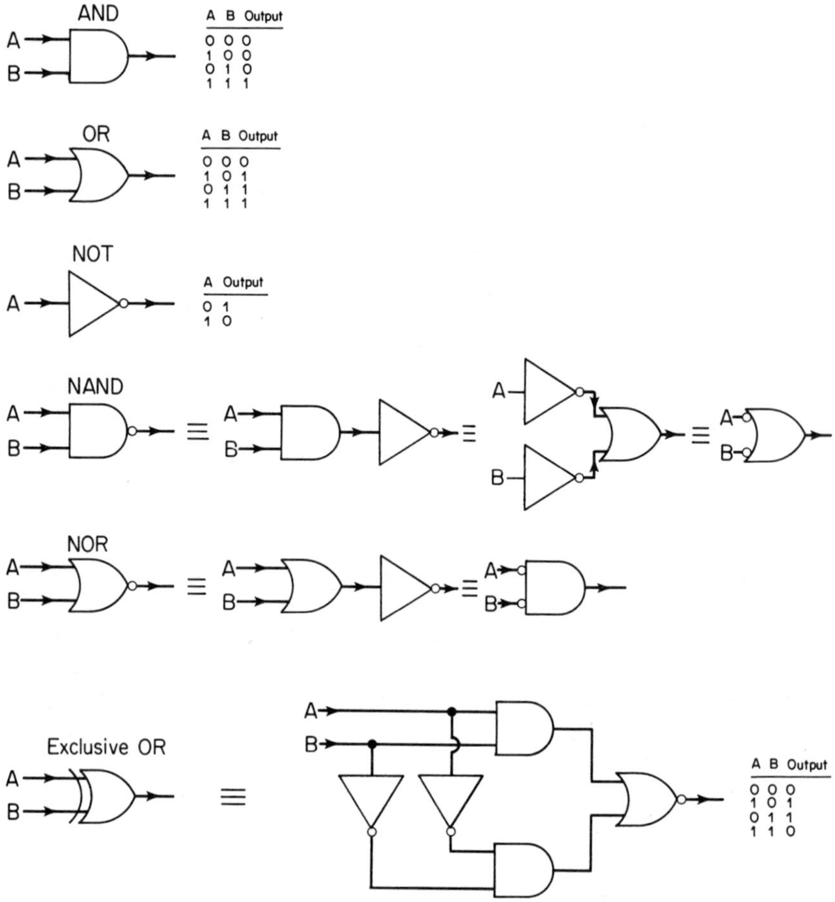

Figure 2.17 Some basic functions of digital logic, their symbols and their truth tables

in, for example, sequential order or, alternatively, at time instants conditional on the progress of other processes.

The two possible values of a binary digit are named as '0' and '1'. The basic logical functions are illustrated in Figure 2.17. In the electrical realization of a binary digit the two possible values take the form of two different voltage or current levels, logical '0' usually realized as a voltage near to zero and logical '1' as a level of several volts positive. The arrival of the logic pulse output from, say, a pulse height discriminator thus changes the output line from its quiescent state of '0', to a value of '1' for a period of the length of the pulse, and it then returns to '0'. It is of course important that any following circuit recognizes the two states '0' and '1' correctly and hence it is necessary to specify the permissible ranges of voltage or current corresponding to each of the two states.

In nuclear instrumentation it is often necessary to interconnect electronic units of different manufacture by means of flexible coaxial cable. In order to ensure compatibility of such units certain standards have been internationally agreed which specify the permissible ranges of the current or voltage representing the '0' and '1' states and these are given in Section 6.1. At the level of individual circuit elements within an electronic unit a similar problem of compatibility of logic signals exists. Most currently available integrated circuit logic elements, with the exception of ECL circuits, utilize '5 V positive logic'; a typical example is illustrated in Figure 2.18, although the values will depend on the circuit design and type. Figure 2.18 shows the range of voltages corresponding to '0' and '1' states to which a circuit must respond and also shows the permissible ranges which a circuit must produce at its output. It will be observed that compatibility is assured as the output voltage ranges lie within the input voltage ranges.

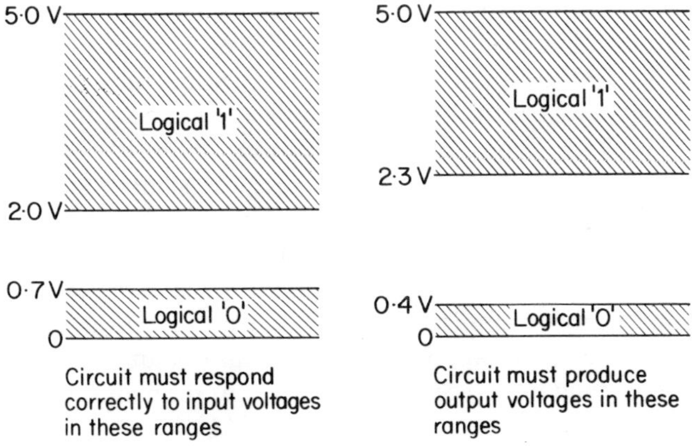

Figure 2.18 Typical example of permissible ranges for input and output voltages corresponding to logical '0' and '1'

2.3.2 *Logic gates*

A large number of arrangements have been used to perform the AND, NAND, OR and NOR functions and it is only possible in this section to mention a brief selection. Many of the recent varieties have been evolved to meet the requirements of economical integrated circuit production and, as the pace of development has been so rapid, the reader is referred to one of the many recent texts on the subject for a fuller description.

Probably the simplest type of gate utilizes diodes. Figure 2.19(a) shows a diode OR gate and it will be seen that logical '1' (a positive potential E of several volts) present at one or more of the inputs causes the respective diode(s) to conduct and a voltage nearly equal to E appears at the output.

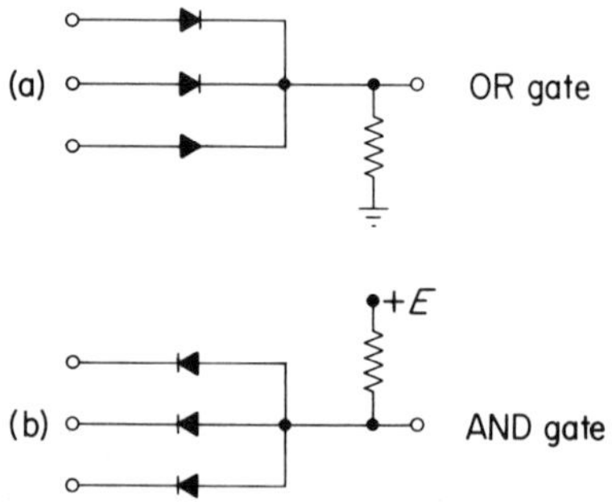

Figure 2.19 Use of diode gates to perform the AND and OR functions

Figure 2.19(b) shows diodes used to perform the AND function. If any input is held at '0' (i.e. near-zero volts) its diode conducts and the output is held at near-zero volts. Only when all the inputs are at '1' do all the diodes cease to conduct and the output goes to '1'. The disadvantage of these simple diode gates is that the output voltage levels corresponding to the '0' and '1' states differ slightly from the '0' and '1' levels at the inputs and this is due to the forward voltage drops across the diodes (0·2–0·4 V for Ge diodes; 0·5–0·8 V for Si diodes). In addition the source of the input signals will have a finite output impedance and there will be a slight attenuation due to ohmic drops. Hence if diode gates are cascaded the progressive shifts in voltage levels will eventually cause malfunction. Diode gates, being extremely rapid, therefore find their main application where only one or two gates are required to be cascaded. For other applications some form of amplification between gates is necessary and until recently this has meant the inclusion of a junction (i.e. bipolar) transistor.

DTL (diode transistor logic) gates are one solution to this. Figure 2.20 shows a typical NAND gate available as an integrated circuit and the analogy

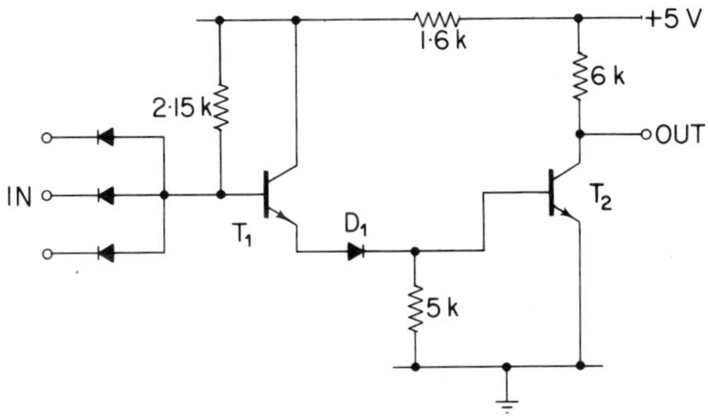

Figure 2.20 DTL NAND gate type MC962 (Motorola Semiconductor Products Inc.)

with the diode gate is obvious. The output can be used to drive a number of other gates, the maximum permissible number being usually termed the 'fan out'. T_1 acts as an emitter follower and T_2 as an amplifying stage. The function of D_1 is to act as a voltage shifting element. DTL gates typically have transmission delay times of 20–30 ns.

TTL (transistor transistor logic) gates give shorter transmission times (~ 10 ns) and Figure 2.21 shows a typical integrated circuit NAND gate. If any input is at '0' the current in R_1 is conducted via the base-emitter diode and negligible collector current flows. Only when all the inputs are at '1' does significant collector current in TR_1 flow, so causing TR_2 to conduct and hence causing the output voltage to fall to '0'. The 'totem pole' output stage may be seen as a push–pull arrangement and serves to maintain a low output impedance in either output state. D_1 is a voltage shifting diode.

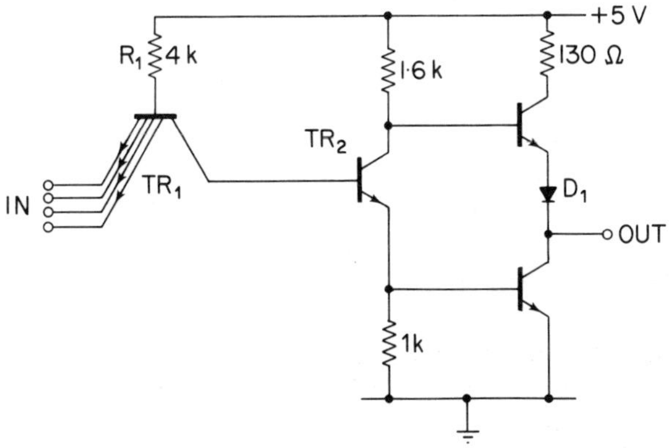

Figure 2.21 TTL NAND gate type FJH 111 (Mullard Ltd)

The speed of switching of a transistor depends on, among other things, the quiescent state before switching. For a common emitter stage using a collector resistance, if the base-emitter voltage is increased slowly the collector current increases to a point and then saturates. At this point V_{ce} has become reduced so as to become comparable with V_{be} so that the base-collector junction becomes forward biased. Further increase in V_{be} causes the extra emitter current to flow predominantly into the base lead. If it is attempted to rapidly switch the transistor from the saturated state, the collector current will switch quite slowly due to the necessity of removing the large number of charge carriers in the base region. It follows that for fast switching it is

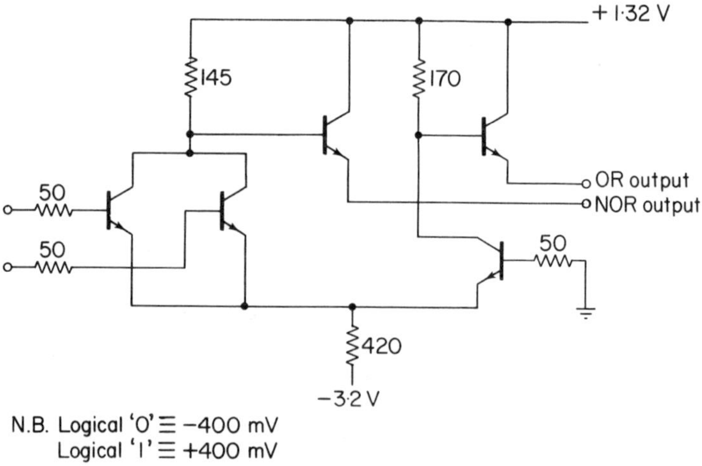

Figure 2.22 ECL gate for OR or NOR function, type ECL2502 (Texas Instrument Inc.)

necessary to prevent the transistor from entering the saturated region. For integrated circuits one current technique is the use of Schottky clamping. In this case a Schottky barrier diode is formed so as to be effectively placed in parallel with the base-collector junction of the transistor. The Schottky diode has a lower forward voltage drop than the base-collector junction so most of the excess base current in the fully conducting transistor is shunted through the Schottky diode. Since minority carriers are not stored in a Shottky diode the switching to the non-conducting state can take place rapidly and Schottky clamped TTL gates give propagation delays of ~ 3 ns.

ECL (emitter coupled logic) is another current approach which avoids the saturation region of the transistors and gives a propagation delay of 2–3 ns. Figure 2.22 shows a gate performing the OR or NOR function according to which output is taken. For this example logical '0' corresponds to − 400 mV and logical '1' to + 400 mV.

2.3.3 *Regenerative trigger circuits*

The elements of regenerative circuits may best be elucidated by a consideration of the Schmitt trigger, Figure 2.23. The amplifier of gain A has a fraction β of the output voltage added to the input so as to constitute positive feedback.

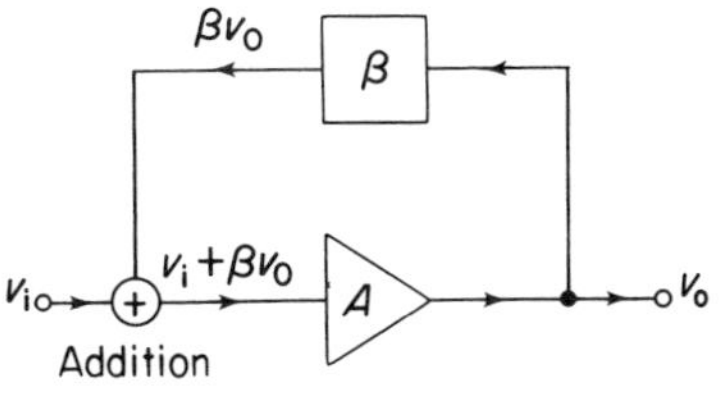

Figure 2.23 Principle of a Schmitt trigger circuit

The overall gain with feedback may easily be seen to be $A/(1 - A\beta)$ for the linear region. The input–output characteristic is plotted in Figure 2.24(a) and assumes the amplifier limits when the output voltage v_o reaches $\pm V_{max}$. For $A\beta > 1$ part of the characteristic has a negative slope and corresponds to an unstable region (see below) so the arrangement has only two stable states, corresponding to $v_o = \pm V_{max}$. If the circuit is in the state $v_o = -V_{max}$ and v_i is gradually increased the output will remain in the same state until v_i exceeds $((A\beta - 1)/A)V_{max} = V_{trig}$ when the output voltage will switch to $+V_{max}$ as shown by the dotted line XY in Figure 2.24(b). In order to switch the circuit back to its original state, by similar reasoning, it is necessary to reduce the input voltage to below $-((A\beta - 1)/A)V_{max} = -V_{trig}$, so the circuit exhibits a hysteresis with respect to input voltage and the magnitude of this hysteresis is $2((A\beta - 1)/A)V_{max}$.

A closer analysis of the behaviour of the circuit needs to take into account the parasitic and other capacitances in the circuit. If we assume these can be lumped together into a single RC filter of time constant τ as in Figure 2.25(a), and assuming the feedback network takes negligible current, we have for the linear region

$$\frac{dv_o}{dt} = \frac{v_o' - v_o}{\tau} \text{ and } v_o' = A(v_i + \beta v_o)$$

$$\therefore \tau \frac{dv_o}{dt} - (A\beta - 1)v_o = Av_i$$

That the circuit is unstable in the linear region may now be simply shown. Assume the initial condition on the output is $v_o = E$ then for the case of $v_i = 0$ the solution of the differential equation gives an output voltage of the form $v_o = E\, e^{(A\beta - 1)t/\tau}$. This solution is plotted in Figure 2.25(b) and it will be seen that for the case $A\beta > 1$ the exponent is positive and the output increases in principle without limit but in practice it will be limited when v_o reaches $\pm V_{max}$.

To consider now the case of switching from the state $v_o = -V_{max}$ to $v_o = +V_{max}$ we have as initial condition $v_o = -V_{max}$ and assume a steady

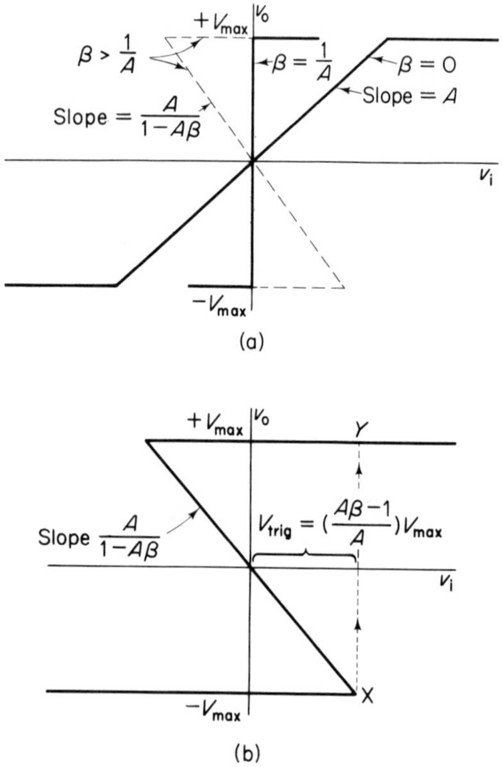

Figure 2.24 Output–input voltage characteristics for a Schmitt trigger circuit (a) with different amounts of feedback, (b) for the case $\beta > 1/A$

voltage $v_i = V_i$ is applied at the input. The solution of the differential equation is now

$$v_o = V_{max}\left[\left(\frac{V_i}{V_{trig}} - 1\right) e^{(A\beta - 1)t/\tau} - \frac{V_i}{V_{trig}}\right]$$

where, as before, V_{trig} is defined as $((A\beta - 1)/A)V_{max}$. Only if $V_i > V_{trig}$ does the circuit switch and representative time courses of the output voltage are shown in Figure 2.26.

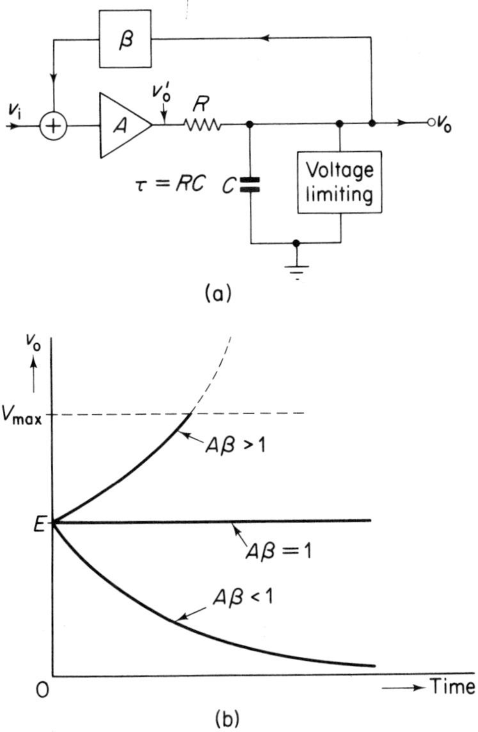

Figure 2.25 (a) Schmitt trigger circuit with the inclusion of a capacitive element prior to the limiting section. (b) Schmitt trigger circuit: time course of output voltage for the initial condition $v_0 = E$ and for various values of $A\beta$

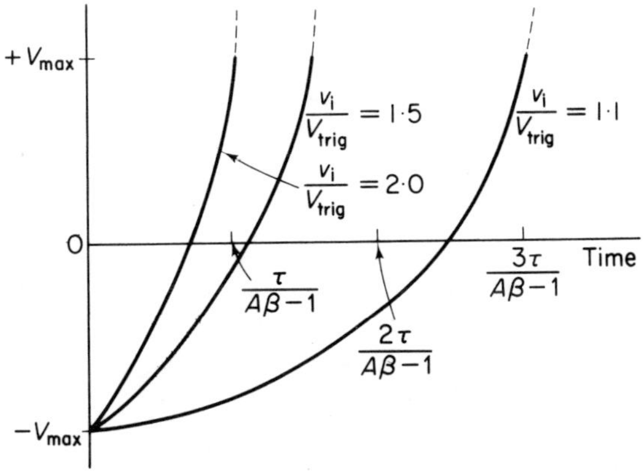

Figure 2.26 Time course of output voltage of Schmitt trigger circuit for various triggering voltages

It will be seen that the speed of switching is increased when $\tau/(A\beta - 1)$ is small and when the input voltage exceeds V_{trig} by a large factor. In conclusion to the above simple treatment it should be noted that the lumping of the parasitic and other capacitances together into a single capacitance before the limiting section of the circuit is a simplification and a fuller treatment should include capacitive components after the limiting action.

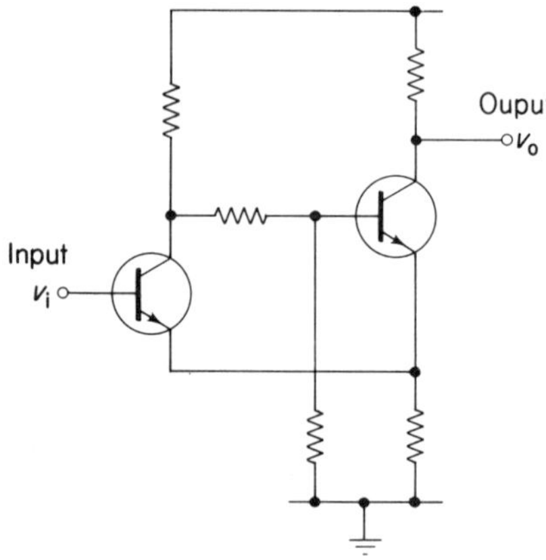

Figure 2.27 Example of a Schmitt trigger circuit

A practical realization of a Schmitt trigger is shown in Figure 2.27. Since both transistors will limit at certain voltages it is necessary to ensure that their biasing is such that their linear regions are compatible. The full design procedure for this type of circuit is given in many texts and manufacturer's publications, for example that of Mullard[5]. For maximum speed of switching it is necessary to prevent the transistors entering their saturated regions and this may be accomplished by the use of clamping or catcher diodes between each collector and an appropriate voltage level.

2.3.4 *Bistable (flip-flop) circuits*

The Schmitt trigger may be seen as a special form of bistable in that the input voltage is summed with the feedback voltage and so can cause a change of stage (toggling). For operation by digital signals other types of bistable are more versatile. Figure 2.28 shows a simplified circuit of an example of a RS (reset–set) bistable. If both the set and reset inputs are at '0' (i.e. near-zero volts) the circuit can exist in two stable states, the first state with T_1 conducting so $\overline{Q} = 0$ and T_2 non-conducting so $Q = 1$ (i.e. several volts positive), and the second state with T_1 non-conducting ($\overline{Q} = 1$) and T_2 conducting ($Q = 0$). If the set input is now taken to '1' (i.e. several volts positive), the

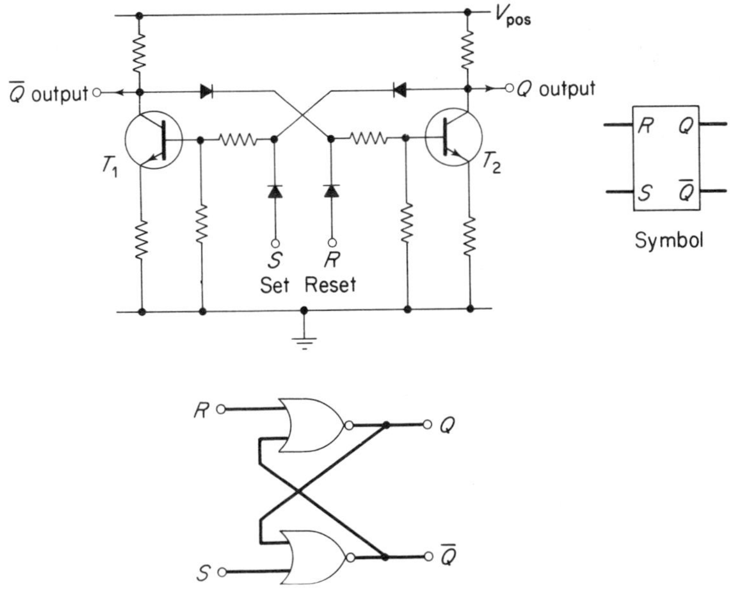

Figure 2.28 An example of a RS bistable, its symbol and its synthesis from NOR gates

reset input still being at '0', then, whatever the prior state, the circuit is now constrained into the state $\overline{Q} = 0$, $Q = 1$. Furthermore the circuit remains in this state if the set input is returned to '0'. In a similar manner the application of a '1' input on the reset input (the set input held at '0') causes the circuit to take up the state $\overline{Q} = 1$, $Q = 0$ and to remain so after the reset input is returned to '0'. These changes are summarized in the truth table of Figure 2.29.

It may be noted that this RS bistable may be fabricated out of logic gates, as in the example of Figure 2.28.

Very often it is useful to have a bistable in which the input conditions can be set up (the bistable 'enabled') beforehand, but the actual instant of

Input conditions		Output states during application of inputs				Output states after returning inputs to $R = 0, S = 0$	
S	R		Q		$\overline{Q}$	Q	$\overline{Q}$
0	0	Either	0	&	1	0	1
		or	1	&	0	1	0
1	0		1		0	1	0
0	1		0		1	0	1
1	1		0		0	Ambiguous	

Figure 2.29 Truth table for the RS bistable of Figure 2.28

toggling controlled by the application of another input, a 'clock' input pulse. To take the case of a clocked RS bistable this could be achieved by the addition of the AND gates shown in Figure 2.30. Change of state can now only occur when the clock pulse arrives, but whether toggling does or does not occur is still conditional on the states of the *R* and *S* inputs during the clock pulse according to the truth table already given. This clocking facility is of great use. For example if many bistables are clocked at a suitable instant from the same clock pulse, any changes in their outputs will all occur at the same instant even though their individual *R* and *S* inputs may not have been applied simultaneously. This therefore offers a means of synchronizing signals which would otherwise have become displaced in time, arising, for instance, from differing propagation delays in previous circuit elements. Although not shown in Figure 2.30 some RS bistables have, in addition, direct set and reset inputs which operate independently of the presence of a clock pulse.

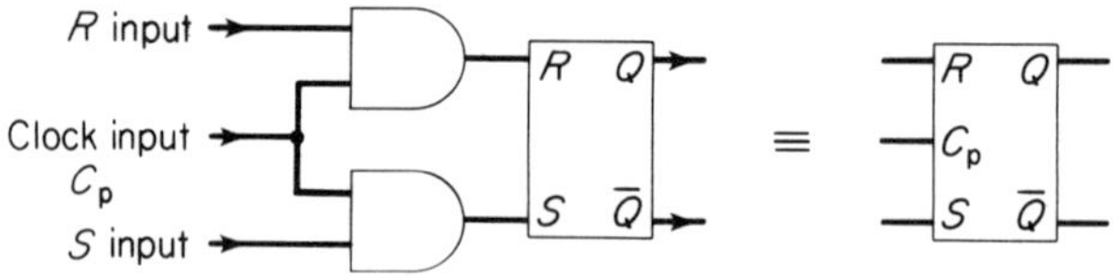

Figure 2.30 Addition of AND gates to give the clocked RS bistable action and the symbol for a clocked RS bistable

Another bistable that is often encountered is the D bistable. In this case, apart from the clock pulse input, there is only one input. Regardless of the previous state of the bistable, on receipt of a clock pulse the *Q* output takes on the same state as the *D* input as shown in the truth table Figure 2.31. Provided the *D* input is constant immediately prior to and during the clock pulse it follows that its value is stored and is available at the *Q* output at least until the next clock pulse. This type of circuit is sometimes known as a 'staticizer' and could be synthesized from a clocked RS bistable by the arrangement of Figure 2.31.

A function that is often required is that performed by a T (triggered) bistable, or scale-of-two bistable, which toggles with each triggering pulse arriving at its input. By cascading such bistables so that the output of any one acts as input for the next it is possible to realize a simple counter for the pulses supplied to the first, the total number of which appears at the bistable outputs in parallel binary code. One way of realizing a T bistable is to use a clocked RS bistable connected as shown in Figure 2.32. The operation of the circuit exploits the inherent delay in the appearance of the new states of the output signals Q and $\bar{Q}$ after the application of the clock or trigger pulse,

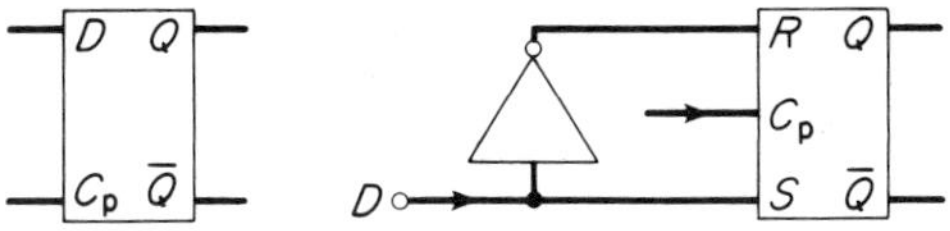

Input	Output after receipt of clock pulse	
D	Q	$\bar{Q}$
0	0	1
1	1	0

Figure 2.31 Truth table for a D bistable, its symbol and a method of synthesizing its operation from a clocked RS bistable

and it is essential that the trigger input is of brief duration compared to this delay. Assuming the bistable is initially in the state $Q = 0, \bar{Q} = 1$ the R and S inputs are enabled to cause toggling on the arrival of the trigger pulse. In the time it takes for the change of state to be completed the triggering input has returned to '0' and the circuit remains in the second state $Q = 1, \bar{Q} = 0$. The bistable is now enabled for a change to the original state which will occur on the arrival of the next triggering pulse. The triggering pulse must be of short duration, otherwise the bistable may toggle several times, and it is usual to pass it first into a CR differentiating circuit. This type of circuit is widely used where discrete components are employed. However another approach is possible which avoids the necessity of a brief triggering pulse and makes use of the JK bistable.

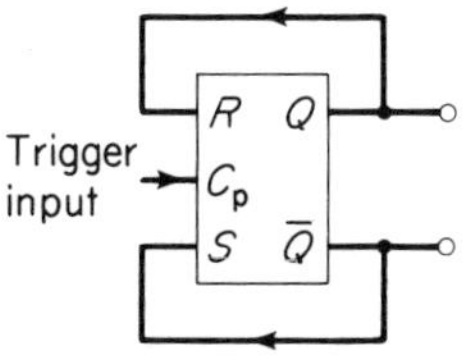

Figure 2.32 Realization of a triggered (T) bistable action by the use of a clocked RS bistable. The operation exploits the inherent delay in the appearance of the new output states and it is necessary for the trigger pulse to be of short duration

The JK bistable is a very versatile element and the type that will be considered here operates on the master–slave principle. The outputs of the bistable change at an instant deriving from the trailing edge of the clock pulse and the truth table, Figure 2.33, shows the state of the outputs after the clock pulse. Subscript n indicates the state prior to the clock pulse and $n + 1$ indicates that after the clock pulse. Some of the versatility of the element may be seen from the truth table. For the input conditions $J = 0$ and $K = 1$, or $J = 1$ and $K = 0$, the input conditions are transferred to the output on receipt of a clock pulse and held there so that the element can be used as a store or register for signals on the J and K lines provided the J and K signals are always arranged, to be mutually complementary. If the J and K inputs are both '0' the clock pulse causes no change, but if they are both '1' the clock pulse causes the bistable to toggle i.e. the clock pulse acts as the triggering pulse in a T bistable. When used in this latter mode however there is no restriction on the duration of the clock (i.e. triggering) pulse. Finally, although these are not shown in the truth table, the JK bistable is often provided with set and reset inputs which operate independently of the clock input and change the Q and $\bar{Q}$ outputs in a manner similar to that of the unclocked RS bistable.

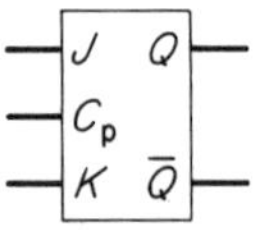

Input conditions		Output states before clock pulse		Output states after clock pulse	
J	K	Q_n	$\bar{Q}_n$	Q_{n+1}	$\bar{Q}_{n+1}$
0	1	A	$\bar{A}$	0	1
1	0	B	$\bar{B}$	1	0
0	0	C	$\bar{C}$	C	$\bar{C}$
1	1	D	$\bar{D}$	$\bar{D}$	D

Figure 2.33 Truth table and symbol for a JK bistable

The master–slave JK bistable uses two component bistables and the sequence of events may be related to the clock pulse waveform Figure 2.34. As the voltage on the clock input rises to point A, the first (master) bistable is isolated from the second (slave) bistable. When the clock input has risen to point B, the information on the J and K inputs is stored in the master bistable. At point C the J and K inputs are disconnected from the master bistable by appropriate gating and at the point D the information from the master bistable is transferred to the slave bistable.

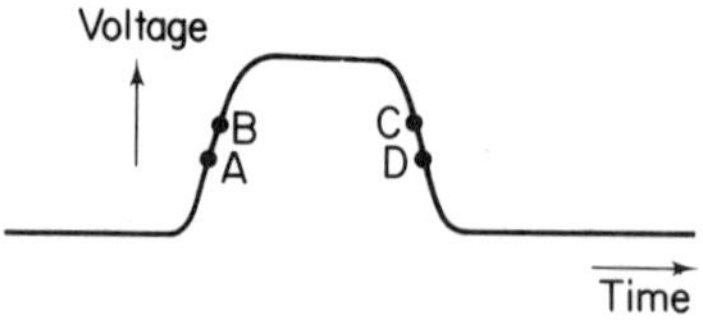

Figure 2.34 Clock pulse

2.3.5 *Monostable circuits*

When an input pulse is applied to a monostable, the circuit changes state but after a certain time it returns to the original state. The arrangement therefore finds application for standardizing the form of logic pulses for use in further circuits. In addition, by utilizing the trailing edge of the output pulse to trigger a following circuit element the effect of the monostable is equivalent to the introduction of a time delay equal to the reset time of the monostable.

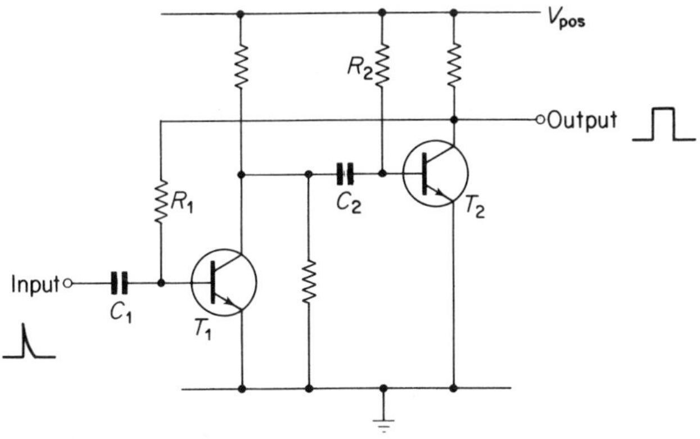

Figure 2.35 Example of a monostable circuit

Many circuit configurations give the monostable action and Figure 2.35 shows one example. Quiescently T_2 is biased so as to be conducting so the output is at '0' and the transistor T_1 is held non-conducting via R_1. On application of a brief positive pulse at the input, T_1 conducts and the voltage change of its collector is transmitted to the base of T_2 via the capacitor C_2 so causing T_2 to cut off and the output to go to '1'. The resistor R_1 provides positive feedback and holds T_1 on, and since the above sequence of events have all occurred rapidly the circuit remains in the second state for some time even if the input triggering pulse falls to zero. After some time the capacitor C_2 has discharged sufficiently for the voltage on base of T_2 to rise to permit partial conduction of T_2, at which time the resultant voltage change on the collector is transmitted via R_1 which, by the positive feedback action, causes the circuit to switch back to the original state. A further time must elapse before the voltage across C_2 can regain its original quiescent value. The time the circuit depends in the metastable state is determined primarily by the discharge time constant of C_2 and equals C_2R_2 approximately. The triggering pulse should ideally be of short duration compared to the duration of the metastable state otherwise its continued presence may affect the resetting of the circuit and prolong the output pulse. A short triggering pulse may usually be assured if C_1 is small enough but this solution is not satisfactory if the triggering pulse has a slow rise since it may fail to initiate the triggering action. For such slowly rising pulses it is therefore preferable to first pass them into a Schmitt trigger circuit and use the output of this to trigger the monostable.

All simple monostable circuits need a certain time to fully recover, after undergoing a transition, and in the circuit of Figure 2.35 this will be primarily due to the necessity for C_2 to regain its quiescent state. During part of this recovery period it will not be possible to trigger the circuit again and, in the latter part of the recovery period, it will only be possible with a larger trigger pulse. This is particularly important for pulses from a nuclear radiation detector which will occur randomly in time as a certain number of these will fall in the recovery period and fail to trigger the monostable. This topic is more fully discussed in the sections dealing with pulse height discriminators.

2.3.6 Tunnel diode circuits

The construction of a tunnel diode differs from that of an ordinary diode in that the doping level is much higher. The charge carrier density is therefore sufficient that the depletion layer is extremely thin, some 100 Å compared to 10^4 Å for an ordinary diode, and when forwardly biased there is a possibility of charge transfer by a quantum mechanical tunnelling process. Arising from the tunnelling effect, superimposed on the normal forwardly biased current–voltage characteristic of the diode, is a peak as shown in Figure 2.36. It will be observed that this leads to a region of negative resistance in the characteristic and it is this feature which enables the device to be exploited in trigger circuits. The other important aspect of the tunnelling process is that, unlike the normal mechanism of forward conduction, the speed with which the current responds to changes in voltage is not limited by charge storage in the depletion layer. Thus, at least in the region of the characteristic up to the valley point, the diode responds extremely rapidly.

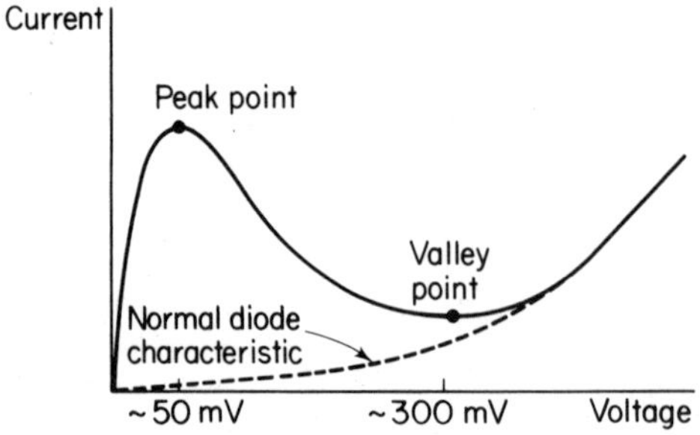

Figure 2.36 Typical current–voltage characteristic for a Ge tunnel diode

Consider first the case of a tunnel diode biased from a constant current source so that the load line intersects the characteristic at points A and B in Figure 2.37. The circuit can exist stably in two states, corresponding to points A and B, and if we assume that it is initially in state A it is possible to switch the circuit into state B by the application of a pulse of current ΔI. ΔI must have a magnitude greater than ΔI_p in Figure 2.37 and during the application of ΔI the operating point will move from A to P and via the dotted line to C and then to D. On removal of the signal ΔI the operating point will fall to B. The circuit has thus been switched into the second state and in order to switch it back into the original state it is necessary to apply a negative current pulse at the input or, alternatively, to temporarily reduce the current from the constant current source.

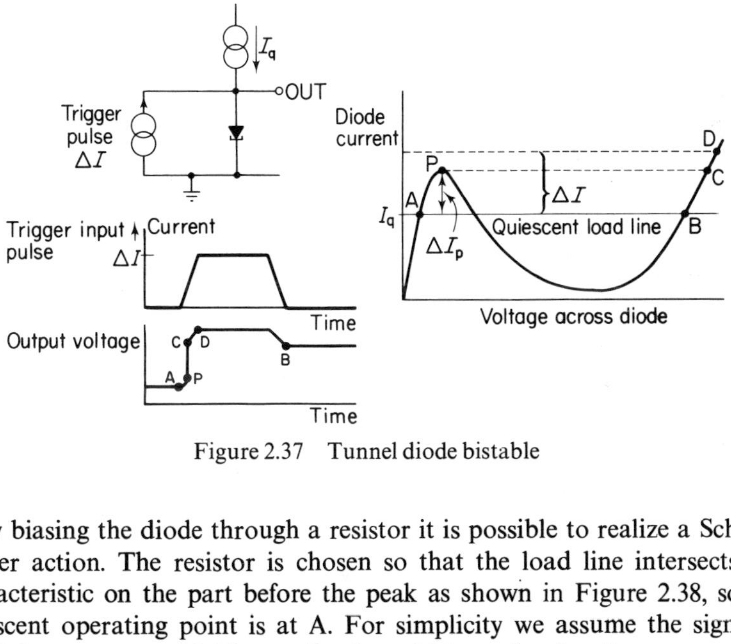

Figure 2.37 Tunnel diode bistable

By biasing the diode through a resistor it is possible to realize a Schmitt trigger action. The resistor is chosen so that the load line intersects the characteristic on the part before the peak as shown in Figure 2.38, so the quiescent operating point is at A. For simplicity we assume the signal is applied as a current ΔI. The application of ΔI may be therefore seen to shift the load line to the new position shown as a dotted line in Figure 2.38 and, provided ΔI is large enough, causes the operating point to move to B. Removal of ΔI causes the operating point to shift back to A. For true triggering action the slope of the load line must be less steep (i.e. more horizontal in Figure 2.38) than that of the negative resistance portion of the diode characteristic. The threshold for triggering is then equal to that magnitude of ΔI which will just shift the load line beyond that point on the negative resistance part of the characteristic where its slope is steeper than the load line. It may be noted that the shape of the output pulse is not perfectly

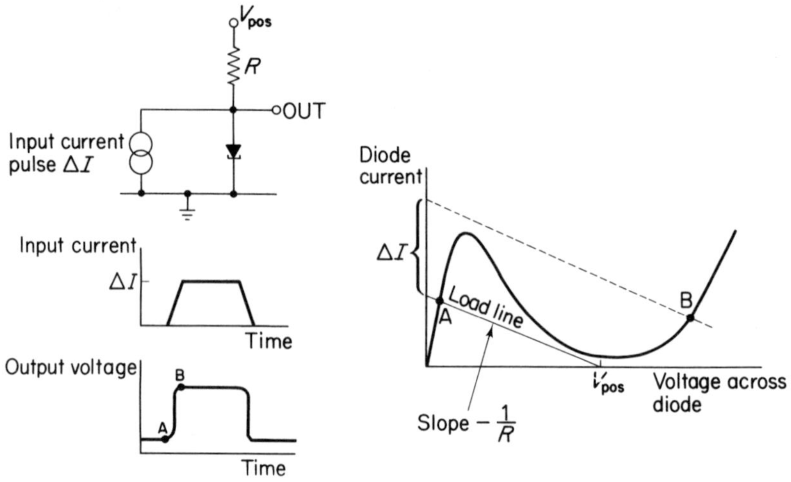

Figure 2.38 Use of a tunnel diode to give a Schmitt trigger action

square and furthermore sub-threshold input pulses will still cause a small output pulse even though triggering does not take place. Hence it is often necessary to follow such a circuit by a second circuit, usually using another tunnel diode to clean up the pulse shape and provide extra discrimination against the action of pulses below the triggering threshold.

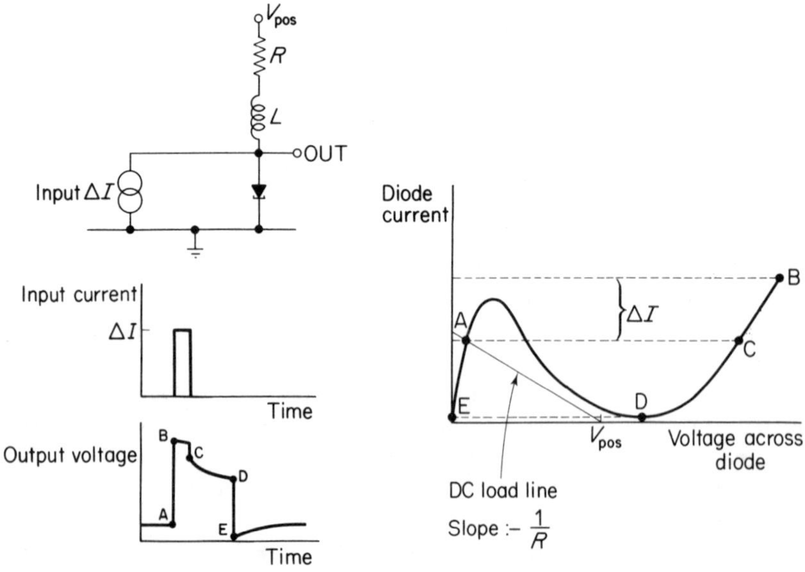

Figure 2.39 Use of a tunnel diode to give a monostable action

The most common way of using a tunnel diode for monostable operation is to use an inductor load as shown in Figure 2.39. If a current triggering pulse ΔI is applied, the current through the inductance cannot change instantaneously and the operating point shifts to B. Assuming the triggering pulse is rapidly terminated, the operating point shifts to C and, following this, the current through the inductance gradually reduces until point D is reached. The circuit then switches to point E and the current through the inductance gradually rises to the quiescent value. Ideally the triggering pulse should be brief, otherwise the metastable state may be prolonged.

2.4 Counting and storage circuits

2.4.1 Counters and scalers

The terms counter and scaler are usually used synonymously; the latter was commonly used in nuclear electronics but is now being superseded by the term counter, which has gained currency in computer terminology. Both denote an arrangement to count up the number of sequentially occurring pulses. Since practically all electronic counting arrangements utilize elements that can exist in two alternative states, there are many different ways that the number of counts can be coded in terms of these elemental states. Of these the simplest is probably that using straight binary code.

A counter operating in binary code may be simply constructed by cascading a series of T bistables. Figure 2.40 shows an arrangement where JK bistables are connected so as to give an action equivalent to T bistables. Assuming that all the bistables are initially reset so that all the Q outputs are at '0' the trailing edge of the first input pulse will cause a toggling of the first bistable so $Q_1 \rightarrow 1$. On receipt of a second input pulse the trailing edge will cause $Q_1 \rightarrow 0$ so causing the second bistable to toggle so $Q_2 \rightarrow 1$. The trailing edge of a third input pulse will now cause the transition $Q_1 \rightarrow 1$, Q_2 remaining at '1'. The table in Figure 2.40 summarizes the states of the outputs after successive input pulses and it will be seen that the Q outputs give the number of pulses in binary code.

It may be noted that if, instead of the Q outputs being used to drive the clock inputs of the following bistables, the $\bar{Q}$ outputs are used then, with each input pulse, the states of the bistables change in the opposite sequence of that in the table in Figure 2.40. This facility may be of use, for instance, when a count has been accumulated with the original mode of connection (i.e. count up mode) and it is then required to subtract from it the counts due to background radiation using the count down mode. For this purpose it is possible to insert gates (AND gates for instance) in all the Q and $\bar{Q}$ outputs and enable the appropriate gates to give either the count up or count down action.

When an input pulse is received in the arrangement of Figure 2.40, the bistables have a finite propagation delay so it is possible that the first bistable may receive a second pulse before the effects of the original input pulse have propagated to the end of the chain. For this reason this type of counter is often termed a 'ripple through' counter. Although the ripple through effect may be of no importance in a simple counter, in other applications it

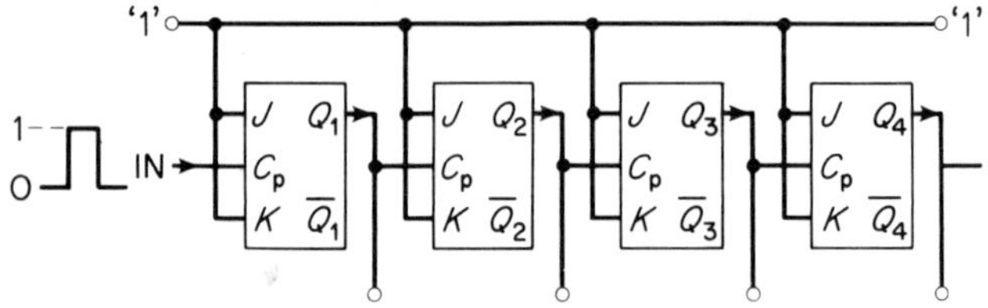

Number of input pulses, n	States of outputs after n pulses Q_4	Q_3	Q_2	Q_1
0	0	0	0	0
1	0	0	0	1
2	0	0	1	0
3	0	0	1	1
4	0	1	0	0
5	0	1	0	1
6	0	1	1	0
7	0	1	1	1
8	1	0	0	0
9	1	0	0	1

Figure 2.40 Asynchronous binary counter assembled from JK bistables

may cause misinterpretation of the counter contents. Such is the case when the contents of the counter are to be sampled by other circuitry while counting is in progress. Similarly it is sometimes necessary to compare the contents of the counter by electronic means with some other preset number and there is a possibility of anomalous parity being achieved. To avoid the problems of the ripple through effect it is necessary to use a synchronous counter in which all the outputs take up their new states simultaneously.

Figure 2.41 shows one method of realizing a synchronous binary counter from JK bistables. It should be noted that the AND gates have a small propagation time and, while this does not affect the synchronous nature of the output signals, following a clock pulse the *J* and *K* inputs take a certain time to settle to their new values. For a long cascade of bistables this puts a limit on the maximum acceptable input pulse rate. For the case of the asynchronous counter the only limitation on input pulse rate is set by the maximum toggling rate of the first bistable.

Although binary counters are often preferred where the data is to be further processed, if the count is to be presented as a visual display it is usually inconvenient to have it presented in binary code. Hence it is often preferable to use counters which count in binary coded decimal (BCD) code. It may be noted that it is a simple matter to convert a number coded in BCD into straight binary, if necessary, but the reverse conversion is not as easy. Many forms of coding can be used to give a BCD code but the one that

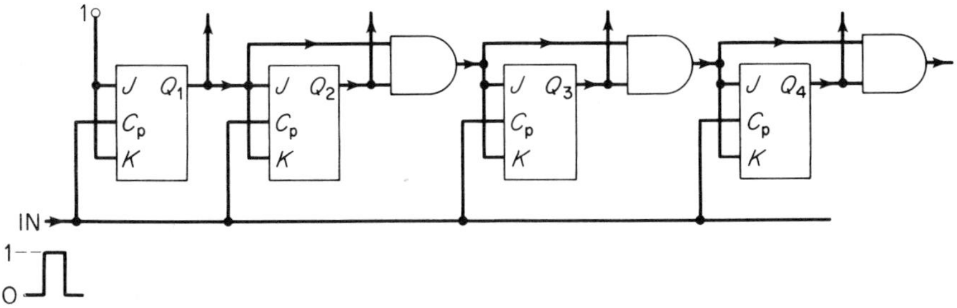

Figure 2.41 Synchronous binary counter assembled from JK bistables

is most commonly used is probably that using the 1–2–4–8 code. This has the advantage that the weights assigned to each digit within a decade are the same as in straight binary and it is this system which will be exclusively considered here. Figure 2.42 shows the BCD code equivalents of some decimal numbers and it will be observed that for numbers 0 to 9 the BCD code is the same as the straight binary, however the states which in straight binary correspond to numbers 10 to 15 do not exist in BCD code.

Many possible arrangements can give a counter operating in BCD and Figure 2.42 shows one example of a synchronous version. Assuming all the bistable Q outputs are initially reset to '0', on receipt of the first eight input pulses the counter behaves in the same way as the binary counter described earlier (Figure 2.41). After the eighth input pulse, however, the $\bar{Q}_4$ output is at '0' and so blocks gate G_1 so that the ninth input pulse causes only the first bistable to toggle to give $Q_1 = 1$. The connection between Q_1 and K_4 now ensures that on receipt of the tenth pulse the fourth bistable toggles to give $Q_4 \rightarrow 0$ and so the four bistables making up the first decade have returned to their initial states.

For the operation of the next decade a similar arrangement of four bistables may be used but with the difference that the J and K inputs of the first bistable are connected to the output of an AND gate whose inputs are derived from Q_1 and Q_4 of the first decade. When the first decade has counted up to nine, both Q_1 and Q_4 are at '1' and the tenth pulse will cause the first bistable of the next decade to toggle.

For the purposes of visual display of the contents of the counter a simple arrangement is to cause the state $Q = 1$ on any bistable to cause an associated lamp to glow so that the decimal count may be inferred. More commonly

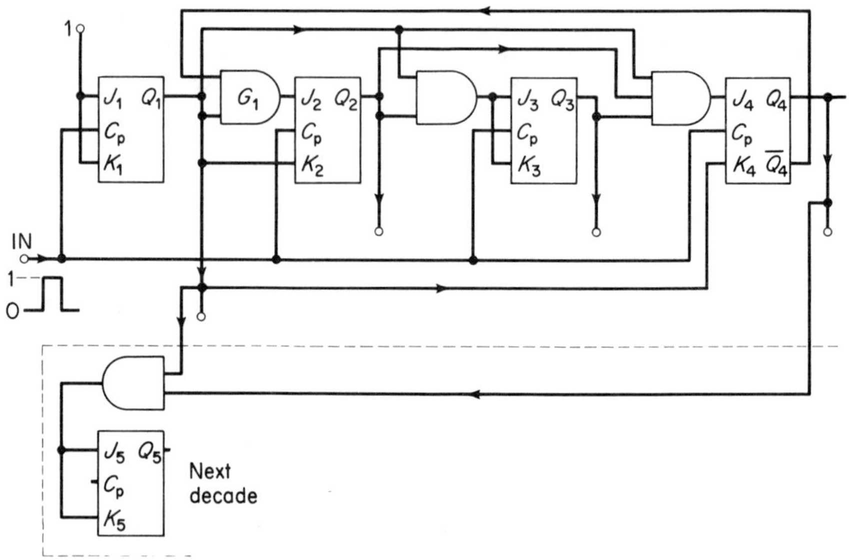

Number of input pulses, n	Output states after n pulses				
	Q_5	Q_4	Q_3	Q_2	Q_1
0	0	0	0	0	0
1	0	0	0	0	1
2	0	0	0	1	0
3	0	0	0	1	1
4	0	0	1	0	0
5	0	0	1	0	1
6	0	0	1	1	0
7	0	0	1	1	1
8	0	1	0	0	0
9	0	1	0	0	1
10	1	0	0	0	0
11	1	0	0	0	1
12	1	0	0	1	0
13	1	0	0	1	1
14	1	0	1	0	0
15	1	0	1	0	1

Figure 2.42 Synchronous BCD counter

the count is displayed as a decimal number by the use of numerical indicator devices. These may take the form of gas filled discharge tubes (Nixie tubes), incandescent lamp devices or solid state devices and to drive these devices the BCD coded number must be converted into a suitable code. The gas filled numerical indicator tubes, for example, require a steady signal to be

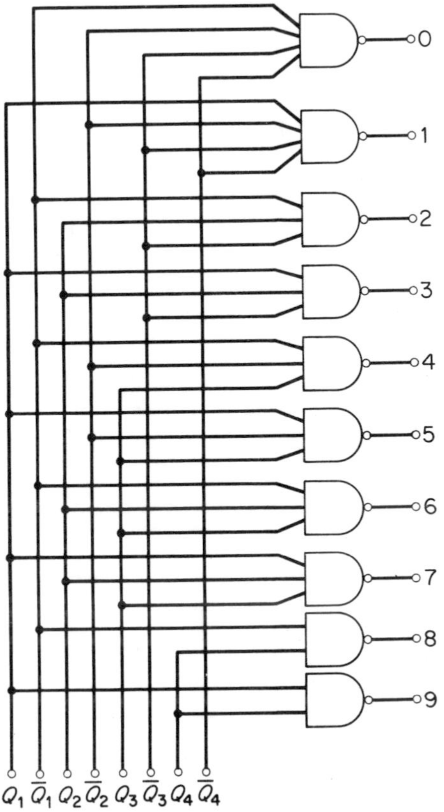

Figure 2.43 Converter from BCD code to one-out-of-ten decimal code

applied to the appropriate one of the ten electrode connections corresponding to the ten numerals and Figure 2.43 shows an arrangement of gates that will decode the BCD number into a signal on the appropriate line. Such circuits are available in a single integrated circuit package.

A rather different approach to counting is used in the ring counter. As its name implies this consists of a ring of bistables and it has the property that on receipt of a clock pulse the state of each Q output is shifted to Q output of the next bistable. We will only consider the situation where one of the bistables

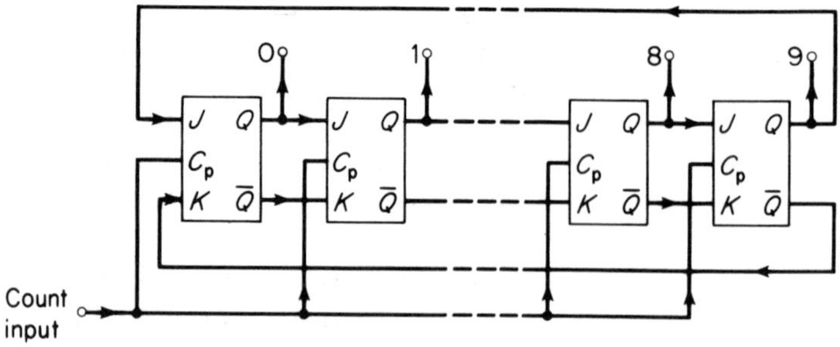

Figure 2.44 Ring counter assembled from JK bistables

has an output of '1', the rest having $Q = 0$. Figure 2.44 shows such a ring counter made up of ten bistables. Since each clock pulse shifts the '1' output to the next bistable it is obvious that the arrangement may be used as a decimal counter. The principle of this circuit is analogous to that used in the Dekatron tube. This consists of ten electrode assemblies in a circular assembly enclosed in a gas filled tube and the conditions of operation are such that any one of the ten main electrodes can act as cathode for the glow discharge. Although this device combines the functions of counting and display it is slow compared to transistor circuits, many types being limited to a maximum

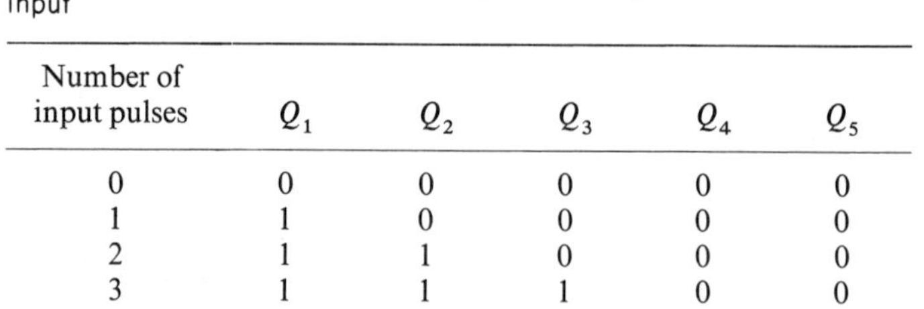

Number of input pulses	Q_1	Q_2	Q_3	Q_4	Q_5
0	0	0	0	0	0
1	1	0	0	0	0
2	1	1	0	0	0
3	1	1	1	0	0
4	1	1	1	1	0
5	1	1	1	1	1
6	0	1	1	1	1
7	0	0	1	1	1
8	0	0	0	1	1
9	0	0	0	0	1

Figure 2.45 Twisted ring counter using JK bistables

counting speed of ~ 50 kHz. Faster devices have been developed using an electron beam in a vacuum tube but for most applications they have been rendered obsolete by the advent of low cost integrated circuits.

The decade ring counter, Figure 2.44, requires ten bistables as against four for a BCD counter. A more economical arrangement is to use the 'twisted ring' or Johnson configuration of Figure 2.45, which requires only five bistables. Assuming all the Q outputs are initially reset to '0' the effect of successive input pulses is shown in the table in Figure 2.45 and it will be seen that the first five pulses progressively fill the bistables with '1's. The next five pulses progressively fill the bistables with '0's. It is a simple matter to decode the outputs of the bistables to give BCD or one-out-of-ten decimal code. With any ring counter there is a possibility of spurious states being accidentally set up which will then circulate so it is necessary to include additional gating to safeguard against this. The attraction of the ring counter over other counters is that each bistable of the twisted ring, for example, only toggles at a frequency one fifth of the input pulse rate so that it can accept a higher input pulse rate.

2.4.2 Registers and memories

Registers and memories perform the same function—the storage of digital data. There is a spectrum of devices available for such storage which vary according to the cost per bit of information stored and the time, the so-called access time, with which the data storing or recalling operation ('writing' or 'reading') may be executed. The fastest type of store with the shortest access time is that based on bipolar transistor bistables, giving access times of the order of 3–100 ns. The cost of such devices usually limits their use to the function of temporary storage, such as storage during arithmetical operations, or as a buffer store pending storage in other slower devices. The term register also tends to be applied to such stores.

For more economical storage of larger amounts of data reasonably fast access can be obtained by the use of devices based on the properties of ferromagnetic materials and, more recently, on semiconductor devices. Ferrite cores are, at the moment, the most commonly used element, giving access times in the region of 0·2 μs to a few microseconds. Rather faster access times can be obtained with the use of thin film and plated wire ferromagnetic devices. Semiconductor storage devices are undergoing rapid development and currently give access times comparable with that of ferrite cores.

For the low cost storage of large amounts of data some form of surface magnetic recording is used. This may be on magnetic disc, drum or tape.

A transistor bistable register can be simply constructed from a number of D or JK bistables. Thus for example, when a counter has been used to accumulate a count it is often necessary to transfer the number to a register in order to free the counter for a further counting operation. This may be accomplished as shown in Figure 2.46. On receipt of a transcribe pulse, the values of the outputs of the count bistables are transcribed in parallel to the outputs of the register where they are stored. By the use of additional gates it is possible to arrange for transfer of data from one of several counters to one of several registers.

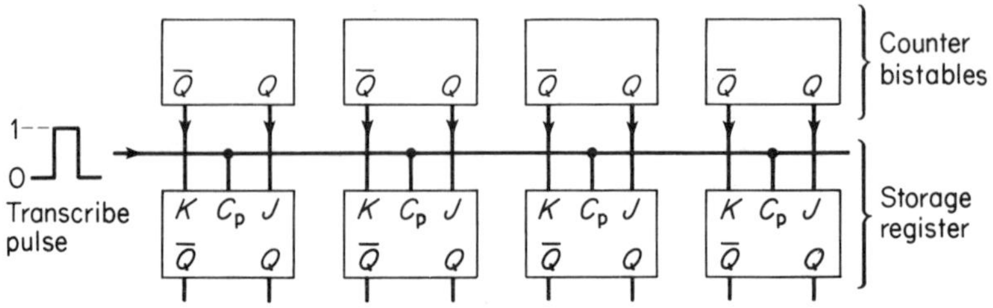

Figure 2.46 Use of a storage register of JK bistables to store the counter contents

Registers also find application as a means of de-randomizing digital data. For example, in the case of a multichannel analyser, each input pulse causes the generation of a digital number proportional to the magnitude of the pulse. Depending on the resolution of the analyser this digital number may be a word of up to 13 bits and this information must be classified and stored. Since the storage process takes some time it is possible temporarily to store the digital word in a buffer register so as to free the analyser to accept another input pulse. The classification and storage of the digital information may then proceed independently. Since the input pulses will occur with all possible time spacings there is a possibility of a second pulse being accepted for analysis before the original contents of the buffer register have been processed, in which case the information from the second event would be lost. To reduce such losses additional buffer registers may be provided. Incoming data occurring at random times is read into the buffer registers in rotation and read out at regular intervals according to the requirements of the following processing circuitry, so the registers act as a de-randomizing agent. The effectiveness in reducing events lost will depend on the ratio of the incoming mean rate to the read-out rate and the number of buffer registers; this is further discussed in Appendix B.5.

In the ferrite core memory element the magnetic medium is fabricated as a toroid of outer diameter 0·5mm to 2 mm through which is threaded a number of wires (Figure 2.47). An essential property of the material is that the direction of magnetization can be reversed by a field H_m whereas a field of magnitude $\frac{1}{2}H_m$ has no significant effect on the state of magnetization.

The direction of magnetization in the core (clockwise or anticlockwise) can be set by the application of a total current $\pm I_m$, corresponding to a field $\pm H_m$, through one or more of the wires. We will arbitrarily assume that a current of $+I_m$ sets the state to (1) and current of $-I_m$ sets the state to '0'. Assuming that the core has been set to the required state, in order to read out the state of the core use is made of a 'sense' wire threading the core.

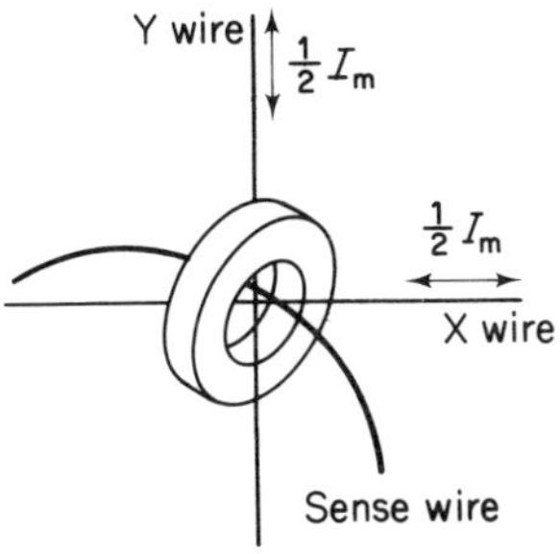

Figure 2.47 Ferrite core element

If a further 'read' current pulse of magnitude $-I_m$ is applied then, regardless of the intial state, the core resets to '0'; however, the voltage induced in the sense wire during the application of the 'read' pulse will differ according to whether switching of the state occurs. If the core was initially in the '0' state no switching would have occurred and only a small voltage would be induced in the sense wire, but if it was initially in the '1' state the large change in flux would cause a larger change in voltage on the sense line. By connecting the sense line to a suitable discriminator circuit it is therefore possible to distinguish between the two alternatives. Since the process of applying the read pulse resets the state to '0', it destroys the information recorded. With any such destructive read-out it is therefore necessary to replace the information (assuming it is to be preserved) and, to achieve this, the information from the sense line, obtained in a read operation, is placed in a register and followed by a write operation where the information in the register is used to set the core back to the original state, the whole operation being termed a read–write cycle.

For most applications where a large number of cores is required, it is impracticable to provide each with separate wires. The cores are therefore arranged in a plane matrix and threaded with common wires running in the X and Y directions, as shown in Figure 2.48, through which the read or write current pulses of magnitude $\pm \frac{1}{2}I_m$ may be passed. If read or write pulses are applied to a pair of such X and Y wires only the core at the intersection of the wires will experience sufficient field to cause a change of state. By suitable gating it is possible to address any core in the matrix and only the signal induced by the change of state of this core will give a significant voltage in the common sense wire. Since it is usually necessary to store a digital word of *n* bits this may be accomplished by stacking *n* plane matrices of cores in a three dimensional array. The digital word is therefore located 'vertically' in the stack, each plane storing one digit of the word at the same XY location. The X and Y wires may be connected in various ways to give addressing of the cores. In the 3D array each X wire of a digit plane is connected to the corresponding X wires of the other digit planes and similarly for the Y wires. On application of currents $\frac{1}{2}I_m$ to the appropriate pair of X and Y wires the digital number is read out in parallel on the *n* sense wires. For the purpose of writing,

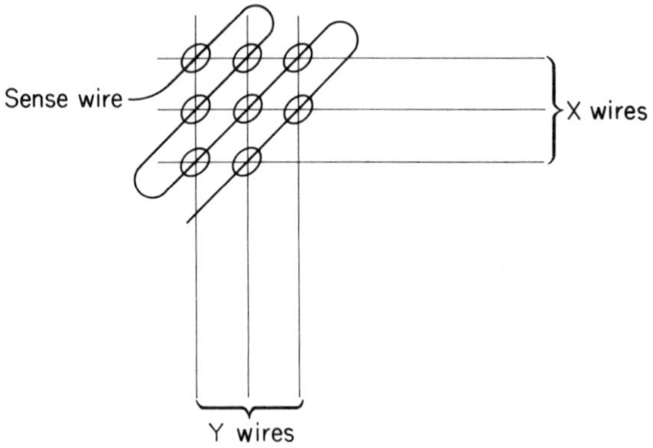

Figure 2.48 Principle of organization of a plane array of ferrite core elements

the fact that the *n* planes have common X and Y wires means that they alone cannot uniquely control the individual cores corresponding to the digits of a number. For writing purposes therefore it is necessary to provide each digit plane with an extra common wire, an inhibit wire, which passes through the cores in the opposite direction to the Y wires. By supplying the inhibit wires of each digit plane with an appropriate current during the application of the write pulse to the X and Y wires it is therefore possible to neutralize the effect of the Y current and so determine whether the selected core in that digit plane is affected.

The 3D array described above, by using X and Y wires common to all planes, places minimum requirements on the driving electronics. In recent years the reduced cost of electronics has led to the use of so-called 2D and $2\frac{1}{2}$D arrays. In the 2D array the cores may be imagined to be arranged in a two dimensional matrix, one of the sides of which has *n* elements, and the *n* digits of the word are written along a line through the matrix. Each core is uniquely defined by the combination of X and Y wires which are consequently of greater number than in the 3D array of the same total capacity. The $2\frac{1}{2}$D array is a compromise between the 2D and 3D array and consists of a stack of digit planes, as in the 3D array, with Y wires common to all digit planes, but with the X wires of each digit plane separately driven. Both the 2D and the $2\frac{1}{2}$D arrays require more driving circuits than the 3D array but require a lesser number of wires threading each core and give faster cycle times.

In thin film and plated wire memories use is made of a thin layer of ferromagnetic material having uniaxial anisotropy and, in the absence of applied field, the two memory states correspond to the two stable states of magnetization parallel to the axis of anisotropy. Switching from one state to the other is achieved by a combination of two fields, one parallel to the axis of anisotropy and the other normal to it. This form of storage is potentially much faster than that using ferrite cores and plated wire storage has found some commercial application, giving access times of the order of 50 ns.

Many types of memory based on semiconductor elements are under development. At the moment the most likely contenders are based on metal-oxide silicon (MOS) technology which can give a very high density (and hence low cost) of devices on a semiconductor chip. The insulated gate field effect transistor, or MOSFET, is a product of this technology and is similar to the junction field effect transistor (Section 5.2) in that the gate voltage can control the channel current. However the MOSFET has an insulating layer of silicon dioxide interposed between the metallic gate electrode and the channel so that very little gate current is drawn. The MOSFETs may be arranged to give a bistable circuit operating in similar fashion to bipolar transistor bistables but give access times somewhat longer. Another approach is possible, however, which reduces the number of components per memory unit and gives an overall cost per memory unit which compares favourably with ferrite core memories. If the gate-channel capacitance of the MOSFET is charged up so as to make the channel conduct, because of the low gate current the charge is retained for considerable time. It is this charge storage that forms the basis of the memory. When charged up, so the channel conducts, the device is in the '1' state and when the charge is low enough to prevent conduction in the channel it is in the '0' state. It is obvious that the charge will eventually leak away and so, before this can significantly occur, it is necessary to refresh the charge by external means. As with ferrite core elements in it is impractical to provide each memory element with individual control and read wires, and various arrays of elements may be used with common X and Y wires. This type of memory currently gives access times in the region of 100–300 ns. The main drawback is that it is volatile, that is, if the power supply should fail the information stored is lost. However, other types of MOS devices are under development which have a hysteresis characteristic and so are non-volatile.

Apart from the random access memories discussed above there are several other types of semiconductor memory devices which are becoming available. Programmable read only memories (ROMs) can be electrically programmed and erased but in normal operation can be rapidly accessed for reading only. This will allow many of the functions now performed by software to be implemented in hardware. Another development of interest in nuclear electronic data processing is the contents-addressable or associative memory. If it is necessary to locate the address of a word, A, stored in the conventional memory it is necessary to search a large number of addresses. A contents-addressable memory however in one operation gives the address or addresses where the word A is stored and this facility is of great importance wherever sorting of data is concerned. Nuclear data applications requiring this facility are discussed in Chapter 9.

For the storage of very large amounts of data the cost of fast memory becomes prohibitive and such data must be placed in a backing store which invariably uses some form of magnetic surface recording. Such stores cannot give random access to the data stored and the access time is therefore limited by the mechanical speed of the device. For a disc store, depending on the speed of rotation and whether fixed multi-track heads or movable heads are used, the average access time is in the range 5 ms to 200 ms. For magnetic tape the access time will obviously depend on how much tape searching is required to locate the address.

2.4.3 Count rate meters

The count rate meter has the function of producing a signal, either as an analogue signal of voltage or current or as a digital signal, proportional to the rate of input pulses. The most common type of rate meter is of the analogue type and operates on the principle of Figure 2.49. The incoming pulses are arranged to be of standardized height and duration (i.e. logic type pulses) and are fed into the parallel RC circuit. Assuming that the input pulses, at a mean rate n per unit time, are supplied from a constant current generator, and that the charge contained in each pulse is Q, then the voltage

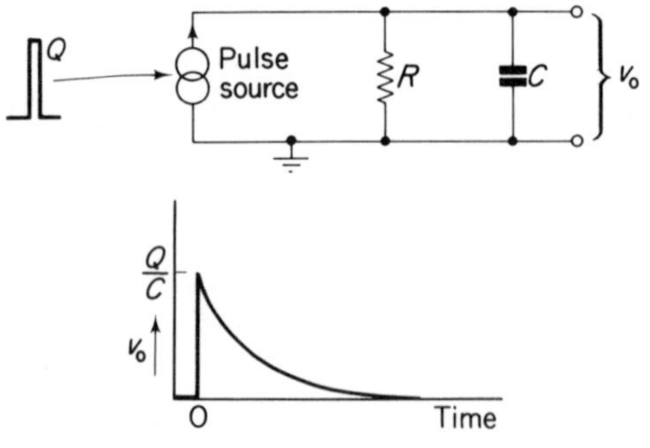

Figure 2.49 Principle of the analogue count rate meter and the form of the output voltage for a single brief input pulse of charge Q

across the capacitor will reach equilibrium when the mean current applied from the input, nQ, equals the current flowing through the resistance $\overline{V}_o/R$. Hence the mean output voltage is $\overline{V}_o = nQR$ and this may be used to operate a meter or chart recorder. The output voltage v_o will of course fluctuate due to the random nature of the incoming pulses and we will assume the root mean square value (i.e. the standard deviation) of the output voltage is V_{RMS}.

Since the circuit is linear the output voltage may be seen as the superposition of the effects of individual input pulses. If the input pulses are brief compared to the time constant, RC, of the circuit, the output voltage due to a single input pulse is of the form

$$v_o(t) = 0 \qquad \text{for } t < 0$$

$$v_o(t) = \frac{Q}{C}\mathrm{e}^{-t/RC} \quad \text{for } t \geqslant 0$$

and this is illustrated in Figure 2.49. For the case of a mean pulse rate n it is possible to find the value of V_{RMS} by using Campbell's theorem (see, for example, Rice[6]) which states

$$V_{RMS}^2 = n \int_{-\infty}^{\infty} v_o^2(t)\,\mathrm{d}t = \frac{nQ^2}{C^2}\int_0^{\infty} \mathrm{e}^{-2t/RC}\,\mathrm{d}t = \frac{nQ^2R}{2C}$$

Hence the fractional standard deviation $V_{RMS}/\overline{V}_0$ is given by $(2CRn)^{-1/2}$. This is the same fractional standard deviation that would be given by a counter counting over a time period $2CR$, and so $2CR$ may be regarded as the effective counting period of the rate meter. By increasing the value of CR the fractional standard deviation may be reduced to any desired value but this will only be obtained at the expense of a sluggish response of the output to changes in the input pulse rate. If, for example, the pulse rate is suddenly increased from n_1 to n_2, by moving the radioactive source nearer the detector for instance, Figure 2.50 shows the time course of the output voltage neglecting the statistical fluctuations. This may be understood by regarding the input current to the circuit as changing from n_1Q to n_2Q and it will be observed that the output voltage is of the form

$$QR[n_1 + (n_2 - n_1)(1 - \mathrm{e}^{-t/RC})].$$

As commonly implemented, the analogue rate meter uses the diode pump circuit of Figure 2.51. On arrival of an input pulse of standardized voltage, the charging current of the capacitor C_1 flows through D_1 into C_2. On termination of the input pulse the capacitor C_1 regains its original state by recharging via D_2. During the input pulse the charge injected into C_2 from C_1 should ideally be constant regardless of the voltage level across C_2 and to ensure this it is essential to make $C_2 \gg C_1$.

As well as the linear rate meter already described it is sometimes useful to have a rate meter giving an output signal proportional to the logarithm of the pulse rate. Several devices are available that can be added on to the linear rate meter circuit and, over several decades of magnitude of voltage, will give an output signal approximately proportional to the logarithm of the voltage. Such devices usually exploit the approximately logarithmic

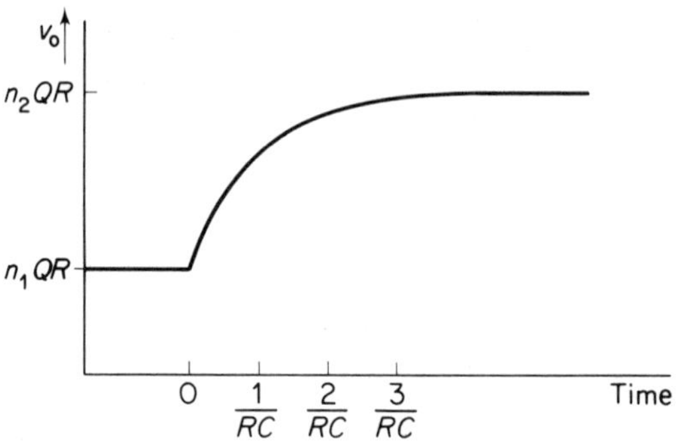

Figure 2.50 Analogue rate meter: time course of output voltage when the mean input pulse rate is suddenly changed from n_1 to n_2. Statistical variations of the output voltage are not shown

characteristic of PN diode junctions and a common technique is to incorporate the junction in the feedback loop of an operational amplifier configuration.

The advent of integrated circuits has increased the popularity of digital rate meters. In essence this is a counter which counts for consecutive constant time periods. At the end of any period the count is read out, probably

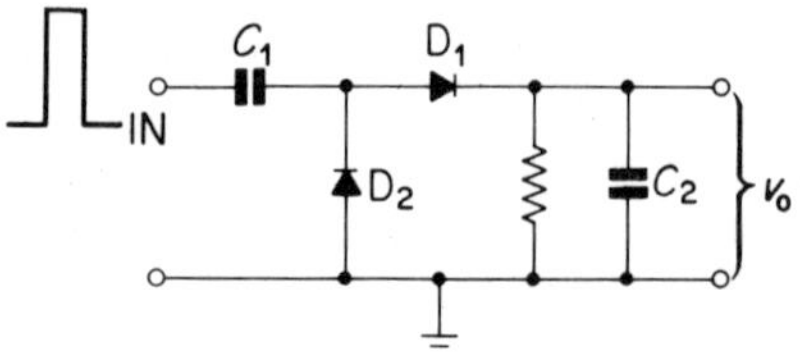

Figure 2.51 Diode pump analogue rate meter

into a buffer register, the counter is reset to zero and counting recommences for the next period. The time necessary to stop, read out and reset the counter is usually small compared to the counting period and so negligible information is lost. The buffer register at any time contains the count of the last period and this may be made available at the output as a digital signal for operation of a digital recorder (e.g. a printer or paper tape punch) or, via a linear or logarithmic digital-to-analogue converter, as an analogue signal to operate a meter or chart recorder.

Many multichannel analysers have facilities to enable them to be used as digital rate meters. For this purpose the memory is made available to store the values of the counts obtained in the successive time periods and these may be read out at the end of the run. Although the amount of data that may be stored is necessarily limited by the memory capacity the technique is valuable for following rapidly changing count rates since many common forms of digital recording (e.g. printers, paper tape punches) cannot cope with the high rate of data output.

References

1. G. E. Valley and H. Wallman, *Vacuum Tube Amplifiers*, McGraw–Hill, 1948.
2. J. Millman and H. Taub, *Pulse and Digital Circuits,* McGraw–Hill, 1956.
3. A. V. Bedford and G. L. Fredendall, *Proc. I.R.E.,* **27**, 277 (1939).
4. B. P. Lathi, *Signals, Systems and Communication,* Wiley, 1965.
5. Mullard Ltd., *Electronic Counting*, London 1967.
6. S. O. Rice, *Bell System Tech. J.*, **23**, 282 (1944) and **24**, 46 (1945).

3

Pulse Shaping Methods for Spectroscopy

3.1 Factors affecting the energy resolution

3.1.1 Pulse height spectra

In all the detectors used for spectroscopy by purely electronic methods the signal appears on the collector as a pulse of current, typically lasting for a period of time in the range 0·1 μs to 10 μs. In most applications the charge contained in this pulse is of primary interest as it is this which is related to the energy deposited in the detector by the ionizing nuclear particle. The most common arrangement is to employ pulse shaping which gives a final output of voltage having a maximum pulse height proportional to the charge in the detector pulse. Voltage pulses produced by such a system can then be analysed on the basis of pulse height. Such a pulse height analyser might select pulses in the height range $v + \delta v$ and these can then be counted in a scaler or register. By making countings for several values of v the pulse height spectrum or histogram may be constructed. Figure 3.1 shows the type of spectrum that might be obtained where the nuclear particles are monoenergetic 5 MeV alphas. The finite width of the peak in the spectrum is due to imperfections in the detection process and the pulse processing system. (We will always assume that the spectra are taken with a small enough value of δv that additional spreading of the peak due to this is negligible). The full width at half maximum (the 'fwhm'), E_{fwhm} in Figure 3.1, is a measure of the resolution of the system and for the 5 MeV alphas in Figure 3.1 may be seen to be 0·5 MeV. The fractional fwhm E_{fwhm}/E is therefore 0·1 or 10%.

The fwhm for monoenergetic nuclear particles is a widely used index of resolution of a spectroscopy system. An alternative index is the standard deviation σ of the heights of the output pulses arriving at the analyser. Where the peak in the pulse height spectrum can be approximated by a Gaussian (i.e. normal) distribution the fwhm, E_{fwhm}, and the standard deviation, σ, are related by

$$\frac{E_{\mathrm{fwhm}}}{E} = 2{\cdot}36\frac{\sigma}{\bar{V}} \tag{3.1}$$

where $\bar{V}$ is the mean pulse height. This relationship is derived in Section 4.2.

It is possible to isolate four components contributing to the overall resolution. One component arises from the nature of the detection process itself. It manifests itself as fluctuations in the magnitude of the charge arriving on the detector collector and has been discussed in Section 1.2. In the absence

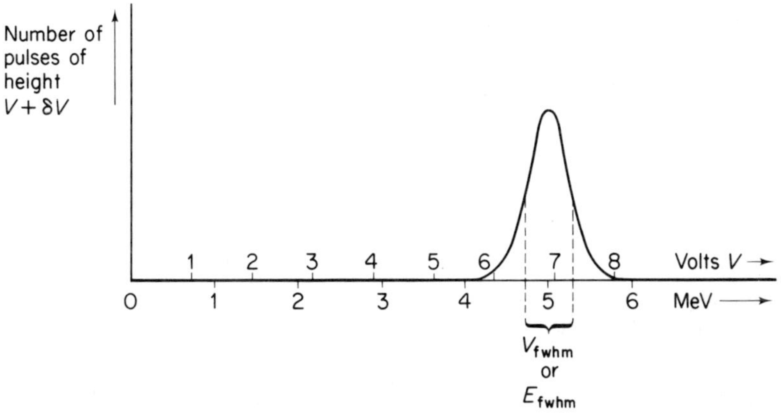

Figure 3.1 Pulse height spectrum; the plot shows the distribution of pulses against the pulse height (and equivalent particle energy)

of any other components we will assume that this would give a standard deviation σ_1 in the heights of the output pulses of the system. The other three components are due to imperfections in the pulse processing system and arise from pulse pile up, noise and variations in the ballistic deficit. We will assume that these three components would give standard deviations in the output pulse heights of σ_2, σ_3 and σ_4 respectively if each were present alone in the absence of all the others. Since the four processes are statistically independent, according to the theorem discussed in Subsection 4.1.1, the overall standard deviation, σ, when all four are present is given by

$$\sigma^2 = \sigma_1^2 + \sigma_2^2 + \sigma_3^2 + \sigma_4^2 \tag{3.2}$$

This result is exact regardless of the nature of the distributions of pulse height produced by the four processes. If each of the four processes alone may be regarded as giving Gaussian (i.e. normal) distributions of pulse height then the resultant distribution due to all four processes together is also a Gaussian distribution and the fwhm may be deduced from (3.1).

The optimum pulse processing system would be that which reduces $\sigma_2^2 + \sigma_3^2 + \sigma_4^2$ to a minimum. Since the requirements of pulse processing necessary to reduce σ_2, σ_3 and σ_4 are, in general, mutually exclusive, some compromise must be found. The pulse processing system chosen will therefore depend on the relative magnitudes of the different components and this in turn will depend on factors such as the type of detector and preamplifier and on the mean pulse rate that the system is required to handle. The nature of the components due to pile up, noise and ballistic deficit variations is discussed qualitatively in the following sections. The quantitative aspect of energy resolution is more fully considered in Chapter 4.

3.1.2 Pulse pile up

Owing to the random nature of nuclear events there is always a finite possibility that two nuclear events may occur in rapid succession. Since the output pulses of a processing system have a finite duration there is a possibility that two such pulses will be superimposed (i.e. will 'pile up') on one another so that the pulse height analyser will not see them as their true height (Figure 3.2). To minimize pile up, the output pulse should ideally be brief and should return cleanly to zero volts. If this is not the case and the output pulses have, for instance, slowly decaying tails as in Figure 3.2(a) the effect on the pulse height spectrum would be to increase the fwhm by distorting the high energy side of the peak. In a similar way output pulses having long undershoots will cause distortion of the low energy side of the peak (Figure 3.2(b)).

The effects of pulse pile up are discussed more fully in Section 4.5; however, it may be noted that the fraction of pulses piling up will depend on the mean

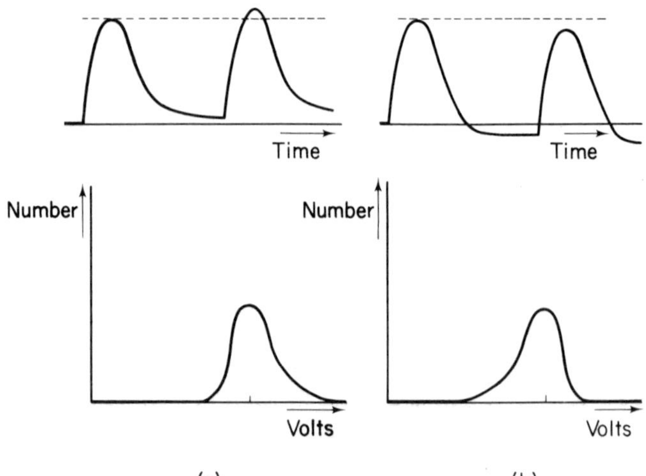

Figure 3.2 Effect of pulse pile up in worsening the resolution for the case (a) of a pulse shape with a long tail, (b) of a pulse shapc with a long undershoot

pulse rate. At low pulse rates the mean spacing between pulses is large and pulse pile up will only slightly broaden the fwhm of the peak so that a pulse shaping system would then generally be chosen rather to minimize the contributions of the other components arising from noise and ballistic deficit variations. Conversely at high pulse rates it will usually be preferable to use a pulse shape which is brief and has a rapid return to zero volts even though this will not generally be optimum for the noise or the ballistic deficit contributions.

The discussion of pulse pile up has centred on the pile up of output pulses. Since most pulse shaping is done in several stages the pulse length in the intermediate stages may be much longer than it is at the output so that pile up may occur at an intermediate stage and this would not be apparent from a consideration of the output pulse shape alone. Provided the system is linear, however, this of no account since any such intermediate piling up of a pair of pulses will be resolved when the pulses are finally shaped prior to the output. Provided any such intermediate pulse pile up does not cause the amplifier to overload it is therefore sufficient to consider pile up of the output pulses only as a contributory cause to loss of resolution. The possibility of amplifier overloading due to pile up is further considered in Section 4.5.1.

3.1.3 Noise

Consider the case of constant height pulses which are arriving at the input of the pulse height analyser. If we ignore the effects of pile up the pulse height spectrum will be a line as shown in Figure 3.3(a). If now noise is present the pulses will be superimposed on the noise voltage and will no longer be of

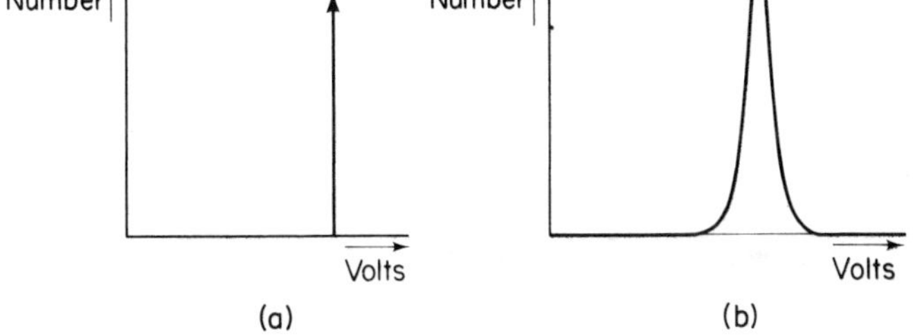

Figure 3.3 (a) Pulses of constant height in the absence of noise; the resulting pulse height spectrum. (b) Effect of the presence of noise on the pulse heights; the corresponding pulse height spectrum

constant height. The noise voltage may be characterized by its root mean square (RMS) voltage (i.e. the standard deviation of the voltage) V_{RMS} and it has a mean voltage of zero. The resultant pulse heights are therefore distributed about a mean equal to the original pulse height with a standard deviation σ_3 equal to V_{RMS}. Since the distribution of the noise voltage is given by a Gaussian distribution (Section 4.1) we have

$$\text{voltage fwhm} = 2{\cdot}36\,\sigma_3 = 2{\cdot}36\,V_{RMS} \tag{3.3}$$

The contribution of noise to the overall fwhm is fully discussed in Chapter 4 and we may only note here that one object of the pulse processing system is to reduce V_{RMS} as much as possible.

Part of the noise originates from the early stages of the preamplifier and the RMS value due to this source alone is partly determined by the bandwidth of the system between preamplifier and output. Where this source of noise is important it is necessary to select a pulse shaping system which limits this bandwidth. Preamplifier noise is usually unimportant in the case of a scintillation counter where minimal amplification of the detector signal is necessary and the inherent energy resolution of the detector is poor. The pulse shaping for this case does not have to be of a type that limits the bandwidth of the system since the inherent bandwidth of the amplifier usually limits the noise sufficiently. DL or $(DL)^2$ shaping, for example, is therefore adequate.

For nearly all the other types of detector however the noise from the early stages of the amplification system is a significant contributor to the overall fwhm and the pulse shaping system must be chosen with this in mind. This precludes the choice of DL or $(DL)^2$ shaping.

3.1.4 Detector pulse shape and ballistic deficit

All the detectors of interest require the application of a polarizing or bias voltage to the collector. The detection of an ionizing particle causes the detector current to momentarily increase by an amount i. If the bias voltage is applied to the collector through a load resistor R_L (Figure 3,4(a)) the signal will appear as a momentary change of collector voltage. For an ionization chamber or semiconductor detector operated in a saturated condition i is independent of collector voltage and therefore behaves as a constant current generator. The approximate equivalent circuit for the signal in this case is shown in Figure 3.4(b) where C_d is the detector capacitance.

For proportional counters the magnitude of i is dependent on the collector voltage V_d which will change slightly during the passage of a signal and so it can be represented by i = constant + V_d/R_d where R_d is a quantity analogous to the anode resistance of a triode or pentode. In a photomultiplier the dynodes are usually held at constant voltage during the passage of a signal by using decoupling capacitors in which case the signal current can be

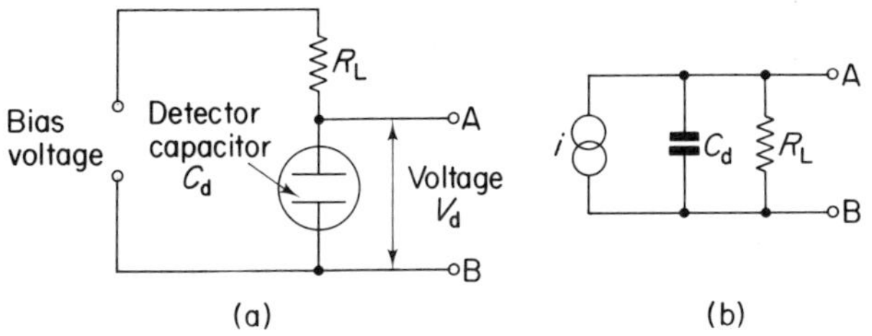

Figure 3.4 Input circuit for the detector; the equivalent circuit for detector current signals of magnitude i

related to collector voltage by a similar expression. As with the anode resistance of a triode the effect of R_d can be taken into account in the equivalent circuit by placing it across R_L in Figure 3.4(b), however this is rarely necessary since R_L is usually much smaller than R_d.

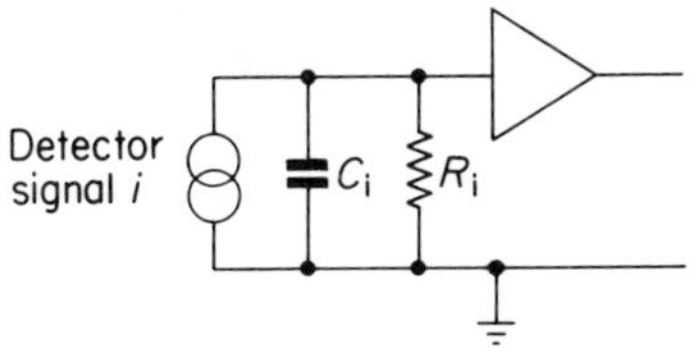

Figure 3.5 Equivalent circuit for detector signal, i, at the pre-amplifier input

As a simple example, the integrating of the detector current pulse to give an output pulse of proportional height may be accomplished by the arrangement of Figure 3.5 where C_i and R_i include the detector capacitance and collector load. If the detector produces a square shaped pulse of current of magnitude i_o and duration T the voltage at the amplifier input is shown in Figure 3.6 and has a peak value given by

$$i_o R_i (1 - e^{-T/\tau_i}) \tag{3.4}$$

where $\tau_i = R_i C_i$. A true integration would give an output pulse proportional to $i_o T$ and (3.4) only approximates to this when $T \ll \tau_i$, i.e. when the input approximates to an impulse of current when the output has a form shown in Figure 3.7. The *ballistic deficit* is a measure of the inefficiency of integration and is the fractional deficit of the output pulse height compared to that for an infinitely short input pulse of the same charge. For the present case the ballistic deficit is e^{-T/τ_i}.

The importance of the ballistic deficit is where the input pulse shape varies due to statistical factors arising from, for instance, different locations and orientations of the ionization track in gas filled and semiconductor detectors. Unless the differently shaped input pulses are each integrated accurately some resolution would be lost. In scintillation counters the pulse shape is practically independent of the track position and, since the resolution is inherently poor, a much larger deficit is acceptable than in other cases.

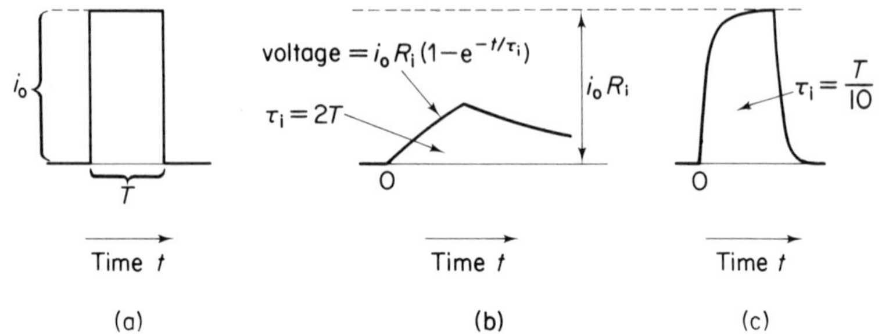

Figure 3.6 (b) and (c) show the voltage waveform observed at the preamplifier input for the detector signal shown in (a). (b) is for the case $\tau_i = 2T$ and (c) is for the case $\tau_i = T/10$

The pulse shaping effect of any system is usually expressed in terms of its impulse response, that is, the response to a unit impulse of current from the detector. To find the output pulse shape for any other shape of detector pulse the convolution method of Subsection 2.2.2 may be used. For a low ballistic deficit the output pulse for an input detector pulse of unit charge should approximate to the impulse response. This has been discussed in Subsection 2.2.2 when it was shown that this implies that the impulse response should have a top which is flat for a length of time comparable to the duration of the detector current pulse.

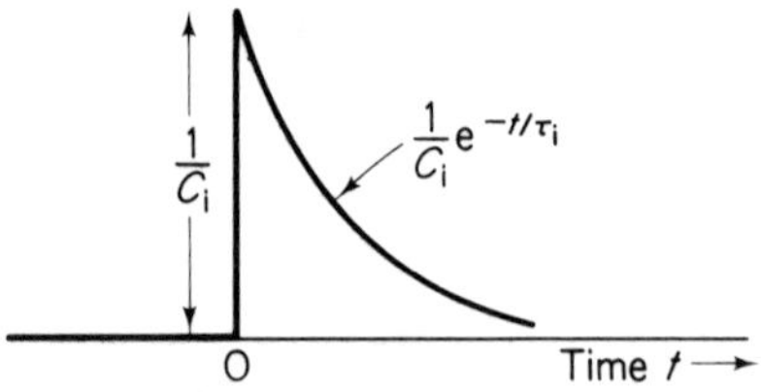

Figure 3.7 Impulse response of the input circuit, i.e. the voltage at the input of the preamplifier when a unit impulse of current is supplied by the detector

Examination of the impulse response (Figure 3.7) of the circuit of Figure 3.5 in the light of this criterion shows that, because of its sharp peak, it is most unsatisfactory. A further disadvantage of the arrangement is that there is no limitation of the high frequency response by the pulse shaping components.

Hence the noise arising from the earlier stages of the amplifier (Figure 3.8) is limited only by the inherent high frequency cut off of the amplifier itself. These disadvantages limit the application of such rudimentary pulse shaping to use with scintillation counters where the ballistic deficit and amplifier noise are unimportant in the context of the inherently poor resolution of the detector. Even for this case the method of pulse shaping is not very satisfactory as it requires the use of a pulse height analyser which can respond rapidly to the sharp peak of the pulse.

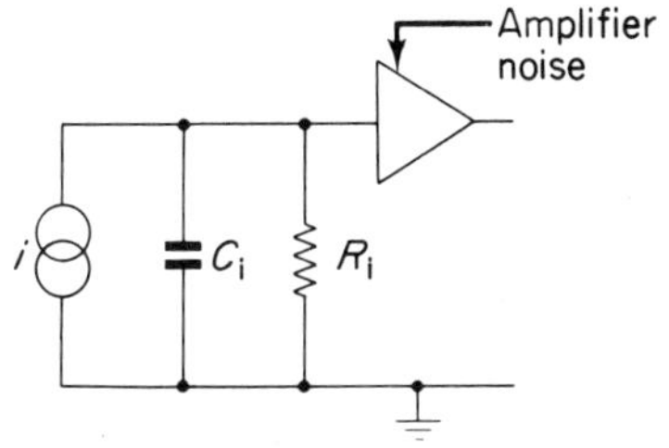

Figure 3.8 Input circuit with noise generated by amplifier

The calculation of the ballistic deficit for commonly used methods of pulse shaping has been done by Baldinger and Franzen[1] and by Hatch[2]. However, the loss of resolution arising from variations in ballistic deficit, which is of particular importance for semiconductor detectors has received little attention.[10]

3.1.5 *Unipolar and bipolar pulse shaping*

Ideally, an amplifier for pulses from a radiation detector should have a frequency response which extends do-vn to DC. However, the problems of using DC coupling between amplifier stages often precludes this. Usually the stages of the amplifier will be CR coupled and the amplifier can transmit no DC component of the signal. It therefore follows that no AC coupled amplifier can transmit strictly unipolar pulses. Since the order of the pulse shaping networks and amplifying stages is of no importance for the purpose of calculating the output pulse shape in a linear system, the effect of the amplifier can be seen more clearly if the pulse shaping is assumed to take place first and is followed by all the amplification.

The effect of CR coupling on pulse transmission has been mentioned previously (Subsection 2.1.2) and assuming the coupling time constant *RC* is large (i.e. the amplifier has a good low frequency response) the effect on pulse shape may be made negligible but the baseline will be shifted. Since the CR coupling can transmit no DC current, from a consideration of charge conservation it follows that for an input of truly unipolar square pulses shown in Figure 3.9(a) and of rate *n* per second the output will have the form shown in Fig. 3.9(b) where the shaded areas above and below the line of zero voltage are equal. In this case the baseline shift is nTV_o. A pulse height analyser would register the apparent pulse heights less that the true height by this amount.

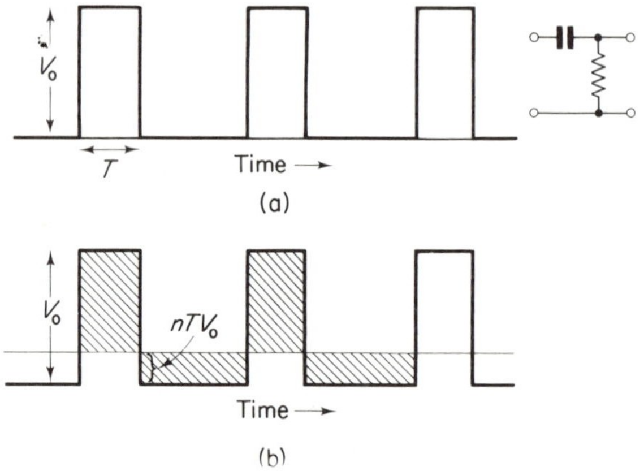

Figure 3.9 Effect of passing unipolar square pulses into a CR high pass filter. *n* is the rate of the pulses and the result is shown in (b)

Pulses from a radiation detector will have different heights and it is plainly undesirable to measure a pulse height spectrum which contains a baseline shift that depends on the nature of the source of pulses in an unknown manner. Unlike the case of periodic pulses of constant height the random arrival of nuclear pulses of varying size means the baseline shift will show fluctuations and this may worsen the resolution. Provided the CR coupling time constants are large enough this loss of resolution can be made negligible (Subsection 4.5.3) but this then introduces the possibility of other slow fluctuations and 50 Hz or 60 Hz interference being amplified and affecting the resolution.

For nominally unipolar pulses the effects of baseline shift and slow fluctuations are usually removed by using a 'baseline restorer' and the mode of action is discussed in Section 3.3. An alternative solution is to use output pulses of bipolar shape (Figure 3.10) having an area balance between positive and negative lobes of the pulse so the pulse has no DC component and no baseline shift would be produced. The bandwidth of the system then does not extend to very low frequencies and this provides a useful attenuation of slow parasitic fluctuations from other sources.

One disadvantage of using a bipolar pulse shape in a linear system is that the signal to noise ratio is nearly always worse than can be obtained with a unipolar pulse shaping method. (This is not necessarily true for non-linear methods employing bipolar pulses (Subsection 3.3.5 and Section 4.7).) On the other hand the use of a baseline restorer with a nominally unipolar pulse shape itself causes some degradation of the signal to noise ratio (Section 4.6). Another disadvantage of a bipolar pulse is that, for the same overall pulse length, the peak of the pulse would have a smaller region of approximate flatness than in a unipolar pulse (Figure 3.10) and so would have a larger ballistic deficit.

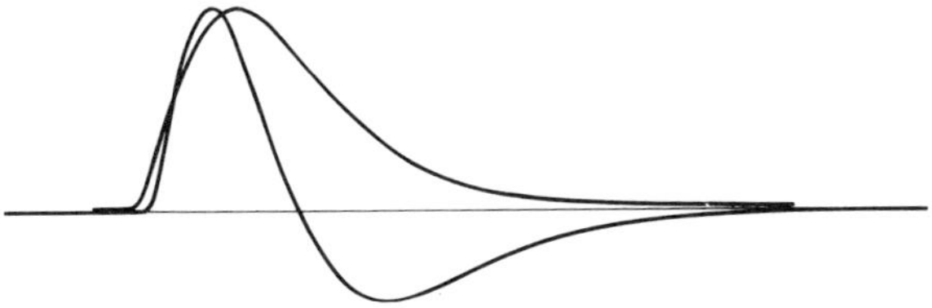

Figure 3.10 Comparison of unipolar and bipolar pulses of the same approximate overall length showing the greater region of flatness of the peak of the unipolar pulse

Theoretically the best solution is to use unipolar shaping with a DC coupled system. This is often not possible or, owing to parasitic baseline fluctuations, practicable. Where an AC coupled system is used, in general nominally unipolar pulses, with or without a baseline restorer, would be preferred for low counting rates and bipolar shaping for high counting rates.

3.2 Linear pulse shaping methods

3.2.1 Conditions at the detector

The equivalent circuit at the input is shown in Figure 3.8 where C_i includes the detector capacitance, amplifier input capacitance and stray capacitance. R_i includes the detector load R_L and any other resistances at the amplifier input. Since the amplifier will generate a component of noise independent of the values of C_i and R_i it follows that to maximize the ratio of the signal to this noise component the impedance $|Z_i|$ of the combination of C_i and R_i should be as large as possible. Thus C_i should be kept to a minimum and R_i made large. This optimum condition is derived more rigorously in Subsection 5.3.1.

The adoption of these conditions on C_i and R_i will have a certain effect on the pulse shape and on the ratio of the signal to *detector* noise. The effect of the *complete* system on these factors is not, however, in any way irrevocably restricted by this choice of C_i and R_i as its effect on these factors can, if necessary, be cancelled by subsequent circuitry further along the amplifier (e.g. by pole zero cancellation, Subsection 3.2.9).

There is an upper limit of the value of R_L, and hence to R_i, due to the necessity of maintaining a reasonable constant bias voltage across the detector at high counting rates. Figure 3.11 shows the form of the detector voltage just after counting has begun. If the mean counting rate is n and the mean charge in each pulse is $\bar{Q}$ the mean DC current flowing in the detector is $n\bar{Q}$ and the mean potential drop across the load R_L is $n\bar{Q}R_L$ (Figure 3.11).

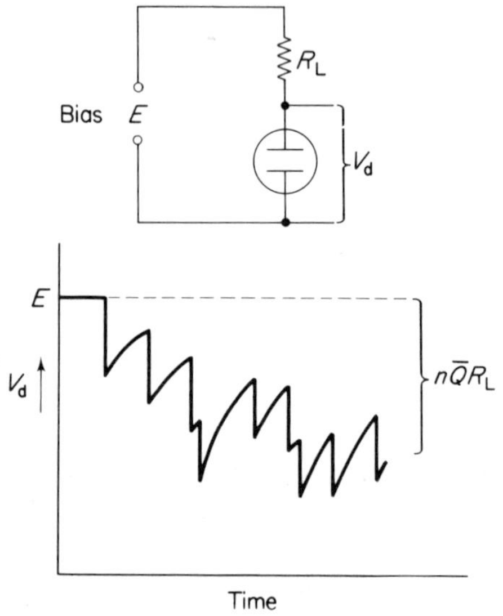

Figure 3.11 Time course of detector voltage just after counting has begun. n is the mean counting rate and $\bar{Q}$ the mean detector charge per count

Although this upper limit is not usually restrictive in practice, it may be circumvented if the load resistor is incorporated in a feedback arrangement around the preamplifier (Subsection 5.4.2). It is then possible to apply the biasing voltage through an equivalent resistance that is much lower without degrading the signal to noise ratio.

In the following discussion on pulse shaping circuits the response of the circuits to an input current pulse of infinitely short duration (i.e. an impulse) will always be considered. This impulse response will be a good approximation to the response to the actual detector current pulse, since, if the ballistic deficit is to be small, the two responses must be similar. In any case, if the impulse response is known it is a simple matter to find the response to the actual input pulse using the convolution method (Subsection 2.2.2). If the overall transfer function for (voltage out)/(current in) is known and has a simple form it is often possible to find the impulse response of the overall system by inspection (Subsection 2.2.1). The overall transfer function will, of course, equal the product of the transfer functions of the input circuit and those of the various pulse shaping circuits. In the following sections it will always be assumed that any amplifier sections have infinite input impedances and zero output impedances. Constant factors due to the gain of the amplifier sections will usually be omitted from the transfer functions. The voltage appearing at the amplifier input (Figure 3.5) when the detector delivers a current impulse containing a charge Q will have the form:

$$\frac{Q}{C_i}\mathrm{e}^{-t/\tau_i} \quad \text{for} \quad t > 0 \quad \text{where} \quad \tau_i = R_i C_i \tag{3.5}$$

and this is shown in Figure 3.12. If R_i is made very large as required by noise considerations the response can be approximated by a step function (Figure 3.12) of the magnitude Q/C_i so the input circuit approximates the

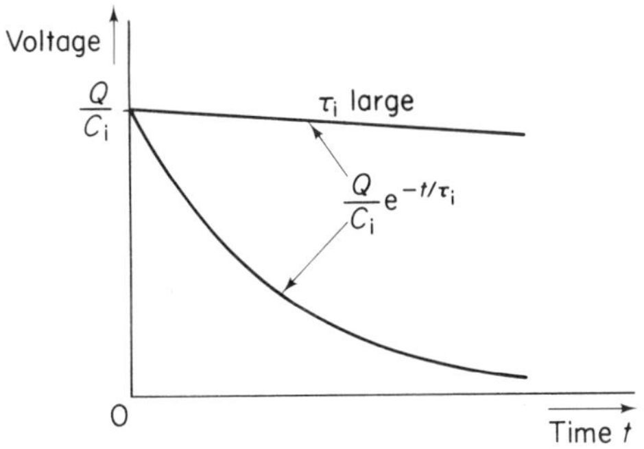

Figure 3.12 Voltage at the amplifier input when an impulse of current containing charge Q is passed by detector

action of an integrator. The transfer function of the input circuit for voltage out/current in is $\tau_i/(1 + s\tau_i)C_i$ so the approximation of $R_i \to \infty$ makes the transfer function approximate to $1/sC_i$. In terms of the pole zero diagram the pole at $-1/\tau_i$ is assumed to be at the origin. This is usually a good approximation but its validity and the means for making it exact are examined in Subsection 3.2.9.

If the signal from the input, assumed to be a perfect step function of magnitude Q/C_i is then passed into a high pass CR filter (Figure 3.13) of time constant τ_1 the output will have the form $(Q/C_i(\mathrm{e}^{-t/\tau_i}$, and this is shown in Figure 3.13 and has the same form as that discussed earlier. Except for use with a scintillation counter this pulse shape is unsuitable as it stands for the reasons given in Subsection 3.1.4. The disadvantages are removed if the signal is further passed through a low pass RC filter (Figure 3.13) of time constant τ_2. This system of pulse shaping is sometimes termed 'single RC differentiation and integration' since the two filters approximate these functions under some circumstances.

This terminology is rather unfortunate as obviously neither filter approximates these functions in the present context. On the other hand the input circuit does approximate to an integrator and this function is not included in the terminology. The term CR–RC shaping will be used in future.

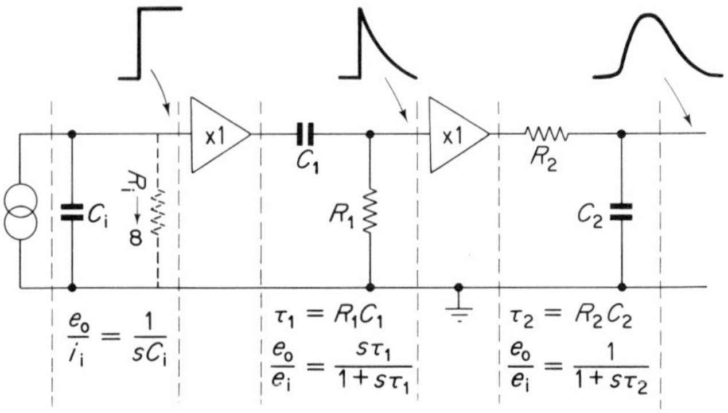

Figure 3.13 Complete equivalent circuit for detector and CR–RC shaping

3.2.2 *CR–RC shaping*

The complete system is shown in Figure 3.13. If R_i is assumed infinite the transfer function for voltage out/current in for the complete system is

$$\frac{\tau_1}{C_i(1 + s\tau_1)(1 + s\tau_2)} \tag{3.6}$$

The impulse response of the system has the form

$$\frac{\tau_1}{C_i(\tau_1 - \tau_2)}(e^{-t/\tau_1} - e^{-t/\tau_2}) \quad \text{for} \quad t > 0 \tag{3.7}$$

It will be seen that τ_1 and τ_2 are interchangeable in the above expressions as far as pulse *shape* is concerned. Figure 3.14 shows the form of the impulse responses normalized to the same height for various pairs of values of τ_1 and τ_2. Figure 3.15 shows the impulse response for pairs of τ_1 and τ_2 selected to give pulse widths approximately the same. Of the various pairs of values it will be seen that $\tau_1 = \tau_2$ gives the best combination of flat pulse top (for low ballistic deficit) and quick return to zero volts (for minimum pile up). Since the condition $\tau_1 = \tau_2$ also corresponds to maximum signal to noise ratio this condition is nearly always used. It may be noted that the expression

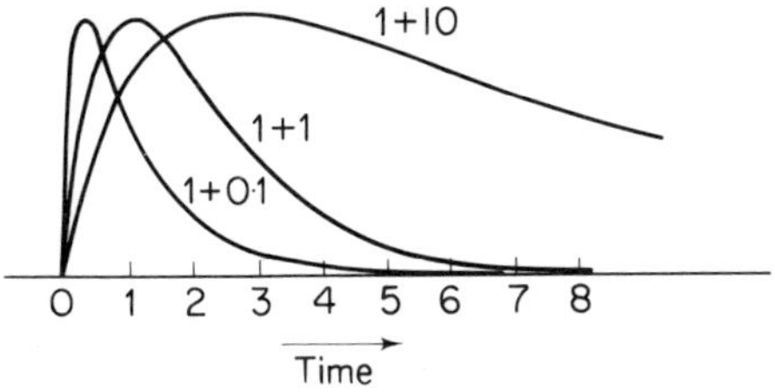

Figure 3.14 Pulse shapes normalized to constant height for CR–RC shaping. Shapes are plotted for various pairs of filter time constants on an arbitrary time scale

for pulse shape given by (3.7) becomes indeterminate for $\tau_1 = \tau_2$ and must be resolved by a procedure such as l'Hôpital's rule which then yields an impulse response of

$$\frac{t}{C_i\tau_1} e^{-t/\tau_1} \quad \text{for} \quad t > 0 \tag{3.8}$$

The actual value of $\tau_1(=\tau_2)$ in a practical system would be set by the requirements of ballistic deficit, pile up and noise considerations. For low noise semiconductor counters at low count rates optimum resolution might be obtained at a value of 2 μs whereas higher counting rates might give optimum resolution at 0·5 μs. Gas filled detectors with slow charge collection require times in the region of 2–10 μs. For scintillation counters the value of τ_1 is not usually critical and a value of 1 μs is typical. Determination of the time constant for optimum resolution under given conditions is best accomplished by trial and error. The simplicity of the system and the ease

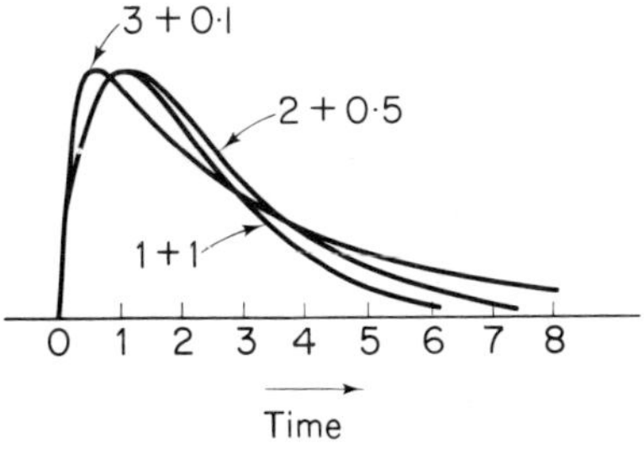

Figure 3.15 Pulse shapes normalized to constant height for CR–RC shaping. The pairs of time constants for which the shapes are shown were chosen to give the same approximate overall duration of the pulses

with which τ_1 and τ_2 may be adjusted makes CR–RC shaping one of the commonest methods of pulse shaping. However, compared to other methods of pulse shaping it does not exhibit a very good signal to noise ratio and the long tail of the pulse means it will give larger pile up errors which become important at high count rates.

The actual location of the filters in the amplifier chain deserves some consideration. If the signal on the detector really were a step function the voltage on it would build up from successive pulses (and it is this that puts an upper limit on the value of R_L). In practice the detector voltage would have the form shown in Figure 3.16(a) and in this form it is unsuitable for much amplification since it would soon overload the amplifier. The CR filtering should therefore be done before the input signal is amplified to a degree that it may overload the amplifier. On the other hand the pulse shaping networks should come after the significant noise producing stages of the amplifier.

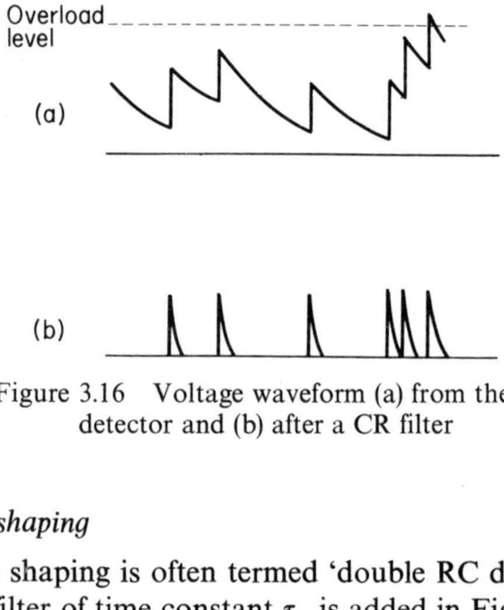

Figure 3.16 Voltage waveform (a) from the detector and (b) after a CR filter

3.2.3 $(CR)^2$–RC shaping

This type of pulse shaping is often termed 'double RC differentiation'. If a second high pass filter of time constant τ_3 is added in Figure 3.17 the pulse will be converted to a bipolar pulse. The transfer function for the whole system has the form

$$\frac{s\tau_1\tau_3}{C_i(1 + s\tau_1)(1 + s\tau_2)(1 + s\tau_3)} \tag{3.9}$$

It will be seen that τ_1,τ_2 and τ_3 are interchangeable with no effect on the form of the transfer function and hence on the shape of the impulse response of the whole system. Figure 3.18 shows the impulse responses normalized to constant height for various combinations of τ_1, τ_2 and τ_3 chosen to give approximately the same width for the initial part of the pulse. Of the various combinations the condition $\tau_1 = \tau_2 = \tau_3$ gives a slightly quicker return to

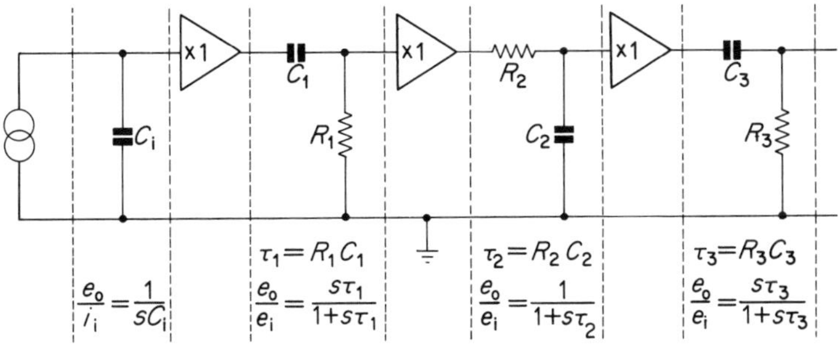

Figure 3.17 Equivalent circuit for $(CR)^2$-RC shaping

near-zero and this arrangement is most commonly used. $(CR)^2$–RC shaping is subject to the usual disadvantages of bipolar pulse shaping compared to unipolar shaping, namely, poorer signal to noise ratio and higher ballistic deficit for the same overall pulse length.

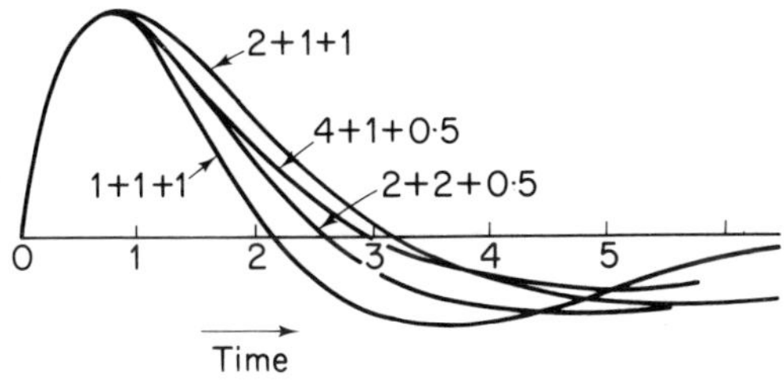

Figure 3.18 Pulse shapes normalized to constant height for $(CR)^2$–RC shaping with the time constants chosen to give approximately the same initial width of pulse

3.2.4 *Delay line pulse shaping*

The essential properties of a perfect delay line may be summarized as follows. When the delay line is terminated at both ends by its characteristic resistance R_o a voltage pulse applied to one end will appear unchanged in shape at the other end but delayed in time by $\sqrt{LC}$ where L and C are the inductance and capacitance per unit length of the line. If one end of the line is terminated by a resistance R the signal will be 'reflected' with a reflection coefficient $(R - R_o)/(R + R_o)$ and will reappear at the input end. In particular if an input pulse is applied at an end terminated with R_o, and the other end of the line is open circuit, the pulse will travel down the line and at the open circuit end will be reflected perfectly and will travel back to the input end. At the input end it will appear superimposed on any part of the original signal that may remain; it will not, however, undergo any further reflection since the input end is terminated with its characteristic resistance. If the far end of the line is short circuited a similar sequence will occur except that the reflection at the shorted end will be perfect but accompanied by a reversal of polarity.

3.2.5 *Single delay line (DL) shaping*

If the step function produced by the input circuit is applied to the end of a delay line through its characteristic resistance and the other end of the line is shorted (Figure 3.19), the voltage resulting at the input end of the line is

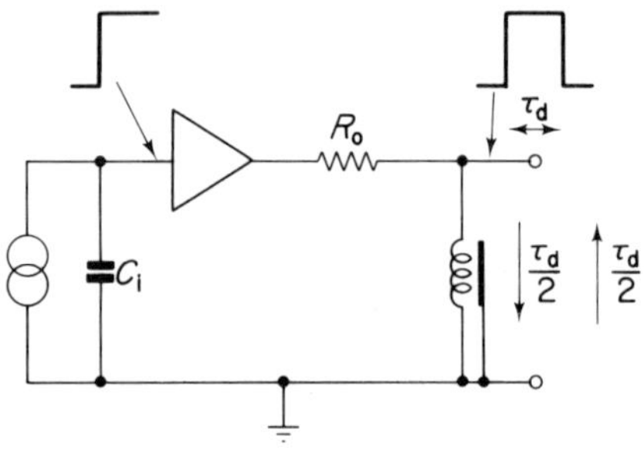

Figure 3.19 Single delay line shaping using a shorted delay line

shown in Figure 3.20 where τ_d is the total delay time of the line (i.e. the time for the signal to travel down and back). This resulting pulse shape is ideal from the point of view of ballistic deficit and has a rapid return to zero volts (i.e. a small pile up error). The transfer function for the delay line section is $\frac{1}{2}(1 - e^{-s\tau_d})$. In terms of frequency the square of the transfer function is $\frac{1}{2}(1 - \cos\omega\tau_d)$ so the delay line itself provides no limitation of the higher frequencies. This means that any noise generated by the amplifier, part of which will be 'white', i.e. covering ali parts of the frequency spectrum, would theoretically be of infinite magnitude at the output. In practice, of course, the high frequency cut off of the amplifier itself would limit the noise to a finite value. In terms of the transient response this is equivalent to saying that a truly square wave shaping could never be realized since the limited high frequency response of the amplifier would degrade the rise and fall times of the output pulse. However, the fact remains that the amplifier noise contribution is limited in an arbitrary fashion by the amplifier high frequency cut off and is much higher than in pulse shaping systems which themselves limit high frequency response. DL shaping (and $(DL)^2$ shaping discussed in the next subsection) are therefore only applicable where the detector signal is above the amplifier noise. This limits its application to the scintillation counter and, owing to the good performance with respect to pile up and ballistic deficit, DL and $(DL)^2$ shaping are often chosen. For use with other types of detector it is necessary to impose limits on the high frequency response of the system much more severe than those afforded by the inherent limitation of the amplifier itself. This is usually accomplished by adding a further low pass filter and this type of shaping is discussed in Subsection 3.2.7.

Unfortunately delay lines do not perform as perfectly as indicated. Fairstein[3] has reviewed the sources of malfunctioning and has examined the properties of several delay lines commercially available in 1962. The following imperfections were noted by Fairstein. The signal is partially attenuated on

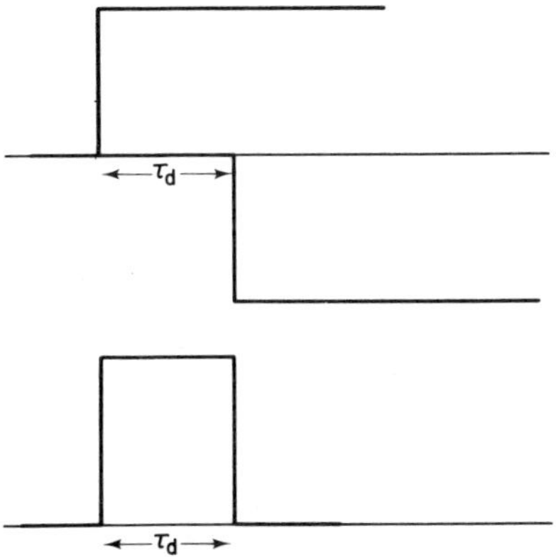

Figure 3.20 Addition of an input step function with the reflected step function to produce a square pulse

passing down the line and so incomplete cancellation with the input step function takes place. There are usually parasitic capacitances that render perfect termination impossible. The end turns of the helix of the line have mutual coupling in one direction only and there is a capacitive discontinuity introduced by external loading. These end effects result in ringing of the final signal waveform and multiple reflections in the line with a typical result shown in Figure 3.21. The end effects can to some extent be compensated out by using a small inductance in series with the line and also by adding other components to the input end. The attenuation of the line can be compensated out by subtracting from the signal a small proportion of the original step function. This involves using either a difference amplifier or an inverter and an adding network.

An arrangement commonly used is to terminate both ends of the delay line with its characteristic resistance as shown, for example, in Figure 3.22. The step function is inverted and applied to the delay line. The delay line output is then added to the original step function at point X in Figure 3.22. The input impedance of the following stage is made very small so that very

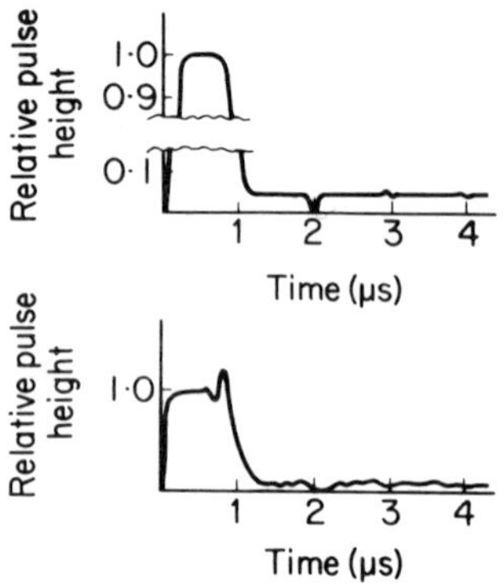

Figure 3.21 Practical examples of pulse shapes produced by shorted delay lines according to Fairstein[3]

little of the signal passing through the delay line finds its way back through to the delay line input and similarly for the undelayed part of the signal that passes through $2R_o'$. In practice a low amplifier input impedance can be easily achieved by the operational amplifier configuration shown in dotted lines in Figure 3.22. In this case the input resistance would be $R_f/(A + 1)$. If the delay line gives no attenuation the value of R_o' required would equal R_o in Figure 3.22 but to compensate for attenuation and the fact that the input is not a pure step function the value would be slightly different from this in practical cases.

In summary, for scintillation counters the advantages of DL pulse shaping are that, in theory, the pulse has a good shape from the point of view of

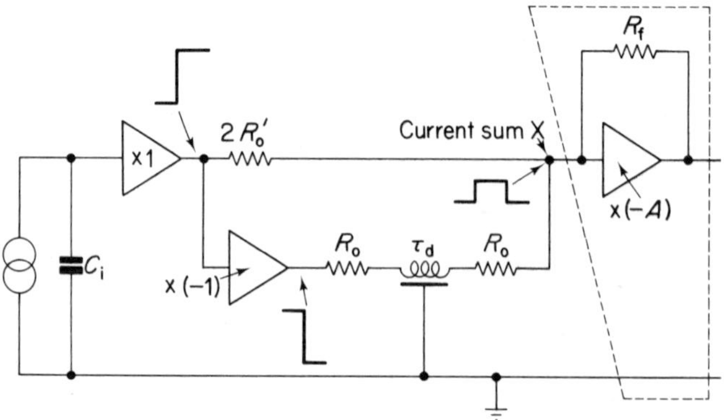

Figure 3.22 Alternative arrangement to produce delay line shaped pulses

ballistic deficit and of rapid return to zero volts (low pile up errors). Its noise performance usually precludes its use for other detector types. The delay line often introduces undesirable artifacts into the pulse and finally, to alter the pulse duration it is necessary to change the delay line length—a more difficult operation than that in the case of changing the time constant of RC filters.

3.2.6 *Double delay line* $(DL)^2$ *shaping*

If a second delay line stage is added having the same delay time as the first, the output pulse will have the bipolar form shown in Figure 3.23(a). As with single DL shaping there is no high frequency limitation due to the delay line

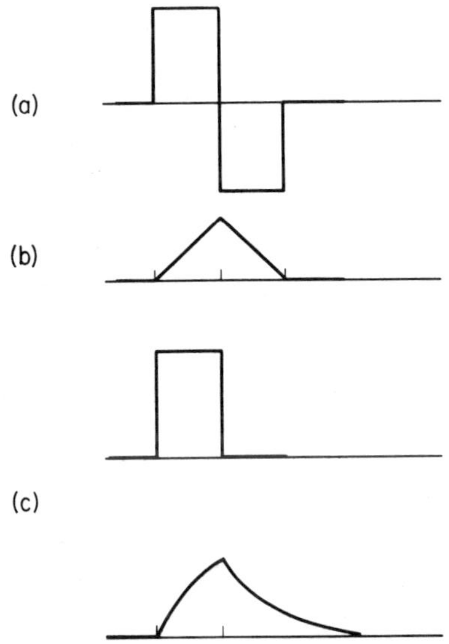

Figure 3.23 Production of a triangular pulse shape (b) by integration of a double delay line shaped pulse (a). Approximation to a triangular pulse obtained by passing a single delay line shaped pulse into a RC low pass filter

action and this confines its application effectively to scintillation counter systems where amplifier noise is of little importance. The advantages of $(DL)^2$ shaping are the same as with DL shaping namely, good pile up immunity and low ballistic deficit; but there is difficulty of adjustment of the pulse width.

3.2.7 *Delay line plus integration (triangular) shaping*

If an integrator circuit is placed after double delay line shaping networks the resultant triangular pulse shape is shown in Figure 3.23(b). This triangular shape is a close approximation to the ideal pulse shape (Figure 4.10) from the point of view of signal to noise ratio. As might be expected this method of pulse shaping therefore gives a better signal to noise ratio than any other methods considered so far. In addition the pulse retains the rapid and complete return to baseline characteristic of delay line shaping. As a consequence this method of shaping is capable of very good performance in some systems, however the ballistic deficit will be higher than for most other pulse shapes of the same width.

To produce a truly triangular pulse it is necessary to use a circuit after the delay lines which approximates to a true integration. A simpler configuration is often used, however, which produces an approximately triangular pulse shape at the expense of a signal to noise ratio only slightly less. This may be achieved by first shaping with a single delay line to give a rectangular pulse and passing this into a RC low pass filter. The time constant of the RC filter should be made equal to 0·96 times the delay time for the optimum signal to noise ratio and the resultant pulse shape is shown in Figure 3.23(c). It will be observed that the tail of the pulse has a slow decay, unlike the truly triangular pulse, and therefore will give a higher contribution of pile up errors.

3.2.8 *Semi-Gaussian shaping*

A truly Gaussian shaped pulse would have a shape given by $e^{-(t/\tau)^2}$ and would therefore extend to infinity on both sides of the lobe. Such a pulse could never be realized in practice since the pulse shaping circuits would take an infinite time to respond. Since the Gaussian pulse shape would give a signal to noise ratio near to the maximum theoretically attainable it is of interest to use pulse shaping circuits which given an approximately Gaussian shape. This may be achieved by a combination of a single CR high pass filter followed by several RC low pass filters. As the number of low pass filters approaches infinity so the pulse shape approximates more truly to the Gaussian shape. In practice four low pass filters are usually considered adequate and the signal to noise ratio for this is better than any other

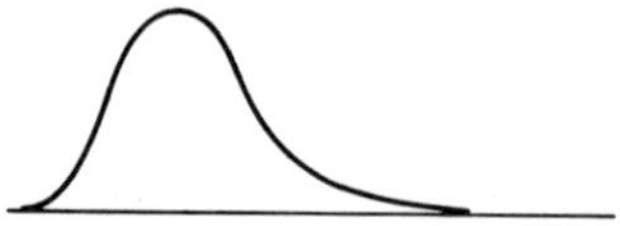

Figure 3.24 Semi-Gaussian pulse shape produced by using one CR high pass filter and four RC low pass filters

practical system considered apart from triangular shaping. This semi-Gaussian shape (Figure 3.24) has a fairly flat top and would be expected to give a smaller ballistic deficit than some other methods, however it does have a slowly decaying tail similar to that obtained with CR–RC shaping and so would be expected to give similar pile up errors.

3.2.9 Pole-zero cancellation

So far it has been assumed that the input circuit (Figure 3.5) has $R_i \to \infty$ which converts the detector current impulse into a voltage step function. The consequence of R_i being finite is that the voltage appearing at the amplifier input has the form $(Q/C_i)e^{-t/\tau_i}$ as shown in Figure 3.7, where $\tau_i = R_i C_i$. If this is now passed through the CR–RC shaping filter system of Figure 3.13 the exact form of the overall transfer function for voltage out/current in is given by:

$$\frac{s\tau_i\tau_1}{C_i(1 + s\tau_i)(1 + s\tau_1)(1 + s\tau_2)}$$

where τ_1 and τ_2, as before, are the time constants of the two filters. The impulse response for $\tau_i = 16\tau_1 = 16\tau_2$ is shown in Figure 3.25 and, as

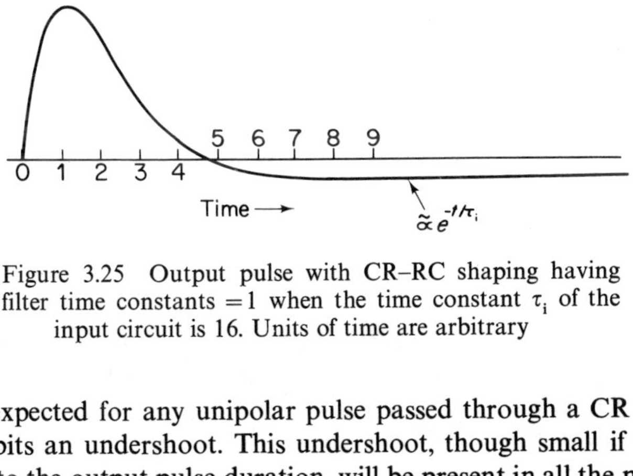

Figure 3.25 Output pulse with CR–RC shaping having filter time constants = 1 when the time constant τ_i of the input circuit is 16. Units of time are arbitrary

might be expected for any unipolar pulse passed through a CR high pass filter, exhibits an undershoot. This undershoot, though small if τ_i is large compared to the output pulse duration, will be present in all the methods of unipolar shaping discussed so far. The undershoot is undesirable under normal conditions (see Subsection 4.5.3) and is even more so when the amplifier passes a large pulse which will temporarily depress the baseline. The property of a rapid recovery of an amplifier from an overloading pulse is important and to prevent baseline shift where limiting is taking place the limiting part of the amplifier should always occur at a point where the pulse is strictly unipolar, as may be seen from the example quoted later.

The problem was solved in a simple way by Nowlin and Blankenship.[4] First it may be noted that the case of $R_i = \infty$ gave a transfer function of

$$\frac{\tau_1}{C_i(1 + s\tau_1)(1 + s\tau_2)} \tag{3.11}$$

which corresponded to an impulse response which was truly unipolar (Figure 3.14). If, therefore, the transfer function of the system can be made to be of this form the problem is solved.

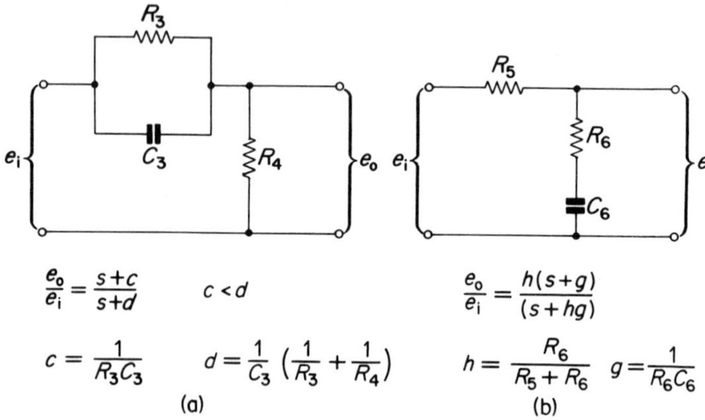

Figure 3.26 Pole-zero cancellation networks according to Nowlin and Blankenship.[4] (Research sponsored by the U.S. Atomic Energy Commission under contract with The Union Carbide Corporation)

The transfer function of the CR high pass filter of Figure 3.13 has the form

$$\frac{s\tau_1}{(1 + s\tau_1)}$$

The filter shown in Figure 3.26(a) has a transfer function $(s + c)/(s + d)$ where $c < d$. If we replace the original high pass CR filter with that in Figure 3.26(a) and make $c = 1/\tau_i$ and $d = 1/\tau_1$ the transfer function of the complete system, Figure 3.27, becomes

$$\frac{\tau_1}{C_i(1 + s\tau_1)(1 + s\tau_2)}$$

which is the desired form for giving a truly unipolar response. This technique is known as 'pole-zero cancellation' since the pole of the input circuit at $-1/\tau_i$ is cancelled by the zero of the filter at $-1/\tau_i$. The technique of pole-zero cancellation may be used to improve the overload characteristic of an amplifier system. In Figure 3.27 it is clear that the pulse at point A has the form e^{-t/τ_i}(for $t > 0$) and is converted by this network to one having the form e^{-t/τ_1}. If now a second pole-zero network is added having a transfer function proportional to $(1 + s\tau_1)/(1 + s\tau_x)$ the pulse shape is further converted to the form e^{-t/τ_x}. Clearly the process could be continued indefinitely and it is therefore possible to specify the decay time of the pulse to be of any value at any part of the amplifier. This is particularly valuable in improving overload recovery if it is arranged that the pulse has a short duration in the amplitude limiting section of the amplifier. For the purpose of lengthening

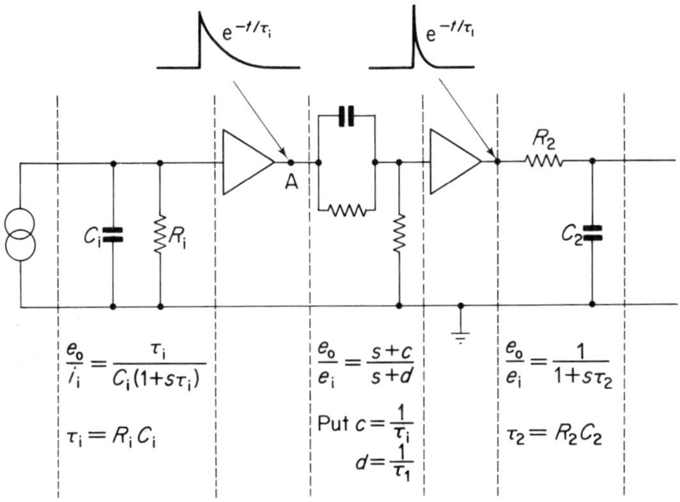

Figure 3.27 Use of pole-zero cancellation to give a truly unipolar pulse in CR–RC shaping. (From Nowlin and Blankenship.[4] Research sponsored by the U.S. Atomic Energy Commission under contract with The Union Carbide Corporation)

the decay time of the pulse a different circuit of the type shown in Figure 3.26(b) is required. To take the case quoted by Nowlin and Blankenship suppose that CR–RC pulse shaping with 10 μs time constant is ultimately required and the pulse at the amplifier input has a decay time τ_i of 1000 μs. This input pulse is first converted to one of decay time 1 μs (Figure 3.28) and later it is converted to a pulse with 10 μs decay time which can then be followed by the 10 μs RC low pass filter. If now the amplitude limiting stage of the amplifier is arranged to be in that part of the amplifier passing the 1 μs pulse, a 500 × overloading pulse would give a deadline of 6·2 μs. This is considerably better than the figure of 62 μs which would be obtained if amplitude limiting took place on a 500 × overloading 10 μs pulse. On the other hand for the simple system of pulse shaping without pole zero cancellation shown in Figure 3.13, not only will the positive part of the 10 μs overloading pulse give a deadtime of 62 μs but in addition the undershoot will account for a further dead time giving a total of 1620 μs although this would possibly be reduced by the use of a baseline restorer. This large contribution from the undershoot illustrates the importance of allowing amplitude limiting to occur only on pulses that are free from long undershoots.

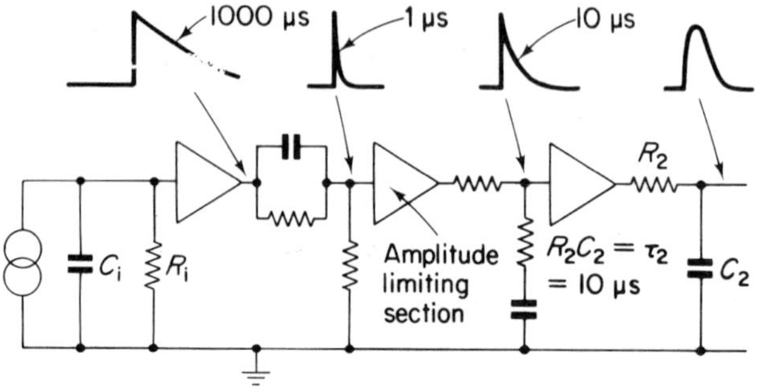

Figure 3.28 Use of two pole-zero cancellation networks in a CR–RC shaping system so as to shift the decay time of the pulse to 1 μs in the amplitude limiting section of the amplifier. (After Nowlin and Blankenship.[4] Research sponsored by the U.S. Atomic Energy Commission under contract with The Union Carbide Corporation)

Pole-zero cancellation can also be applied to shift the decay time of the pulse prior to delay line shaping with a consequent improvement of amplifier deadtime. In Subsections 3.2.5, 3.2.6 and 3.2.7, delay line shaping was only considered for a step function input. An input of the form shown in Figure 3.29(b) would, after a single delay line shaping, give the output shown which does have an undershoot. However this undershoot may be eliminated by altering the proportion of delayed and undelayed signal which are differenced as in Figure 3.29(c) and this may be simply accomplished, for example by adjustment of R_0' in Figure 3.22. However it is clear that the resultant waveform will not have a square shape unless the incident pulse has a decay time long compared to the delay time.

In practice it will not be possible to make the poles and zeros cancel exactly. Thus the first pole-zero filter in Figure 3.28 should ideally have a transfer function proportional to $(1 + s\tau_i)/(1 + s\tau_x)$ but in practice would have a transfer function $(1 + s\tau_i')/(1 + s\tau_x)$ where $\tau'_i \simeq \tau_i$. For the output of this filter to be unipolar it has been shown[4] that either τ_i or τ_x should be greater than τ_i'. To ensure this the value of τ_i' should therefore err on the low side of τ_i. In practice the value would probably be 'trimmed' by viewing the pulse of an oscilloscope until its shape just becomes unipolar.

Finally, it may be noted that a bipolar output pulse may be obtained in the usual ways after pole-zero cancellation has been used. The unipolar pulse is taken and passed through a high pass CR circuit for $(CR)^2$—RC shaping or in the case of double delay line shaping the unipolar pulse is passed into a second delay line stage.

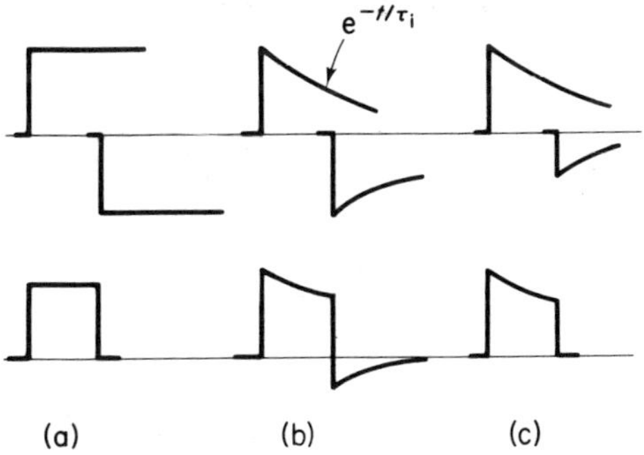

Figure 3.29 (a) Delay line shaping with a perfect step function from the input stage. (b) Delay line shaping of the input pulse shape when τ_i is finite, showing undershoot in the output pulse. (c) As in (b) but with different amplitudes of delayed and undelayed pulses to eliminate undershoot

3.3 Baseline restorers

3.3.1 Purpose of baseline restoration

As mentioned earlier AC coupled amplifiers cannot transmit truly unipolar pulses. Figure 3.30(a) shows the output of a CR coupling network when the input is a train of unipolar square pulses. The essential problem is to restore the baseline of the output to zero volts. Although bipolar pulses need no baseline restorer under ideal conditions, the output of the amplifier will often contain slowly changing voltages from several sources such as 50 Hz or 60 Hz ripple, drifts and long pulse tails from imperfect pole zero cancellation together with statistical fluctuations arising from the random nature of the signals (Subsection 4.5.3) so that the use of a baseline restorer with bipolar pulses often effects an improvement in resolution. A baseline restorer is usually placed immediately before the pulse height analyser to which, of course, it must be DC coupled.

3.3.2 Simple diode restorer

Consider a train of positive pulses applied to the circuit of Figure 3.30(a). Assuming the diode conducts perfectly when point A is negative the output waveform is shown in Figure 3.30(b) on the assumption that the diode is suddenly connected into the circuit at the instant T_1. During the negative going voltage at time T_2 the diode will conduct strongly and the capacitor will charge up with the polarity shown in Figure 3.30(b) until it acquires a potential V_B across it when conduction ceases. In a more realistic case where the signal has a finite source resistance and the diode has a finite forward resistance, together constituting a resistance r, the capacitor will not quite

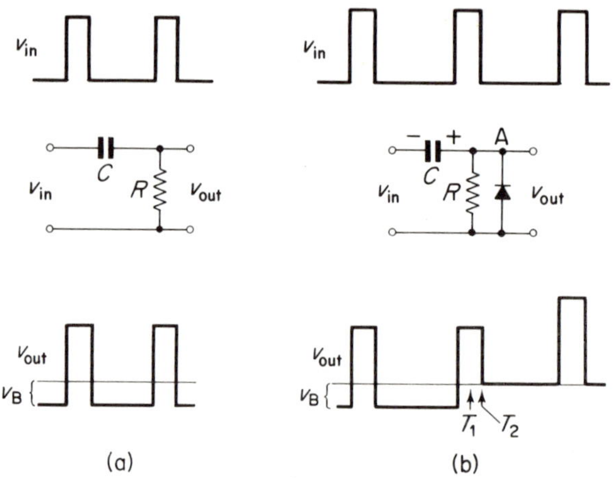

Figure 3.30 (a) Output waveform of a CR filter with a square pulse input. (b) Output waveform when a diode is connected in the circuit at instant T_1

acquire the voltage V_B. The capacitor will slowly discharge between pulses via R with a time constant RC and this charge will be replaced by the next pulse with a time constant rC. Thus the diode will clamp the baseline voltage to practically zero if RC is large compared to the time interval between pulses.

For the circuit to function $R \gg r$ so that the charging and discharging rates of the capacitor are widely different and the restoring action is not symmetrical with respect to polarity. The essential problem with the simple diode restorer is that it cannot operate properly if the pulses are of differing height and have overshoots or if there is noise present. Consider the behaviour when the two pulses shown in Figure 3.31(a) are presented to the restorer that was previously quiescent with the output at zero volts. At point A in Figure 3.31(a) the diode conducts and the restoring action begins with the time constant rC until point B. Figure 3.31(b) shows the restorer output. After point B the signal becomes positive-going and the diode ceases to conduct so that it is not the baseline but the trough of the undershoot that has been restored to zero. A smaller pulse may fail to bring the voltage low enough for the diode to conduct at all so the restoring action is lost until the capacitor has partially discharged through R with a long time constant RC.

Similar considerations apply when there is noise present. Thus the essential problem with the simple diode restorer is that once an undershoot has been restored to zero the restorer cannot reset the baseline (or subsequent smaller undershoots for that matter) except by the charge leaking away through R.

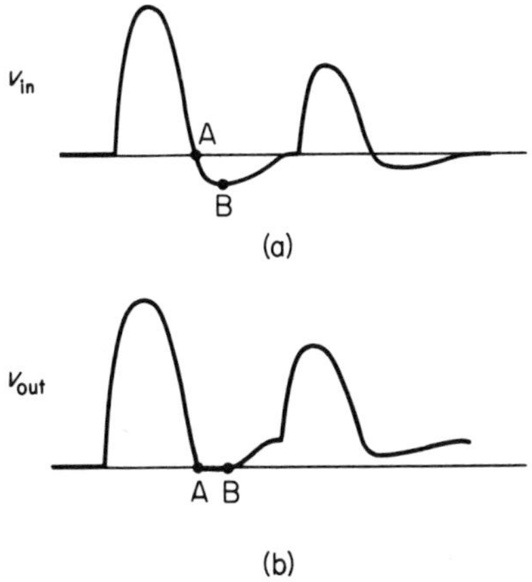

Figure 3.31 Simple diode restorer. Effect of undershoot causing the restoring action to be lost for the following pulse

3.3.3 *The Robinson restorer*

The problem with the simple diode was that it restored in one direction only. The Robinson restorer[5] restores equally well for positive-going or negative-going signals. In Figure 3.32 diodes D1 and D2 each conduct quiescent currents I_0 supplied by a constant current source $2I_0$ and, for D1, taken away by a constant current sink I_0. To understand the functioning of the circuit it is useful to examine the current/voltage characteristic of point A in Figure 3.32.

If a small current i were injected into the circuit at point A the current would flow through D1 and D2 to ground, decreasing the current in D1 to $I_0 - i$ and increasing the current in D2 to $I_0 + i$. The voltage at A will therefore be $2ir$ where r is the forward resistance of each diode, assumed to be constant. If the voltage at A is now increased i will increase until it becomes equal to I_0 when the total current through D1 becomes zero and D1 becomes non-conducting. Any further increase in the voltage at point A will drive D1 further into the non-conducting region and the current i will not increase above the value I_0. A similar situation occurs for negative voltages at point A except it is D2 that becomes non-conducting and the current i cannot decrease below the value $-I_0$. The current/voltage characteristic is therefore

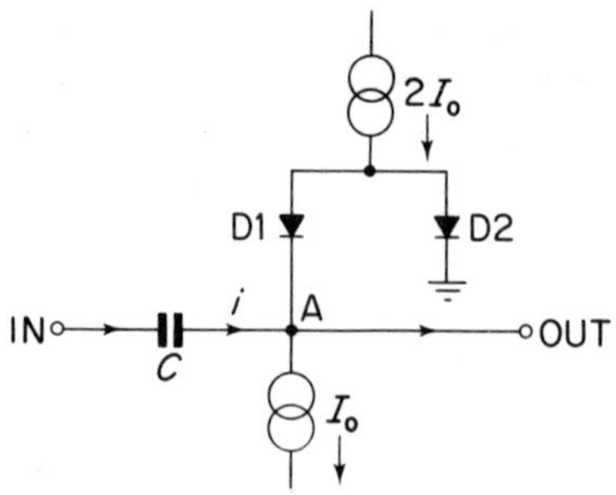

Figure 3.32 The baseline restorer of Robinson[5]

plotted in Figure 3.33 and, as shown, the terms 'linear' and 'saturated' will be used to describe the different regions of the characteristic.

Consider the application of the pulse shown in Figure 3.34 to the complete restorer initially in quiescent state. Immediately before the pulse the diodes are in the linear part of their characteristic. If v_i is the input voltage at any instant and C is the coupling capacity the current flowing into the restorer will be $C(dv_i/dt)$ since r is small. If $C(dv_i/dt) > I_0$ the diodes will enter the saturated region of their characteristic. For the signal of Figure 3.34 we will assume this occurs almost immediately after the leading edge of the pulse. With the diodes saturated the output voltage will therefore be given by $v_0(t) = v_i(t) - I_0Ct$ and this is shown in Figure 3.34 together with a plot of $C\int i dt$ on the same scale. Eventually $v_0(t)$ reduces so the diodes re-enter the linear region (point A in Figure 3.34) and the diodes may saturate in the

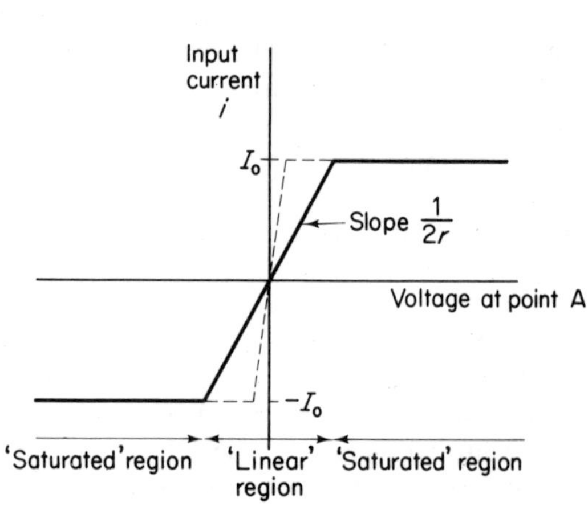

Figure 3.33 Current/voltage characteristic of the Robinson restorer. The dotted line is the characteristic of the Chase–Poulo restorer

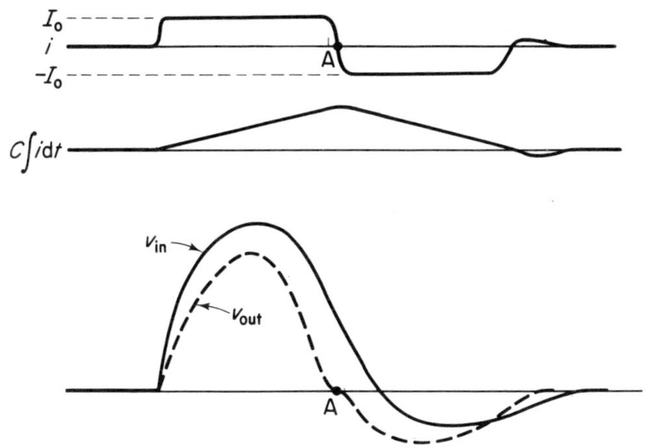

Figure 3.34 Operation of the Robinson restorer. $v_0(t) = v_i(t)$
$C\int i\,dt$

opposite direction, depending on whether $C(dv_i/dt) < -I_0$. This is obviously the case in Figure 3.34 and the output pulse acquires a brief undershoot before returning to zero volts.

The factor I_0/C determines the rate at which the restorer returns to the baseline. For the input pulse of Figure 3.34 a small value of I_0/C gives a shallow overshoot and a large value gives a deep overshoot although there is little difference in the lengths of the overshoots. If the input pulse has a long shallow tail as in Figure 3.35(a) the restorer may be used to shorten the tail. For bipolar input pulses the restorer functions equally well as shown in Figure 3.35(b). A restorer is useful even for bipolar pulses to remove slow baseline fluctuations and it is also useful after an overloading pulse in which the area balance is lost with consequent baseline shift. Various modifications of the Robinson restorer have been described. Chase and Poulo[6] used the arrangement shown in Figure 3.36 and this configuration reduces the

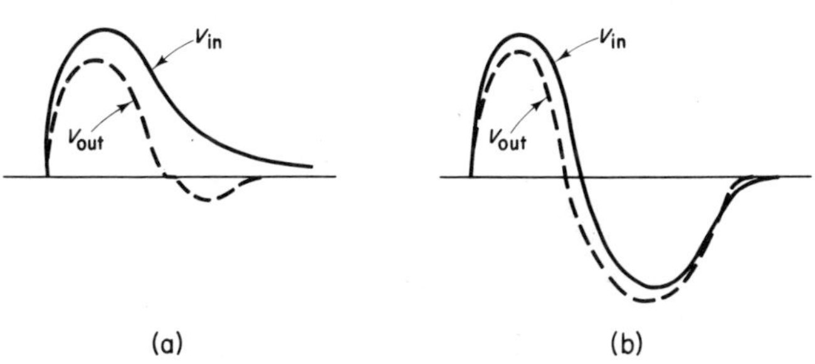

Figure 3.35 Action of the Robinson restorer on (a) a pulse with a long tail (b) a bipolar pulse

effective forward resistance of the diodes to yield a characteristic which is shown in Figure 3.33. Other 'active' restorers are in use[7, 9] and these differ from the Robinson circuit in that the restoration rate is adjustable or the characteristic is asymmetric with respect to voltage.

The use of any of these restorers in an existing system usually worsens the signal to noise ratio (Section 4.6). This can be offset or eliminated by adjusting the pulse shaping circuits to give a longer pulse but this will result in higher pile up errors.

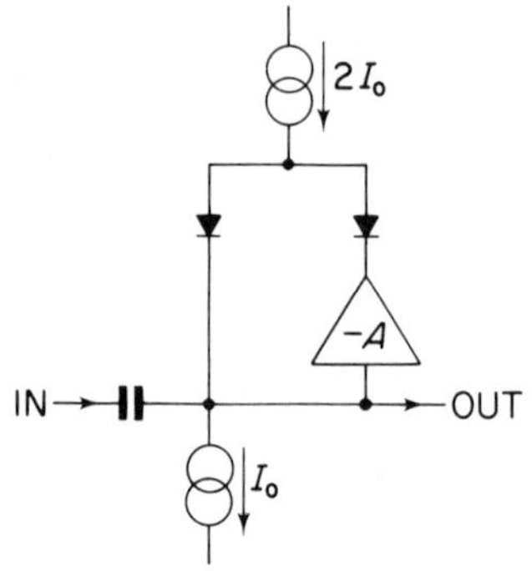

Figure 3.36 Modification by Chase and Poulo[6] of the Robinson restorer

3.3.4 *The Gere–Miller restorer*

The Robinson restorer restores the baseline to zero and enables the pulse height analyser to measure pulse heights in the normal way. The Gere–Miller[8] restorer, Figure 3.37(a), restores the pulse peak (or its undershoot) to zero and so the pulse height analyser measures the peak to undershoot voltage.

To understand the operation of the circuit consider the voltage/current characteristic of point A in Figure 3.37(b) where the diode is assumed to

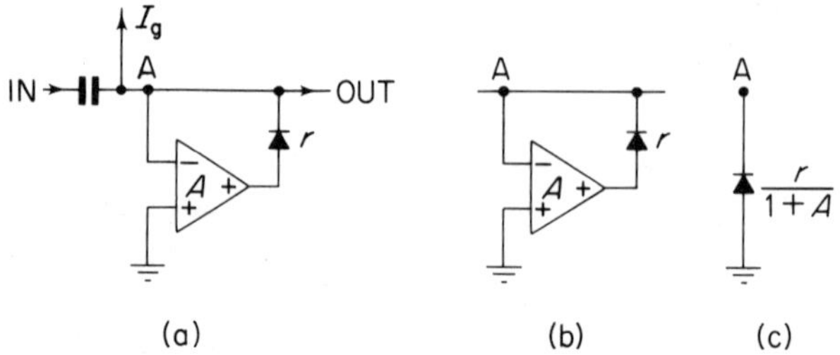

Figure 3.37 (a) Gere–Miller restorer.[8] (c) Equivalent circuit for the circuit of (b)

have a forward resistance of r and the amplifier has a gain of $-A$. If the voltage at A, V_A, is negative a current $(1 + A)V_A/r$ flows while for positive values of V_A the diode becomes non-conducting. Therefore the arrangement of Figure 3.37(b) is equivalent to a diode of low forward resistance $r/(1 + A)$ between point A and ground as shown in Figure 3.37(c).

In the complete restorer, Figure 3.37(a), the gating current I_g flows so the equivalent diode is held conducting and presents a low resistance $r/(1 + A)$ to incoming signals and negligible transmission results. Naturally I_g must be large enough to maintain conduction for all signals, i.e. $I_g > C(dv_i/dt)_{max}$. On removing I_g the restorer behaves like a simple diode restorer. Consider the situation when the signal and gating signal shown in Figure 3.38 are applied to the restorer. At the instant T_1 before the gating signal arrives the equivalent diode is held conducting by I_g and the output is essentially zero.

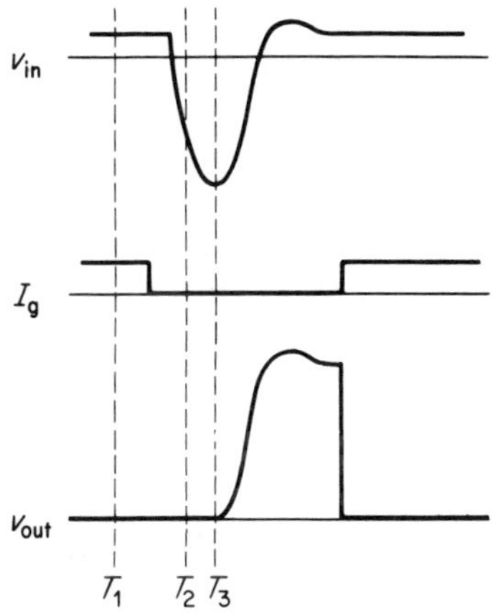

Figure 3.38 Operation of the Gere–Miller restorer

At time T_2, although the gating current is now zero the equivalent diode is still held conducting by the negative-going signal until the signal becomes positive-going at time T_3. The equivalent diode then becomes non-conducting until the end of the gating signal when the re-establishment of I_g forces the equivalent diode to conduct and produces a zero output.

The gating signal associated with each pulse must be generated by separate circuitry and must begin before the main peak and end after the undershoot (or overshoot). This requirement of a separate gating signal is a disadvantage of the Gere–Miller restorer. With nominally unipolar pulses the signal to noise ratio is usually worsened but with bipolar pulses it is sometimes possible to improve the signal to noise ratio (Section 4.6).

3.3.5 Other methods

Bertolaccini and Bussolati[9] have pointed out that, with pulses having a large undershoot, the action of the Robinson type of restorer and its variants may actually prolong the undershoot (Figure 3.39) if the restoration rate is low.

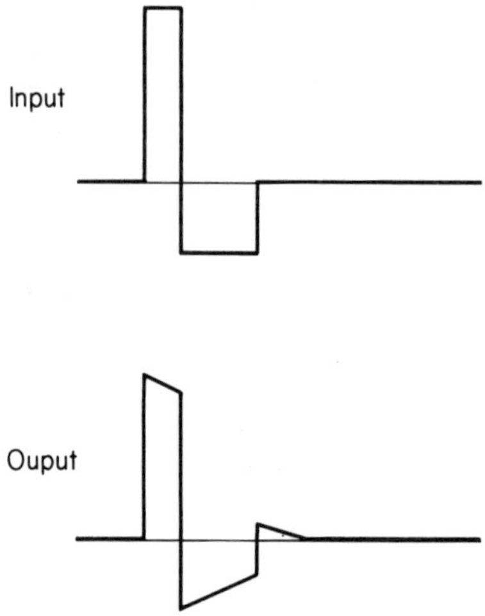

Figure 3.39 Effect of a baseline restorer of the Robinson type on a large undershoot, according to Bertolaccini and Bussolati[9]

They therefore suggested the use of a restorer using an element having the current/voltage characteristic shown in Figure 3.40, where the parameters R, V_1 and V_2 are all adjustable. The restorative action is therefore only operative when the output voltage is between V_1 and V_2, and the rate of restoration is controlled by the magnitude of R. With this circuit the undershoot cannot be prolonged as the restorative action only takes over when the undershoot has approached zero (i.e. within V_1 or V_2).

As with other types of restorer the magnitude of R also affects the signal to noise ratio and its optimum value depends on the magnitude and form of the noise. The optimum values of V_1 and V_2 depended on the pulse shape, pulse height distribution and the amplitude of the noise pulses. Although it was not possible to generalize on the rules for choosing the optimum values of R, V_1 and V_2, it was claimed that in all cases examined it was possible to find a combination of values which gave a performance better than that of other types of baseline restorer which were tested.

The addition of any baseline restorer to a linear pulse shaping system converts it into a non-linear system. Many other types of non-linear pulse

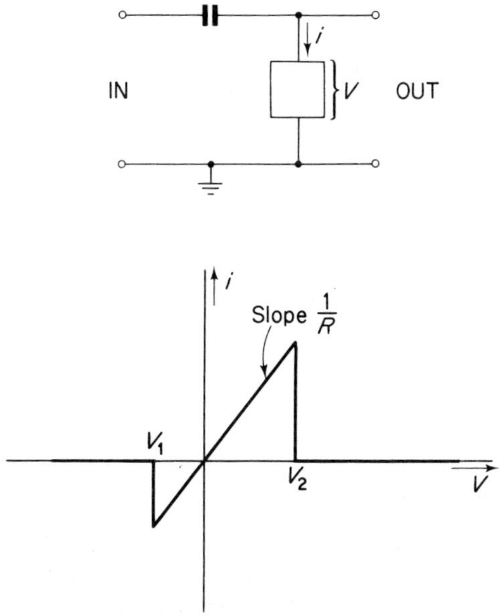

Figure 3.40 Principle of the baseline restorer of Bertolaccini and Bussolati[9]

processing systems have been described and in most cases the object is to realise a system which handles bipolar pulses, and is therefore AC coupled, but which gives a better performance with respect to signal to noise ratio and pile up than that obtained by linear methods. These topics are fully discussed in Section 4.7 and it may only be noted here that baseline restoration methods are a special case of a more general class of non-linear processing.

References

1. E. Baldinger and W. Franzen, *Advances in Electronics and Electron Physics*, **8**, 255 (1956).
2. K. Hatch, *IEEE Trans. Nucl. Sci.*, NS **15**, No. 1, 303 (1968).
3. E. Fairstein, in *Nuclear instruments and their uses* (Ed. A. H. Snell), Vol. 1, Wiley, 1962.
4. C. H. Nowlin and J. L. Blankenship, *Rev. Sci. Instr.*, **36**, 1830 (1965).
5. L. B. Robinson, *Rev. Sci. Instr.*, **32**, 1057 (1961).
6. R. L. Chase and L. R. Poulo, *IEEE Trans. Nucl. Sci.*, NS **14**, No. 1, 83 (1967).
7. C. W. Williams, *IEEE Trans. Nucl. Sci.*, NS **15**, No. 1, 297 (1968).
8. E. A. Gere and G. L. Miller, *IEEE Trans. Nucl. Sci.*, NS **14**, No. 1, 89 (1967).
9. M. Bertolaccini and C. Bussolati, *Nucl. Instr. and Meth.*, **100**, 349 (1972).
10. M. G. Strauss, R. N. Larsen and L. L. Sifter, *Nucl. Instr. and Meth.*, **46**, 45 (1967).

4

Resolution in Spectroscopy Systems

4.1 Noise

4.1.1 Statistical nature of noise

The signal from the detector will be contaminated by noise generated by the detector itself and the amplifier. In order to construct an optimum filter to maximize the signal to noise ratio it is necessary to take account of the form and properties of both signal and noise. In this section certain statistical properties of electrical noise will be considered. These properties are dealt with in the many standard texts on the subject.[1, 2, 3]

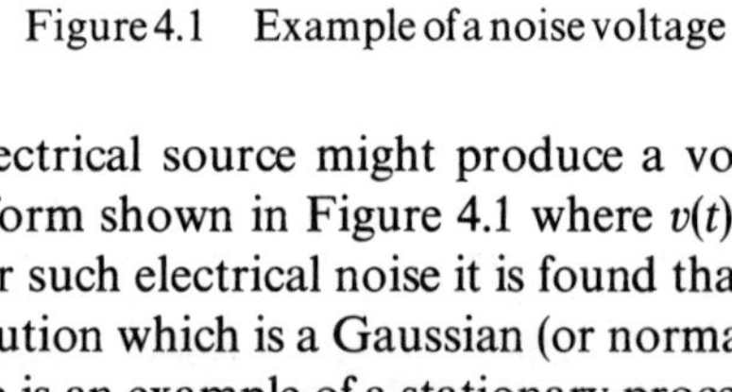

Figure 4.1 Example of a noise voltage

Noise from an electrical source might produce a voltage at the point of interest having the form shown in Figure 4.1 where $v(t)$ is the instantaneous voltage at time t. For such electrical noise it is found that the value of $v(t)$ has a probability distribution which is a Gaussian (or normal) distribution about zero. Electrical noise is an example of a stationary process, that is, one that is temporally homogeneous, and its properties are unaffected by a shift in the time origin. The probability of observing a voltage in the region v to $v + \mathrm{d}v$ is

$$\frac{1}{\sigma(2\pi)^{\frac{1}{2}}} \mathrm{e}^{-(v^2/2\sigma^2)} \, \mathrm{d}v$$

where σ^2 is the variance, or in this context, the mean square voltage, and so σ is the standard deviation or root mean square voltage. This function is shown in Figure 4.2. Such a noise source will be completely specified if either its autocorrelation function $\phi(\tau)$ or its power spectrum $F(\omega)$ is known. The definition and interrelationship of these two quantities will be given later. Before this it is necessary to review some elementary properties of probability distributions.

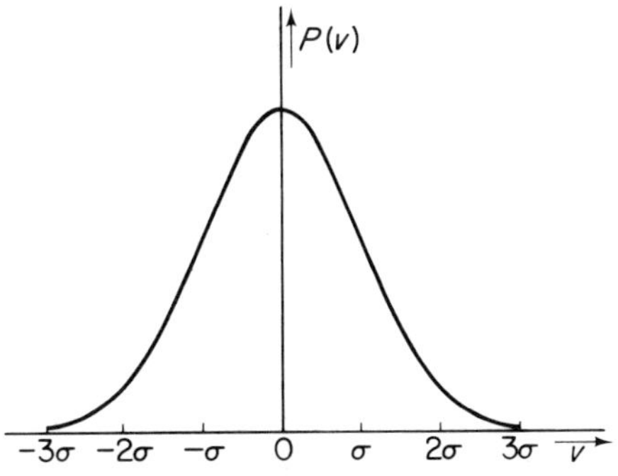

Figure 4.2 Gaussian distribution giving the probability $P(v)\,dv$ of observing a noise voltage in the range v to $v + dv$ at any instant

If x and y are any two variates, that is, quantities possessing probability distributions, then it is shown in elementary textbooks on statistics that the mean or expectation value of $x + y$ is equal to the sum of the means of x and y. The variates x and y are said to be statistically independent if the probability that x will assume a value in a prescribed range $x + dx$ does not depend on the value assumed by y and vice versa. Voltages produced by two independent noise sources are examples of independent variates. For the special case of two such independent variates the variance of $x + y$ equals the sum of the individual variances of x and y. A practical example of this is if two or more independent noise voltages are added, when the resulting mean square voltage is equal to the sum of the mean square voltages each noise source would produce on its own.

If, in addition to being independent, x and y each have Gaussian probability distributions of variances σ_x^2 and σ_y^2 then the distribution of $x + y$ is also given by a Gaussian distribution, and of variance $\sigma_x^2 + \sigma_y^2$. For independent electrical noise voltages added together, it follows that the resultant voltage has a Gaussian distribution about zero and the resultant mean square voltage equals the sum of the mean square voltages produced by the individual sources on their own. Since mean square voltages from such sources are additive, it is usual to characterize noise in mean square voltage or the square root of this quantity, the root mean square (RMS) voltage. Similar considerations apply also to noise currents.

For two independent variates the mean value of the product of the variates is equal to the product of the means of the two variates, i.e. zero for independent noise voltages. If now we consider a single noise source and select as our pair of variates the voltages $v(t)$ and $v(t + \tau)$ at any two instants separated by a constant time or lag τ (Fig. 4.3) it cannot be assumed that $v(t)$ and $v(t + \tau)$ are independent so the value of the product averaged over many values of t will not, in general, be zero. The value of this product averaged over infinite time is termed the autocovariance function or, more commonly, the autocorrelation function and is given by

$$\phi(\tau) = \lim_{T\to\infty} \frac{1}{T} \int_{-T/2}^{T/2} v(t)\, v(t + \tau)\, dt \tag{4.1}$$

$\phi(\tau)$ is a function of τ only and not of t since we have assumed the process is stationary. For a stationary Gaussian noise source $\phi(\tau) = \phi(-\tau)$, and so a knowledge of $\phi(\tau)$ for all positive values of τ completely characterizes it.

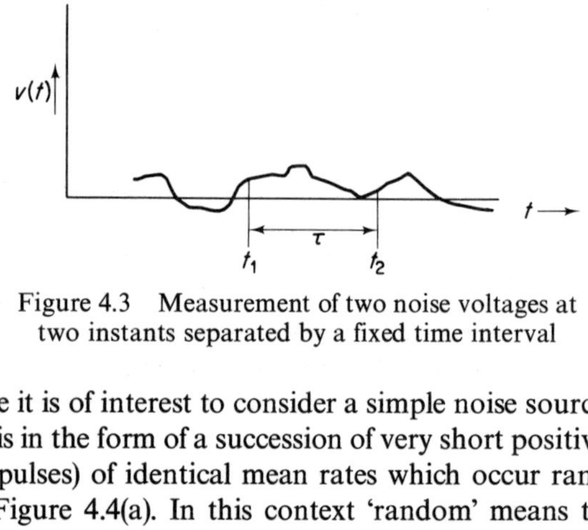

Figure 4.3 Measurement of two noise voltages at two instants separated by a fixed time interval

At this stage it is of interest to consider a simple noise source. Suppose the noise voltage is in the form of a succession of very short positive and negative pulses (i.e. impulses) of identical mean rates which occur randomly in time as shown in Figure 4.4(a). In this context 'random' means the probability that an impulse occurs in the time span t to $t + dt$ is constant and is therefore independent of t. In such a case $\phi(\tau)$ would be zero for all values of τ except $\tau = 0$ and would have the form of a delta function shown in Figure 4.4(b).

If the noise voltage is now passed through a low pass RC filter of time constant τ_1, it will have the form shown in Figure 4.4(c) and $\phi(\tau)$ will not be zero for $\tau \neq 0$. $\phi(\tau)$ will have the form shown in Figure 4.4(d), that is, a curve having a shape (1) given by $e^{-|\tau|/\tau_1}$.

The formal procedure for finding the autocorrelation function, $\phi_2(\tau)$, at the output of a circuit when the input has an autocorrelation function $\phi_1(\tau)$ is as follows. If $h(t)$ is the impulse response function of the circuit we can define an autocorrelation function of the circuit as

$$\rho(\lambda) = \int_{-\infty}^{\infty} h(t)\, h(t + \lambda)\, \mathrm{d}t \qquad (4.2)$$

$\phi_2(\tau)$ is then given by

$$\phi_2(\tau) = \int_{-\infty}^{\infty} \rho(\lambda)\, \phi_1(\tau - \lambda)\, \mathrm{d}\lambda \qquad (4.3)$$

The autocorrelation function is a method of characterizing the noise in the time domain. An alternative method is to characterize the noise in the frequency domain. By means of the Fourier transform it was shown that a single pulse could be characterized in the frequency domain (Appendix A).

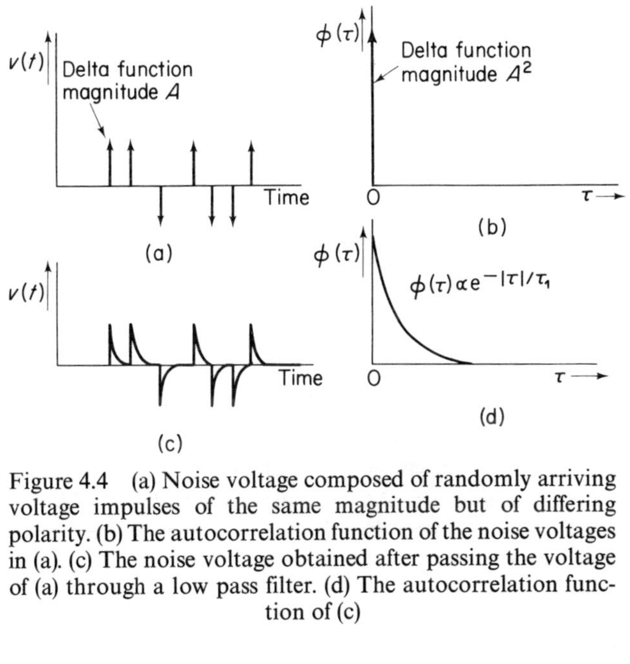

Figure 4.4 (a) Noise voltage composed of randomly arriving voltage impulses of the same magnitude but of differing polarity. (b) The autocorrelation function of the noise voltages in (a). (c) The noise voltage obtained after passing the voltage of (a) through a low pass filter. (d) The autocorrelation function of (c)

For a succession of pulses occurring randomly in time it may be assumed that the contributions from the individual pulses to a given frequency component would be in random phase. All frequency components would therefore add in a similar manner and it would be expected that the resultant spectral density would be the same as for a single pulse. Thus one might guess that the spectral density of the noise in Figure 4.4(a) would be the same as that of an impulse, that is having all frequencies present in equal proportion. Such noise is termed 'white' noise by analogy with white light.

Since the transfer function of a circuit is usually given in terms of its frequency dependence it is usually more convenient to characterize the noise in terms of its spectral density or power spectrum $F(\omega)$. Using this formulation the mean square voltage is given by

$$\int_0^\infty F(\omega)\,\mathrm{d}\omega \tag{4.4}$$

where ω is the angular frequency ($=2\pi f$). If this voltage is applied to, say, a low pass RC filter having a transfer function $1/(1 + s\tau_1)$, the noise at the output has a spectral density of $F(\omega)/|(1 + \mathrm{j}\omega\tau_1)|^2$, that is, $F(\omega)/(1 + \omega^2\tau_1^2)$ and the mean square output voltage is given by

$$\int_0^\infty \frac{F(\omega)}{1 + \omega^2\tau_1^2}\,\mathrm{d}\omega \tag{4.5}$$

The relationship between the spectral density $F(\omega)$ and the autocorrelation function $\phi(\tau)$ is given by

$$\phi(\tau) = \int_0^\infty F(\omega)\cos\omega\tau\,\mathrm{d}\omega \tag{4.6}$$

$$F(\omega) = \frac{2}{\pi}\int_0^\infty \phi(\tau)\cos\omega\tau\,\mathrm{d}\tau \tag{4.7}$$

so that $F(\omega)$ and $\phi(\tau)$ are a Fourier transform pair. The formal justification for this procedure is too long to include here but is dealt with in most textbooks on the subject of noise. This is in accord with the tentative conclusion arrived at earlier for the example of Figure 4.4(a) where the form of $F(\omega)$ is clearly a Fourier transform of the corresponding autocorrelation function in Figure 4.4(b). It may be noted that the mean square voltage is given by $\phi(0)$ and if we put $\tau = 0$ in (4.6) the expression reduces to (4.4) for the mean square voltage.

4.1.2 Johnson or thermal noise

Johnson noise arises from the thermal motion of the charge carriers in a conductor where the number of carriers is large enough for the fluctuation in the numbers of carriers to be comparatively unimportant.[4] It is therefore predominant in majority carrier devices such as resistors and field effect transistors and has a square voltage spectral density given by

$$\mathrm{d}\overline{v^2} = F(\omega)\,\mathrm{d}\omega = \frac{2kTR\,\mathrm{d}\omega}{\pi} \tag{4.8}$$

where k is Boltzmann's constant, T the absolute temperature and R the resistance. It will be seen that Johnson noise is white noise and if it is followed by an amplifier or other circuit of bandwidth Δf, the mean square noise voltage is equivalent to $4kTR\Delta f$. The resistor may therefore be replaced by a noiseless resistor R in series with a noise voltage generator of this magnitude. Alternatively the noise can be represented as a current noise source in parallel with a noiseless resistor R and having a mean square current magnitude of $4kT\Delta f/R$.

4.1.3 *Current shot noise*

Pure shot noise arises from fluctuations in the numbers of charge carriers available for conduction which are assumed to behave independently, rather than from fluctuations in the carrier velocities as in the case of Johnson noise. It is therefore important in, for instance, low density electron beams and in minority carrier devices. For a thermionic valve operated in a region where space charge effects are negligible the electrons arrive on the anode randomly in time and the situation is similar to the example shown in Figure 4.4(a). For this case it is easily shown that the spectral density is given by

$$\mathrm{d}\overline{i^2} = F(\omega)\,\mathrm{d}\omega = \frac{Ie}{\pi}\,\mathrm{d}\omega \tag{4.9}$$

where I is the mean anode current and e the electronic charge. The effects of space charge may have a sufficient smoothing effect that the electrons do not behave independently and this reduces the noise below that given by (4.9) for pure shot noise.

In semiconductor devices utilizing minority carriers the number of carriers is usually controlled by the probability of carriers surmounting a potential barrier and by the effects of trapping and recombination.[5, 6, 7] For a reverse biased PN junction, including semiconductor detectors, the effects of trapping and recombination are usually negligible[7, 8] and the charge carriers may be assumed to be independent so that the noise approximates to the ordinary shot noise,

$$\mathrm{d}\overline{i^2} = \frac{I_1 e}{\pi}\,\mathrm{d}\omega$$

where I_1 is the reverse (or leakage) current. For a forwardly biased diode the noise again approximates to shot noise but at low forward bias an additional component arises from recombination in the depletion layer. For a junction transistor in the common emitter mode Robinson[7] has summarized the main noise components. In the low frequency region the noise arises predominantly from shot noise from the base and collector currents. At higher frequencies the situation is more complicated and a significant additional component arises from the induction of a noise voltage in the base lead from the collector current fluctuations.

4.1.4 Flicker or excess noise

In addition to Johnson noise and shot noise which are white noise sources, there is often a further component of noise in many devices which predominates at lower frequencies. This flicker noise has a spectral density $F(\omega)$ proportional to $1/\omega^n$ where n is usually in the region 0·8 to 1·5. Flicker noise is present in a wide variety of different devices and no single mechanism can explain its presence in such diverse types of device. In thermionic valves its magnitude is a property of the cathode material and construction. In semiconductor devices it appears to be highly dependent on surface conditions and may be reduced by careful construction. In all cases the origin of flicker noise is poorly understood and its magnitude varies considerably in different samples of the same type of device.

4.2 Signal to noise ratio, fwhm and the equivalent noise charge (ENC)

The fwhm as a measure of the resolution in a spectroscopy system has already been defined in Subsection 3.1.1. In the next few sections we are interested in the contribution of noise to the fwhm of a system and we will assume that the overall fwhm is due solely to the noise contribution, the effects of pulse pile up and ballistic deficit variations being neglected. We will assume that, in the absence of noise, the pulses at the output have a constant height V.

If, now, the noise at the output has a voltage v_n at any instant the signal pulses will be superimposed on this noise voltage to give a resultant pulse height v where $v = V + v_n$ (Figure 4.5). If V_n^2 is the mean square noise voltage (i.e. the variance) then the resultant distribution of pulse height will be the Gaussian distribution

$$P(v) = \frac{1}{V_n(2\pi)^{\frac{1}{2}}} e^{-(V-v)^2/2V_n^2} \tag{4.10}$$

which has a mean value of V and standard deviation of V_n. The values of v for which $P(v)$ falls to half its maximum value (Figure 4.5(d)) are given by

$$e^{-(V-v)^2/2V_n^2} = \tfrac{1}{2} \qquad \text{i.e.} \quad v = V \pm V_n(2\log_e 2)^{\frac{1}{2}} = V \pm \frac{V_{fwhm}}{2}$$

The fwhm is therefore given by

$$V_{fwhm} = 2V_n(2\log_e 2)^{\frac{1}{2}} = 2{\cdot}36V_n \tag{4.11}$$

The signal to noise ratio, η, is defined as the ratio of the mean signal pulse height V to the RMS noise voltage V_n. Thus we have

$$\frac{V_{fwhm}}{V} = \frac{E_{fwhm}}{E} = 2{\cdot}36\frac{V_n}{V} = \frac{2{\cdot}36}{\eta} \tag{4.12}$$

The signal to noise ratio in a given system is dependent on both the magnitude of the noise and also on the signal pulse magnitude and hence on the energy of the nuclear particles being detected. For a figure of merit of resolution in a system which is independent of the energy of the nuclear particles it is possible to use E_{fwhm}. An alternative index is the equivalent noise charge, ENC. This is the quantity of charge which, appearing on the

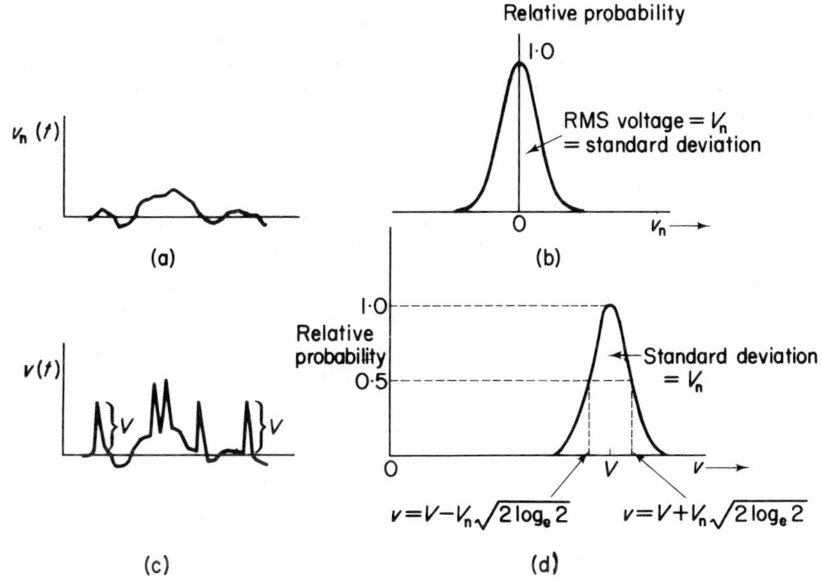

Figure 4.5 (a) Noise voltage only. (b) Probability distribution of the noise voltage in (a). (c) Noise voltage plus pulses of constant height V. (d) The probability distribution of pulse height in (c)

detector, would give an output pulse of height equivalent to V_n volts. Thus it may be seen that if the nuclear particle of energy E produces a charge Q on the detector

$$\frac{ENC}{Q} = \frac{V_n}{V} \qquad \left(= \frac{E_{\text{fwhm}}}{2{\cdot}36E}\right) \tag{4.13}$$

For a silicon detector the value of E/Q (i.e. the w value) is equivalent to about 3·6 eV per carrier pair (i.e. per unit electronic charge of Q). This leads to the following relationship

$$E_{\text{fwhm}}\,(\text{in eV}) = \frac{2{\cdot}36 \times 3{\cdot}6}{1{\cdot}602 \times 10^{-19}} \times (ENC \text{ in coulombs})$$

$$= 5{\cdot}30 \times 10^{19} \times (ENC \text{ in coulombs}) \tag{4.14}$$

For germanium, assuming a w value of 2·9 eV per pair, this leads to a relationship

$$E_{\text{fwhm}}\,(\text{in eV}) = 4{\cdot}27 \times 10^{19} \times (ENC \text{ in coulombs}) \qquad (4.15)$$

4.3 Optimization of signal to noise ratio

4.3.1 General case

Before considering the case of signals from radiation detectors it is useful to consider a more general class of signal. The following discussion closely follows the treatment of Baldinger and Franzen.[9] We will assume that a single signal only is present together with white noise and the signal has a known form. The problem is to measure the magnitude of the signal as accurately as possible. First let us consider the signal having the form shown in Figure 4.6(a), that is, a pair of brief pulses spaced a known time τ_d apart, and such that the ratio of the heights of the pulses is a known quantity γ. We will assume the signal is mixed with white noise of mean square voltage V_n^2 and we wish to measure the signal amplitude A.

Figure 4.6 (a) Elementary signal. (b) Signal plus noise

For such a rudimentary signal the only measurable quantities are the heights of the two pulses plus noise, A_1 and A_2 (Figure 4.6(b)), and if we assume the processing system to be linear the most general measurement obtainable is $A_1 + \beta A_2$. The problem of optimizing the signal measurement is therefore reduced to finding the optimum value for the weighting factor β. It may be noted that a practical method of forming a pulse of maximum height $A_1 + \beta A_2$ is shown in Figure 4.7(a). Since A_1 is the sum of A and noise, the mean square variation in A_1, i.e. the variance of A_1, will be entirely due to noise and equals V_n^2. Similarly the variance of βA_2 will be $\beta^2 V_n^2$. Now, since the noise is white, the noise voltages at the two instants of measurement are statistically independent. This is because white noise has an autocorrelation function which is zero for any non-zero value of τ. The variance of $A_1 + \beta A_2$ is therefore $V_n^2 + \beta^2 V_n^2$. If the signal to noise ratio of the output is η then

$$\eta^2 = \frac{(\text{Signal})^2}{\text{Variance}} = \frac{A^2(1+\gamma\beta)^2}{V_n^2(1+\beta^2)} \tag{4.16}$$

For optimum resolution η should be a maximum and this occurs for $\beta = \gamma$ when

$$\eta^2 = \frac{A^2}{V_n^2}(1+\gamma^2) \tag{4.17}$$

It may be noted that this value cannot be improved upon by making further measurements at other instants since such a measurement would only

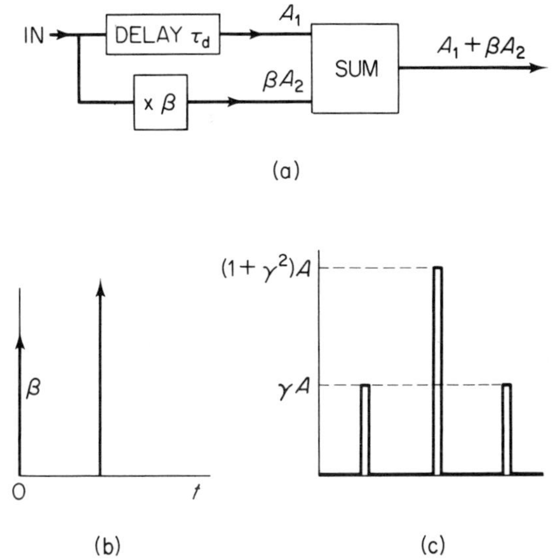

Figure 4.7 (a) Practical arrangement for processing the waveform of Figure 4.6(b) to give an output pulse having a maximum height of $A_1 + \beta A_1$. (b) The impulse response of the circuit in (a). (c) The form of the output pulse when $\beta = \gamma$

measure a noise voltage and this would be statistically independent of the noise voltages responsible for the variances of A_1 and A_2 and so could not yield any information which would reduce these variances. It can further be shown that the signal to noise ratio in (4.17) is the maximum for any system even if non-linear and time variant systems are considered.[9, 10]

It will be observed that since the optimum value of β is equal to γ, the optimum processing system weights A_1 and A_2 according to the magnitude of the signal contained in them. An ideal processing system would yield an output pulse with a height proportional to this weighted sum and the system of Figure 4.7(a) has this property. For more general types of input signal as shown in Figure 4.8(a), the same conclusions apply; the voltage of the signal-plus-noise at each instant should be weighted by a factor proportional to the magnitude of the signal contained at that instant.

In the optimum processing system (Figure 4.7(a)) for the input signal of Figure 4.6 it can easily be seen that its impulse response has the form shown in Figure 4.7(b) and is an exact mirror image of the input signal. The output signal has the form shown in Figure 4.7(c) and has a height $(1 + \gamma^2) A$. For the more general input signal of Figure 4.8(a) the same is true; the optimum processing system has an impulse response which is a mirror image of the input signal shape, as shown in Figure 4.8(b) and, as before, measurement of the height of the resultant output pulse will yield the best signal to noise ratio.

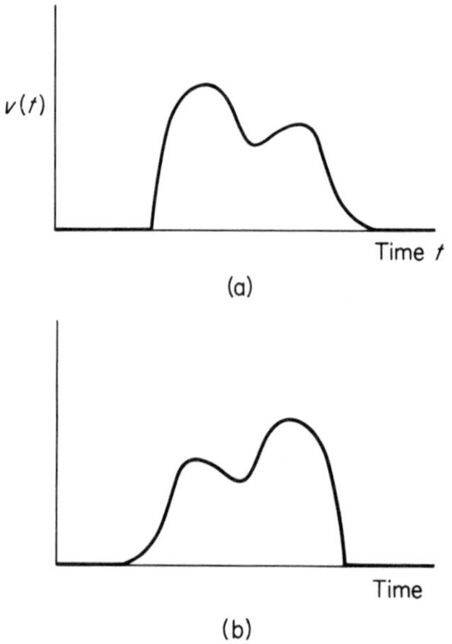

Figure 4.8 (a) More general type of signal. (b) Impulse response of the optimum filter for the signal of (a) in the presence of white noise

This can be seen by examining the convolution integral (Subsection 2.2.2) which gives the output pulse when a known input pulse is applied to a system with a known impulse response. For the case of a mirror image input signal and impulse response the output signal pulse is symmetrical; the peak of the output pulse occurs when the convolution integral is evaluated for the case when the input signal and impulse response lie on 'top' of each other. Clearly the value of the peak is an integration of the ordinates of the input signal plus noise weighted by the signal ordinates.

4.3.2 Radiation detector systems

The preceding discussion was for the case of a signal mixed with white noise. Where the noise is not white the optimum processing system may be derived after passing the signal-plus-noise through a filter which converts the noise to white. For the case of a radiation detector the noise originates partly from the detector itself and partly from the first stage of the preamplifier. This is fully discussed in Chapter 5 where it is shown that, for any detector–preamplifier combination, the optimum conditions at the input are when the input capacitance C_i is kept to a minimum and the input resistance is made as large as possible. For a detector signal approximating to an impulse of current containing a charge Q the signal at the input then approximates to a voltage step function of magnitude Q/C_i. The noise is equivalent to a noise source at the input of spectral density

$$\mathrm{d}\overline{v^2} = F(\omega)\,\mathrm{d}\omega \qquad \text{where} \quad F(\omega) = a^2 + \frac{b^2}{\omega^2} + \frac{c^2}{\omega} \tag{4.18}$$

The first term, the so called A noise, arises from the first transistor (i.e. channel noise for an FET) in the preamplifier. The second term, the B noise, arises from the detector, the Johnson noise of the input resistance and the gate shot noise of the FET or its equivalent. The third term represents the flicker noise of the first transistor and initially we will assume this is negligible. The noise can then be represented as a source at the input of spectral density

$$F(\omega) = a^2\left(1 + \frac{1}{\omega^2\tau_c^2}\right) \tag{4.19}$$

where τ_c is equal to a/b and is termed the 'noise corner time'. The corresponding noise corner frequency $\omega_c = 1/\tau_c$ is therefore that frequency at which the contributions of the A and B noise to $F(\omega)$ are equal.

The noise given by (4.19) may now be converted to white noise by passing through a CR high pass filter of time constant equal to τ_c. This 'pre-whitening' filter has a transfer function of $H_1(\mathrm{j}\omega) = \mathrm{j}\omega\tau_c/(1 + \mathrm{j}\omega\tau_c)$ and the resultant noise spectral density $F'(\omega)$ is

$$F'(\omega) = a^2\left(1 + \frac{1}{\omega^2\tau_c^2}\right)\frac{1}{(1 + 1/\omega^2\tau_c^2)} = a^2 \tag{4.20}$$

The signal is converted by this filter into that shown in Figure 4.9(a), that is having a form

$$\frac{Q}{C_i}\mathrm{e}^{-t/\tau_c} \qquad \text{for} \qquad t > 0 \tag{4.21}$$

For the optimum signal to noise ratio the signal of Figure 4.9(a) must be processed by a second filter whose impulse response $g_2(t)$ has a form that is a

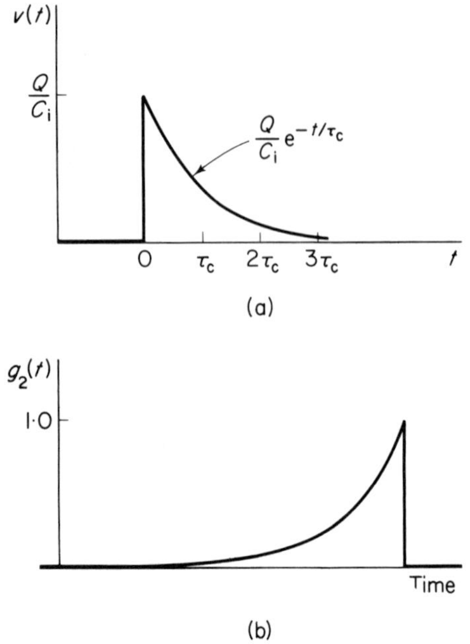

Figure 4.9 (a) The detector waveform after the pre-whitening filter. (b) The impulse response of the optimum filter

mirror image (Figure 4.9(b)) of the signal. The final output pulse shape may be found by evaluating the convolution integral (Subsection 2.2.2) between $g_2(t)$ and the signal of Figure 4.9(a) and the resulting cusp shaped pulse is shown in Figure 4.10. It should be noted that Figure 4.10 is plotted with the time origin shifted to the peak of the pulse and on this time scale the pulse has a voltage time course of

$$\frac{Q\tau_c}{2C_i}\,e^{-|t|/\tau_c} \tag{4.22}$$

The complete optimum processing system is shown in Figure 4.11. To calculate the noise voltage at the output we must find the effect of the second filter on the noise of spectral density a^2. The frequency response $H_2(j\omega)$ of this second filter may be obtained by taking the Fourier transform of the impulse response $g_2(t)$ shown in Figure 4.9(b):

$$H_2(j\omega) = \int_{-\infty}^{\infty} g_2(t)\,e^{-j\omega t}\,dt = \lim_{T\to\infty}\int_0^T e^{(t-T)/\tau_c}\,e^{-j\omega t}\,dt = \frac{1}{1/\tau_c - j\omega} \tag{4.23}$$

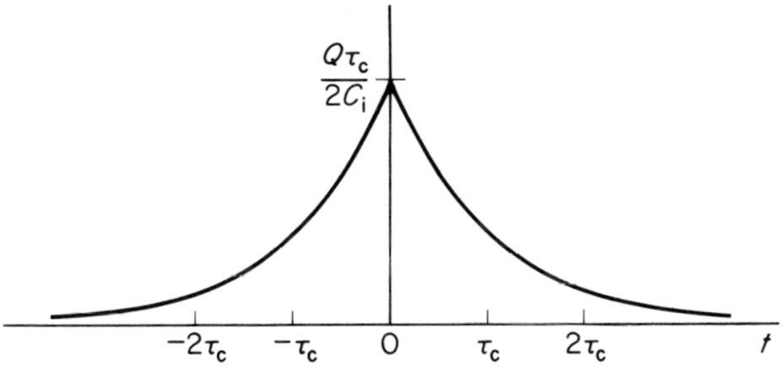

Figure 4.10 Output pulse for the optimum radiation detector system (the time origin is arbitrary)

The noise at the output therefore has a spectral density

$$a^2\,|H_2(j\omega)|^2 = \frac{a^2\tau_c^2}{1+\omega^2\tau_c^2}$$

If V_n is the RMS noise voltage at the output then

$$V_n^2 = \int_0^\infty \frac{a^2\tau_c^2}{1+\omega^2\tau_c^2}\,d\omega = \frac{\pi a^2\tau_c}{2} \tag{4.24}$$

Thus the signal to noise ratio, defined as the ratio of the signal pulse peak amplitude to the RMS noise voltage is

$$\eta_\infty = \frac{Q}{C_i}\left(\frac{1}{2\pi ab}\right)^{\frac{1}{2}} \tag{4.25}$$

and the equivalent noise charge is

$$(ENC)_\infty = C_i(2\pi ab)^{\frac{1}{2}} \tag{4.26}$$

If the noise is predominantly from the amplifier (i.e. the A noise is larger) then τ_c will be large and the optimum processing system would have an output pulse shape with a long mean duration. If the noise is mostly from the detector

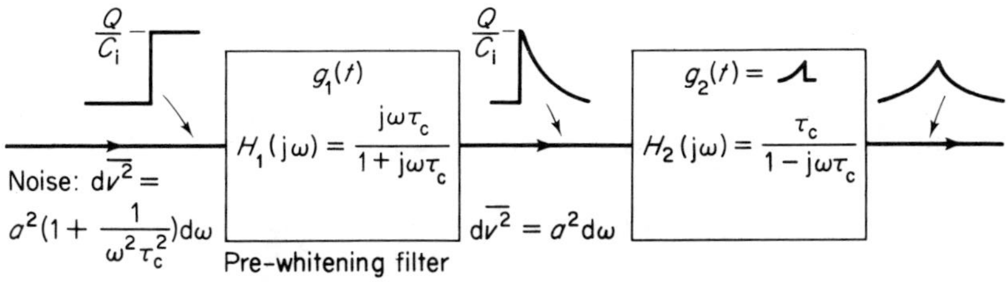

Figure 4.11 Hypothetical optimum radiation detector system

(i.e. the B noise is large) the optimum output pulse has a mean duration that is short. This is roughly the situation for a scintillation counter. It is also the situation in any detector system where the pulses of interest are accompanied by many other background pulses. It is possible to view these background pulses as an additional source of detector noise which increases the contribution of B noise. At high background counting rates this will reduce the effective τ_c and a pulse shape of shorter mean duration would be optimum. This is really only another way of looking at the pile up contribution to the overall resolution. At higher counting rates the pile up contribution becomes more important and a shorter mean pulse is desirable to give the lowest overall fwhm.

The above treatment has been for the optimum *linear* processing system. It has been further shown by Bertolaccini *et al.*[10] that it still yields the best signal to noise ratio of any type of system even if non-linear and time variant systems are considered. (4.25) therefore gives the theoretically maximum signal to noise ratio attainable and is therefore invaluable as a standard of comparison for all other systems.

4.3.3 Optimization under additional constraints

The ideal output pulse shape of Figure 4.10 has an infinite length and the shaping circuits would take an infinite time to respond so such a system could never be realized in practice. It is of interest to know the ideal output pulse shape under the additional constraint of a finite pulse duration. This is of importance in a practical system in order to limit the effects of pulse pile up on the loss of resolution. For a linear system where the output pulse is specified to be of overall length τ_w Arecchi *et al.*[11] have shown the optimum output pulse shape to be of the form $v(t) \propto \sinh(|(\tau_w/2 - t)/\tau_c|)$ for $|t| \leqslant \tau_w/2$ as shown in Figure 4.12 and the signal to noise ratio η_f is given by

$$\frac{\eta_f}{\eta_\infty} = (\tanh(\tau_w/2\tau_c))^{\frac{1}{2}} \tag{4.27}$$

where η_∞ is the signal to noise ratio for the infinite cusp given by (4.25). This ratio is plotted in Figure 4.13 for various values of τ_w/τ_c. It may be seen that

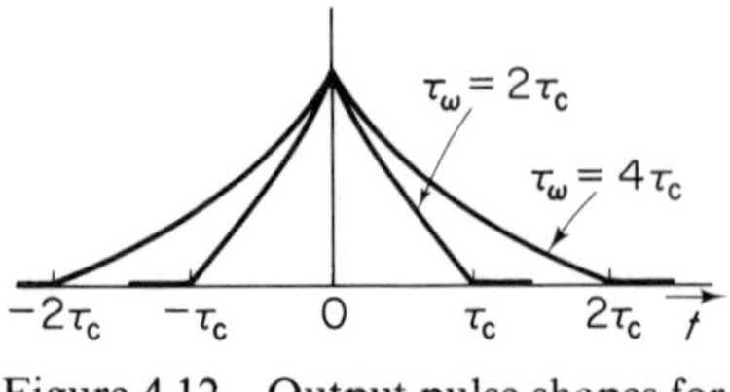

Figure 4.12 Output pulse shapes for the optimum system under the additional constraint of a pulse having a finite duration τ_w

η_f is $<15\%$ less than η_∞ when $\tau_w \geqslant 2\tau_c$. Since shaping by double delay lines followed by integration gives a triangular shaped pulse (Subsection 3.2.7) and this is a close approximation to the finite cusp of Figure 4.12, the signal to noise ratio for such pulse shaping is also included in Figure 4.13. It will be observed that the signal to noise ratio differs very little from that of the finite cusp and so triangular pulse shaping may be regarded as a practical realization of the finite cusp.

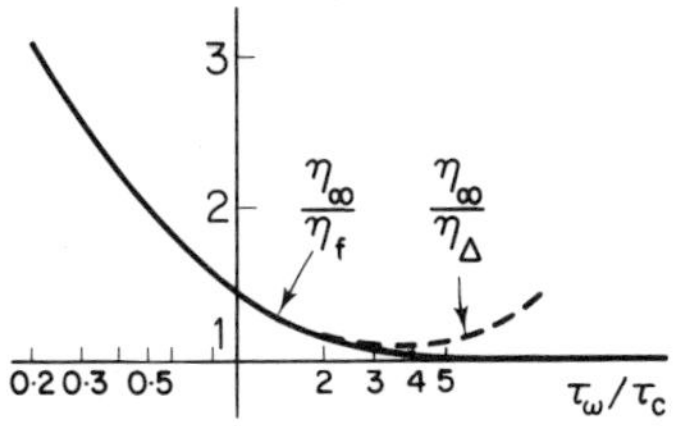

Figure 4.13 The signal to noise ratio η_f for finite duration cusp pulse plotted against the pulse duration τ_w. The signal to noise ratio is expressed as a fraction of the signal to noise ratio η_∞ of the ideal infinite cusp shaping (equation (4.26)). The broken line shows the relative signal to noise ratio for a triangular pulse

Bertolaccini *et al.*[10, 12] have further shown that the signal to noise ratio yielded by the finite cusp is the optimum for a pulse of fixed duration even if non-linear and time variant systems are considered. Although the finite cusp shape is clearly optimum for the maximum resolution as far as the noise contribution is concerned, the sharp peak of the cusp is not very desirable from the point of view of loss of resolution arising from ballistic deficit variations. From this point of view the ideal pulse shape should have a top which is flat for a period of time comparable to the collection time of the detector (Subsection 3.1.4). Arecchi *et al.*[11] therefore considered the case of a pulse shape which, in addition to having a finite duration τ_w, was also constrained to have a top which was flat for a duration $r\tau_w$. The optimum shape was then shown to be of the same cusp form but with the top truncated as shown in Figure 4.14. For the truncated cusp the signal to noise ratio η_t is given by

$$\frac{\eta_t}{\eta_\infty} = \left[\frac{2\sinh(\tau_w(1-r)/2\tau_c)}{2\cosh(\tau_w(1-r)/2\tau_c) + (r\tau_w/\tau_c)\,e^{-\tau_w/\tau_c}}\right]^{\frac{1}{2}} \tag{4.28}$$

where η_∞ is the signal to noise ratio for the infinite cusp (4.25). (4.28) can be approximated to within 10% for all values of τ_w and r by the expression

$$\frac{\eta_t}{\eta_\infty} \simeq \left[\tanh\left(\frac{1-r}{2}\right)\frac{\tau_w}{\tau_c}\right]^{\frac{1}{2}} \tag{4.29}$$

If $\tau_w(1 - r) \geqslant 2\tau_c$ then from (4.28) it may be seen that η_t is $<15\%$ less than η_∞.

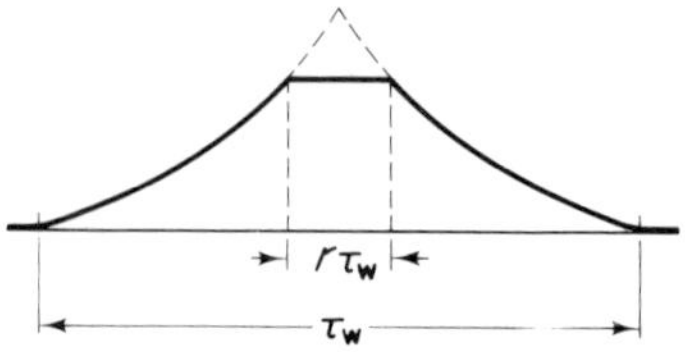

Figure 4.14 The output pulse shape, according to Arecchi *et* al.,[11] of the optimum system under the additonal constraints of finite duration τ_w and having a top flat for a duration $r\tau_w$

It should be noted that all the preceding discussions have been for the case of a single pulse only. Some attempt was made to take account of the effects of pulse pile up by constraining the pulse to have a finite duration. In such a system pulse pile up could then be completely eliminated if a pile up rejector circuit is used which inspects the pulses and rejects any pair separated by a duration less than half the overall pulse width. A rather different approach was taken by Bertolaccini *et al.*[13] who considered a more general class of pulse shaping designed to avoid pile up. Rather than constrain the output pulse to a constant duration they considered the case of an adaptive processing system which sensed the time positions of all the pulses and for each pulse provided the optimum filtering in the time available between the preceding and following pulses. In this way pile up was avoided and their results may be taken as giving the theoretically optimum resolution (neglecting the contribution from ballistic deficit variations) at any counting rate.

Their results are shown in Figure 4.15. Pulses having a close spacing contribute disproportionately to the reduction of the signal to noise ratio and a considerable improvement may be made if such closely spaced pulses are rejected by an additional circuit. Figure 4.15 therefore includes some of their results when pulses are rejected if they have less than a certain critical time spacing. The results in Figure 4.15 were obtained on the assumption that the critical time spacing was adjusted so that a constant fraction of counts were rejected at any counting rate. By way of comparison the corresponding signal to noise ratios are also included in Figure 4.15 for the case of finite cusp shaping together with a circuit rejecting pulses spaced less than $\tau_w/2$ where τ_w is the width of the finite cusp pulse (Figure 4.12). To make the comparison valid the value of τ_w was assumed to be adjusted so as to give a constant fraction of counts rejected at any counting rate. It will be observed that the adaptive processing system plus rejector circuit gives a considerable improvement in the signal to noise ratio at high counting rates over that obtainable with the finite cusp plus rejector circuit. It may be noted again that since there is no additional component as such due to pile up the adaptive system plus rejector gives the optimum signal to noise ratio at any counting

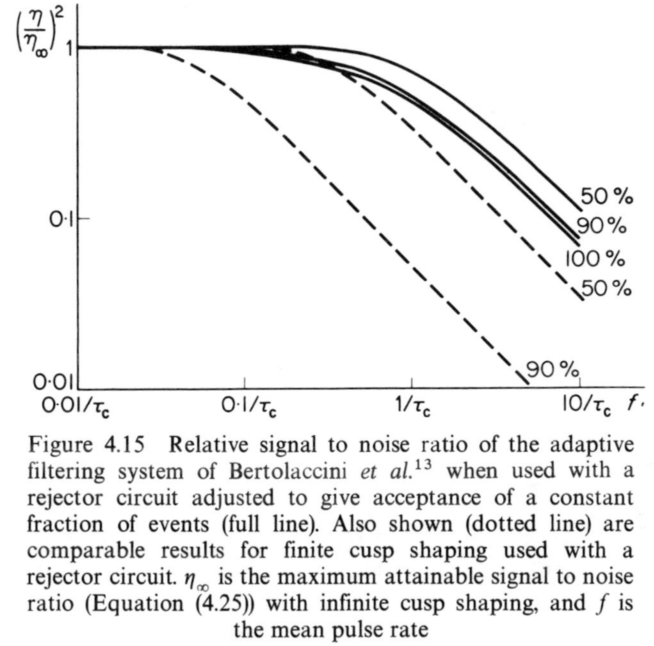

Figure 4.15 Relative signal to noise ratio of the adaptive filtering system of Bertolaccini *et al.*[13] when used with a rejector circuit adjusted to give acceptance of a constant fraction of events (full line). Also shown (dotted line) are comparable results for finite cusp shaping used with a rejector circuit. η_∞ is the maximum attainable signal to noise ratio (Equation (4.25)) with infinite cusp shaping, and f is the mean pulse rate

rate in the absence of ballistic deficit variations. The adaptive system gives a peak in the pulse height spectrum which is not a true Gaussian distribution and has shallow tails on either side of the peak. The fwhm is not accurately calculable from the signal to noise ratios using (4.11), (4.12) or (4.13) and these equations would yield a fwhm slightly larger than the true value.

4.4 Signal to noise ratio for practical systems

4.4.1 CR–CR shaping

This case has been considered by Gillespie[14] and by Baldinger and Franzen.[9] The transfer function for the pulse shaping sections of Figure 3.13 is

$$H(s) = \frac{s\tau_1}{(1 + s\tau_1)(1 + s\tau_2)} \tag{4.30}$$

where τ_1 and τ_2 are the time constants of the CR high pass and RC low pass filters respectively. The output pulse is given by

$$v(t) = \frac{Q}{C_i}\frac{\tau_1}{\tau_1 - \tau_2}(e^{-t/\tau_1} - e^{-t/\tau_2}) \qquad \text{for} \qquad t \geqslant 0 \tag{4.31}$$

and the form of this for various values of τ_1 and τ_2 was illustrated in Figure 3.14 and 3.15. The output pulse has a maximum value of

$$V_{max} = \frac{Q}{C_i}\lambda^{\lambda/(1-\lambda)} \tag{4.32}$$

where $\lambda = \tau_2/\tau_1$.

As before we will assume the noise is equivalent to a mean square voltage spectral density at the input of

$$d\overline{v^2} = F(\omega)\, d\omega = \left(a^2 + \frac{b^2}{\omega^2} + \frac{c^2}{\omega}\right) d\omega$$

The mean square noise voltage V_n^2 at the output is therefore

$$V_n^2 = \int_0^\infty \left(a^2 + \frac{b^2}{\omega^2} + \frac{c^2}{\omega}\right)\left(\frac{\omega^2\tau_1^2}{(1 + \omega^2\tau_1^2)(1 + \omega^2\tau_2^2)}\right) d\omega$$

$$= \frac{\pi a^2}{2\lambda(\lambda + 1)\tau_1} + \frac{\pi b^2 \tau_1}{2(\lambda + 1)} + \frac{c^2}{\lambda^2 - 1}\log_e \lambda \tag{4.33}$$

If η is the signal to noise ratio then $1/\eta^2 = V_n^2/V_{max}^2$. If we put $\tau_1\tau_2 = K$ then, keeping K constant, we can examine the effect on $1/\eta^2$ of varying λ. The results are shown in Figure 4.16. It will be seen that, for $\lambda = 1$, amplifier, detector and flicker noise all give minimum contributions to $1/\eta^2$ and so the condition $\tau_1 = \tau_2 = \tau$ is a clear optimum. In this case from (3.8) we have

$$V_{max} = \frac{Q}{C_i}\frac{1}{e} \tag{4.34}$$

and from (4.33) we have

$$V_n^2 = \frac{\pi a^2}{4\tau} + \frac{\pi b^2 \tau}{4} + \frac{c^2}{2} \tag{4.35}$$

or, in terms of the equivalent noise charge, from (4.13) we have

$$(ENC)^2 = e^2 C_i^2 \left(\frac{\pi a^2}{4\tau} + \frac{\pi b^2 \tau}{4} + \frac{c^2}{2}\right) \tag{4.36}$$

It will be seen that increasing τ does not affect V_{max} but affects V_n^2 or $(ENC)^2$ so as to decrease the contribution from A noise, to increase the B noise contribution and to leave unaffected the contribution from flicker noise. We can now optimize τ for a minimum V_n or ENC and this occurs when $\tau = a/b = \tau_c$ i.e. when the contributions to V_n^2 or $(ENC)^2$ of the A and B noise are equal. Then

$$(ENC)^2 = \frac{e^2 C_i^2}{2}(\pi ab + c^2) \tag{4.37}$$

$$\eta^2 = \frac{V_{max}^2}{V_n^2} = \frac{2Q^2}{e^2 C_i^2(\pi ab + c^2)} \tag{4.38}$$

For the case of no flicker noise, $c = 0$, the theoretically optimum processing system (infinite cusp shaping) gave a signal to noise ratio η_∞ given by (4.25). It will be seen therefore that

$$\frac{\eta_\infty}{\eta} = \frac{e}{2} = 1{\cdot}359 \tag{4.39}$$

so that the fwhm resolution given by CR–RC shaping will be worse than the theoretically optimum by the same factor.

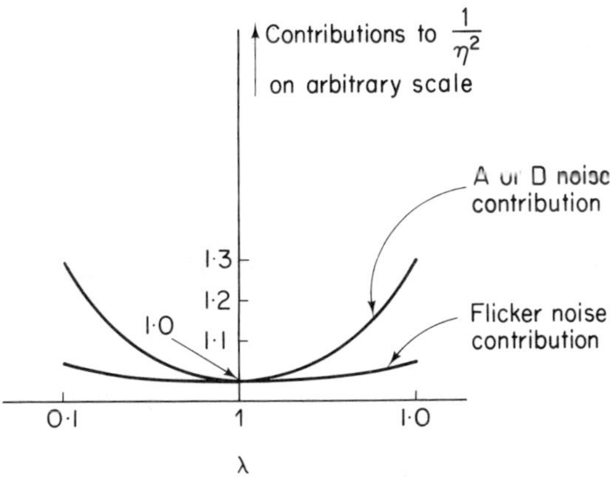

Figure 4.16 The plot shows the effect of $\lambda(=\tau_2/\tau_1)$ on the contributions from A, B and flicker noise to the overall value of $1/\eta^2$ (taken from Baldinger and Franzen[9])

4.4.2 *Various systems*

The output pulse of, for example, a Gaussian pulse shaping system has the form $v(t) = e^{-(t/2\tau)^2}$ so that parameter τ controls the time scale of the pulse but not the shape nor the amplitude scale. We now wish to examine the effect on the signal to noise ratio of the value of such a time scaling factor for the general case. We will therefore consider a pulse shaping system producing an output pulse $v(t) = Qs(t/\tau)$, where Q is the charge on the detector, so that $s(t/\tau)$ is the impulse response of the overall system (i.e. the output voltage for an impulse of current of unit charge in the detector). Before doing this it is useful to consider the case of the system with an impulse response $s(t)$; this system has a transfer function (voltage out/ current in) which may be obtained by taking the Fourier transform of $s(t)$. Since the overall system includes the

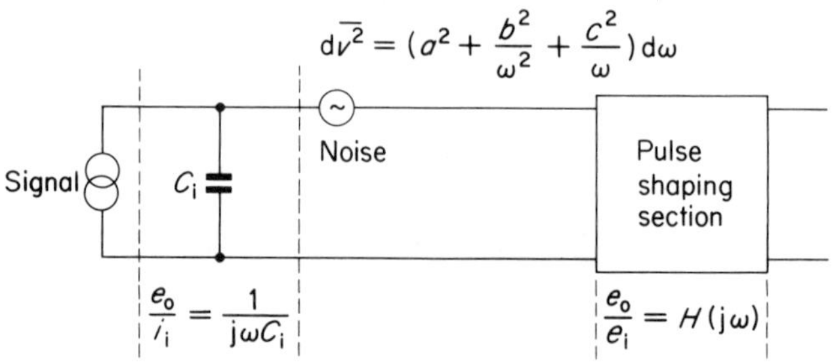

Figure 4.17 General form of a pulse shaping system

integrating action of C_i which has a transfer function of $1/j\omega C_i$, it follows that the transfer function $H(j\omega)$ of the pulse shaping part in Figure 4.17 alone is

$$H(j\omega) = j\omega C_i \int_{-\infty}^{\infty} s(t)\, e^{-j\omega t}\, dt \tag{4.40}$$

If we now consider the case of the system having the impulse response $s(t/\tau)$, the transfer function $H'(j\omega)$ of the amplifying and pulse shaping part is now

$$H'(j\omega) = j\omega C_i \int_{-\infty}^{\infty} s\left(\frac{t}{\tau}\right) e^{-j\omega t}\, dt = j\omega C_i \tau \int_{-\infty}^{\infty} s(x)\, e^{-j\omega\tau x}\, dx = H(j\omega\tau) \tag{4.41}$$

where $x = t/\tau$.

Thus, changing the time scale of the output pulse by $1/\tau$ is equivalent to changing the frequency scale of the amplifying and pulse shaping section by τ.

To calculate the noise at the output we will assume, as before, the noise is equivalent to a magnitude at the input of

$$d\overline{v^2} = \left(a^2 + \frac{b^2}{\omega^2} + \frac{c^2}{\omega}\right) d\omega$$

The output noise voltage will therefore be given by

$$V_n^2 = \int_0^\infty \left(a^2 + \frac{b^2}{\omega^2} + \frac{c^2}{\omega}\right)(|H(j\omega\tau)|)^2 \, d\omega$$

$$= \frac{a^2}{\tau}\int_0^\infty (|H(j\omega\tau)|)^2 \, d(\omega\tau) + b^2\tau \int_0^\infty \frac{(|H(j\omega\tau)|)^2}{\omega^2\tau^2} \, d(\omega\tau)$$

$$+ c^2 \int_0^\infty \frac{(|H(j\omega\tau)|)^2}{\omega\tau} \, d(\omega\tau) = \frac{a^2}{\tau} I_a + b^2\tau I_b + c^2 I_c \qquad (4.42)$$

It should be noted that the integrals I_a, I_b and I_c are independent of *a*, *b*, *c* and τ and are therefore only dependent on the form of the function *H* and hence on the form of the function *s*.

Since the signal output pulse has a peak height which it is independent of τ it follows that the signal to noise ratio η is given by

$$\frac{1}{\eta^2} = \frac{K^2}{Q^2}\left(\frac{a^2}{\tau} I_a + b^2\tau I_b + c^2 I_c\right) \qquad (4.43)$$

and

$$(ENC)^2 = K^2\left(\frac{a^2}{\tau} I_a + b^2\tau I_b + c^2 I_c\right) \qquad (4.44)$$

where *K* is independent of *a*, *b*, *c* and τ and is constant for a given type of pulse shaping system. The three terms in (4.43) and (4.44) represent the contributions from the A, B and flicker noise sources respectively and it will be observed that increasing τ, i.e. lengthening the output pulse, decreases the contribution from A noise, increases that from B noise and does not affect the flicker noise contribution. If (4.43) or (4.44) is minimized with respect to τ we find that the optimum value, τ_{opt}, is given by

$$\tau_{opt} = \frac{a}{b}\left(\frac{I_a}{I_b}\right)^{\frac{1}{2}} = \tau_c \left(\frac{I_a}{I_b}\right)^{\frac{1}{2}} \qquad (4.45)$$

When this value is substituted in (4.43) or (4.44) we see that the contributions from the A and B noise are equal. For a given type of pulse shaping I_a and I_b are fixed and the value of τ_{opt} is proportional to the noise corner time τ_c as shown by (4.45). If the noise is predominantly from A noise τ_{opt} will be large, i.e. an output pulse of long duration would be optimum and, for the case of predominantly B noise, a shorter duration would be optimum.

It should be noted that the foregoing discussion has been quite general and is valid for any linear pulse shaping system. The expressions for η or the *ENC* may be regarded as partly dependent on the value of τ selected and partly on the form of the output pulse, the latter dependence being implicit in the evaluation of I_a, I_b, I_c and K in (4.43) and (4.44). We will now confine our attention to the case of negligible flicker noise, which often approximates to the real case (see Chapter 5), since we have derived the theoretically optimum processing system (Subsection 4.3.2) for this case. The theoretically optimum ENC was found (4.26) to be

$$(ENC)^2_\infty = C_i^2(2\pi ab) \tag{4.46}$$

To put (4.44) into a form comparable to this we may rewrite it (assuming no flicker noise, $c = 0$) as

$$(ENC)^2 = (ENC)^2_\infty \frac{F^2}{2}\left(\frac{\tau_{opt}}{\tau} + \frac{\tau}{\tau_{opt}}\right) \tag{4.47}$$

where

$$F^2 = \frac{K^2(I_a I_b)^{\frac{1}{2}}}{\pi C_i^2}$$

F^2 is independent of a, b and τ and can be regarded as a factor of 'demerit' inherent in the type of processing system; if τ is made equal to its optimum value τ_{opt} (4.47) reduces to

$$(ENC) = F(ENC)_\infty \tag{4.48}$$

so that the ENC and hence the fwhm resolution is seen to be F times that obtainable by the theoretically optimum system (4.46). For non-optimum values of τ the $(ENC)^2$ is increased by a further factor $\frac{1}{2}(\tau_{opt}/\tau + \tau/\tau_{opt})$ and this is plotted in Figure 4.18. It will be observed that it is a fairly slowly changing function of τ in the region of τ_{opt}. If log $(ENC)^2$ is plotted against

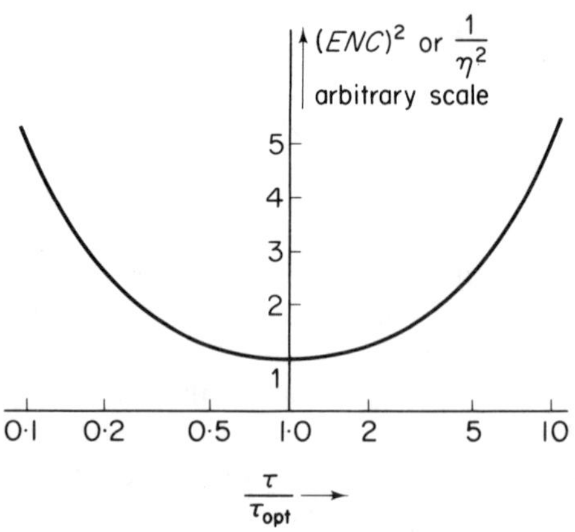

Figure 4.18 Variation of $(ENC)^2$ or $1/\eta^2$ with τ

log τ (Figure 4.19) it will be seen that the curve is asymptotic to two lines of slope 1 and -1 which represent the contributions of B and A noise respectively. This type of plot is further considered in Subsection 5.4.1 for several practical cases.

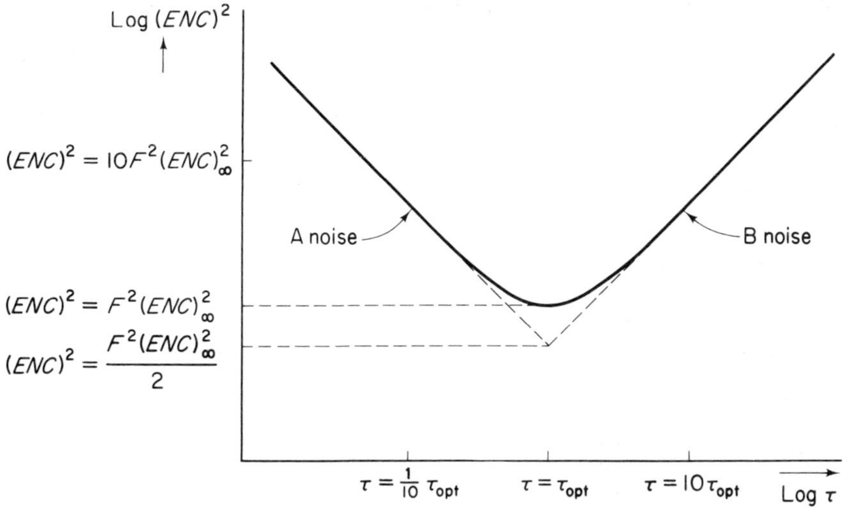

Figure 4.19 Variation of $\log(ENC)^2$ with $\log \tau$

The factors of demerit F for different types of processing system have been given by several workers and these are collected together in Table 4.1. Where a filter consists of components having two time parameters (e.g. the delay time and the RC filter time constant for the case of DL–RC shaping), the ratio of the two has assumed to be optimum as shown in column 1. For each type of shaping τ is defined in the first column and its optimum value is given in column 2. The shaping systems are arranged in descending order of merit as F becomes larger. It will be observed that triangular and DL–RC shaping give noise performances near to the theoretically optimum (infinite cusp), but it should be noted that the sharp peak of the pulse is the least satisfactory from the point of view of resolution loss through ballistic deficit variations (Subsection 3.1.4). Semi-Gaussian shaping may be regarded as a compromise in this respect (see below). CR–RC shaping, although giving a poor noise performance compared to other unipolar shaping methods is better from the point of view of ballistic deficit and owes its popularity to its simplicity and the ease of adjustment of τ. The two bipolar pulse shaping methods give noise performances rather worse than the unipolar pulse shaping systems.

The factor of demerit, F and the optimum value of τ are all based on the presence of A and B noise only. The presence of flicker noise causes an

Table 4.1 Performance of various methods of pulse shaping with respect to noise. For each method the optimum value of τ, as defined in the second column, is given in the third column. The fourth column gives the factor of demerit which is proportional, in the absence of flicker noise, to the *ENC*, the fwhm resolution and the noise to signal ratio on the assumption that the optimum value of τ is used in each case. The sixth column gives the flicker noise component on an arbitrary scale according to Hatch[19]

	Definition of τ	τ_{opt}	F	Reference	RMS flicker noise[19] (arbitrary scale)
Infinite cusp (Subsection 4.3.2)	—	—	1·00	(9, 14)	—
Triangular	Total duration $= \tau$	$2\sqrt{3}\,\tau_c$	1·075	(15, 17, 19)	1.665
DL–RC	Total delay time $= 1{\cdot}036\,\tau$ RC time constant $= \tau$	$1{\cdot}29\,\tau_c$	1·098	(15, 17, 19)	1·626
Gaussian			1·120	(16, 18, 19)	1·773
Semi-Gaussian CR–(RC)^4	CR time constant $= \tau$ RC rime constant $= \tau$	$0{\cdot}378\,\tau_c$	1·165	(16, 19, 20)	1·810
CR–(RC)^2	CR time constant $= 1{\cdot}06\,\tau$ RC time constant $= \tau$	$0{\cdot}57\,\tau_c$	1·215	(15, 16, 19)	1·847
CR–RC (Subsection 4.4.1)	CR time constant $= \tau$ RC time constant $= \tau$	τ_c	1·359	(9, 14)	1·992
(CR)^2–(RC)^4			1·380	(20)	
(CR)^2–RC	CR time constant $= 1{\cdot}38\,\tau$ RC time constant $= \tau$	$1{\cdot}40\,\tau_c$	1·410	(15, 20)	—

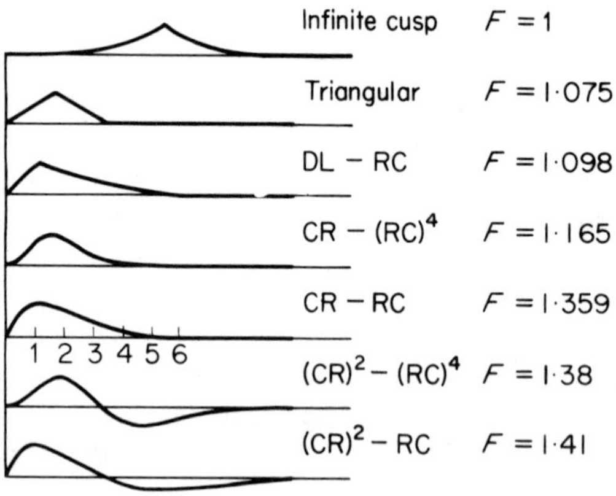

Figure 4.20 Pulse shapes, normalized to constant height, for various systems when the time scale of each (i.e. the value of τ) is optimized for the case of a noise corner time τ_c equal to 1. The curves shown are adapted from those given by Konrad[20]

additional component to be added to the $(ENC)^2$ given by (4.44). This component is independent of the value of τ and, according to Hatch,[19] has a relative magnitude given in the sixth column of Table 4.1. It appears that the flicker noise component may be significant in some FET preamplifiers.[19] However, the flicker noise magnitudes in Table 4.1 do not differ greatly and in any case they follow the factors of demerit F. As better FETs become available it appears that the flicker noise component will become negligible (see Subsection 5.2.2).

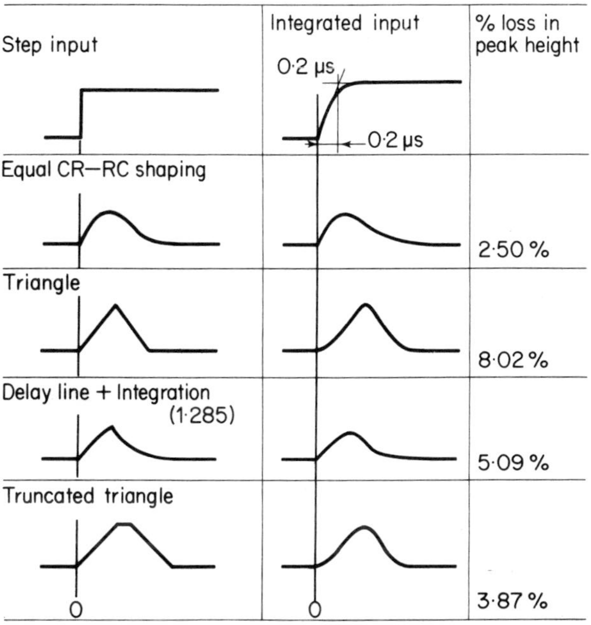

Figure 4.21 Comparison of the output pulse shapes and the loss in peak height due to the finite rise time of the detector signal, according to Hatch[19]

The relative merits of the different pulse shaping systems with respect to ballistic deficit variations are difficult to assess. The pulse shapes corresponding to some of the systems in Table 4.1 are shown in Figure 4.20 after Konrad[20] and it is possible to assess subjectively their relative merit as regards flatness of the pulse top. More quantitative information has been presented by Hatch[19] and is shown in Figures 4.21 and 4.22. The pulse shapes in Figure 4.21 are, with the exception of the truncated triangle, those of Table 4.1 on the assumption that in each case τ is optimized and $\tau_c = 1$ μs. Column 2 of Figure 4.21 shows the pulse shape when the preamplifier input is not a perfect step function but has a rise time of 0·2 μs arising from the finite collection time of the charge carriers in the detector. As expected the sharp peaked pulse shapes give large ballistic deficits (column 3) and CR–RC shaping appears to be best in this respect. Figure 4.22 shows plots of ballistic deficit for three types of shaping for different rise times of the input signal. Since the collection time in

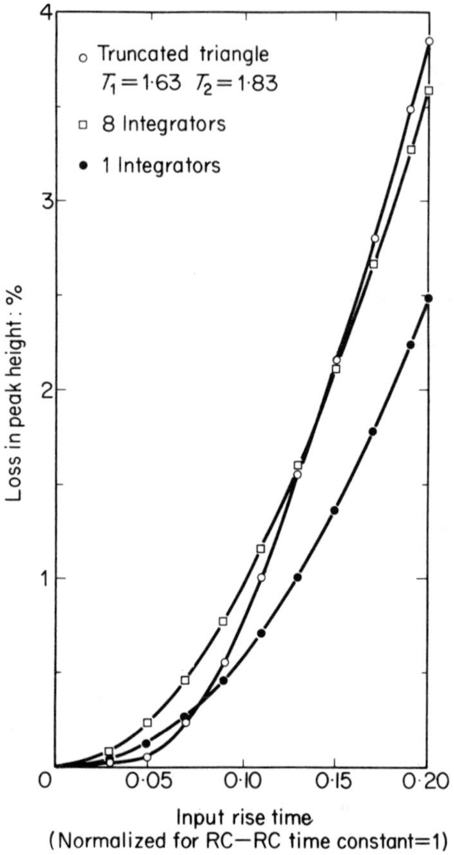

Figure 4.22 Loss in peak height of the output pulse against the detector signal rise time, for three type of shaping, according to Hatch[19]

lithium drifted semiconductor detectors may be of the order of 100 ns and pulse shaping of 1 μs time constant is typical, it appears that ballistic deficit variations in this case may significantly degrade the resolution. Strauss *et al.*[37] have estimated this resolution loss to amount to ~0·03% for a 1 cm Ge(Li) detector used with 1 μs CR–RC shaping and so it is comparable with the other factors determining the resolution at high energies. Apart from this estimate and also some evidence by Radeka[38] there has been little consideration in the literature on the magnitude of ballistic deficit variations.

4.5 Effects of pile up

4.5.1 Overloading

If pulses each of shape $v(t)$ arrive randomly in time at a mean rate n, we can apply Campbell's theorems (see for example Rice[2]) for the mean voltage $\overline{V}$ and root mean square fluctuations V_{RMS} giving

$$\overline{V} = n \int_{-\infty}^{\infty} v(t)\,\mathrm{d}t \tag{4.49}$$

$$V_{\mathrm{RMS}}^2 = n \int_{-\infty}^{\infty} v^2(t)\,\mathrm{d}t \tag{4.50}$$

The voltage appearing at the preamplifier input would have the form shown in Figure 4.23(a) due to the superimposition of individual pulses having a form

$$v(t) = \frac{Q}{C_{\mathrm{i}}} \mathrm{e}^{(-t/R_{\mathrm{i}}C_{\mathrm{i}})} \qquad \text{for} \quad t \geqslant 0$$

$$v(t) = 0 \qquad \text{for} \quad t < 0$$

where R_{i} is the input resistance and C_{i} the input capacitance. From (4.49) we therefore have $\overline{V} = nQR_{\mathrm{i}}$ and the same result could be obtained by considering the mean voltage drop across R_{i} due to the mean counting current of nQ. The RMS fluctuation is given by

$$V_{\mathrm{RMS}} = Q\left(\frac{nR_{\mathrm{i}}}{2C_{\mathrm{i}}}\right)^{\frac{1}{2}} \tag{4.51}$$

Where the pulses are of varying size then the appropriate form of Campbell's theorem causes Q to be replaced by the RMS charge $(\overline{Q^2})^{\frac{1}{2}}$.

If an AC coupled preamplifier with large coupling time constants is used then the input signal of Figure 4.23(a) is converted to that of Figure 4.23(b) with a mean of zero. In this case only the fluctuations are available to cause overloading. It is difficult to decide the maximum voltage excursion the circuits should be able to handle from the RMS value given by (4.51) since the distribution of voltage is not necessarily Gaussian. An upper limit is given by Tchebychef's theorem (see reference 36, for example) which states the probability of v deviating more than λV_{RMS} from the mean must be less than $1/\lambda^2$. This is an extremely conservative estimate since for a Gaussian distribution the probability of v deviating more than $3V_{\mathrm{RMS}}$ from the mean is about 0·25%. Provided the mean rate is such that $n \gg 1/R_{\mathrm{i}}C_{\mathrm{i}}$ the distribution of v does begin to approximate to Gaussian but it is difficult to apply this to a practical case. Before amplification is complete the signals would, if left in the same form, eventually cause overloading and it is necessary to shorten the pulses before this occurs so that pile up is lessened. This shortening (Figure 4.23(c)) may be done by the methods of pole-zero cancellation of Subsection 3.2.9 or alternatively by placing part of the overall pulse shaping at the appropriate point. For the latter alternative, inspection will usually immediately indicate which pulse shaping elements should precede which

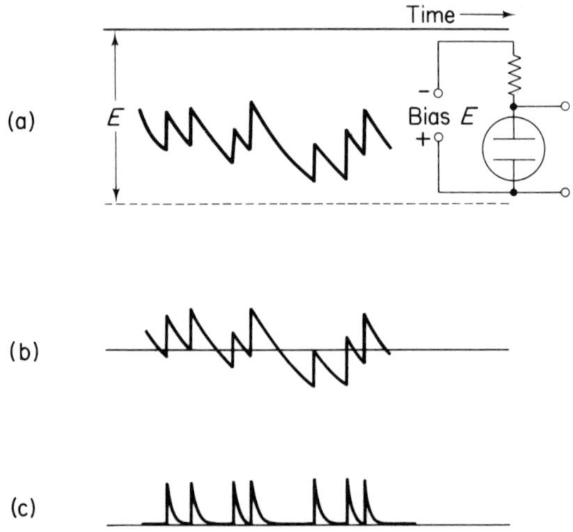

Figure 4.23 (a) Voltage waveform across the detector. (b) Voltage waveform after AC coupling. (c) Voltage waveform after pulse shortening (clipping)

others. For instance, in the case of DL–RC shaping the delay line shapings should precede the RC filter. Finally it may be noted again that, as the overall pulse shaping serves to limit the bandwidth, the pulse shaping elements should not be placed so near the input that they fail to reduce significant noise from later stages of the amplifier

4.5.2 Estimation of loss of resolution

At high counting rates loss of resolution due to pile up becomes important. We will assume that the impulse response of the system is $s(t)$ so the output pulse from a charge Q on the detector has the form $Qs(t)$. According to Campbell's theorems the mean voltage and the RMS voltage at the output are given by

$$\overline{V} = n\overline{Q} \int_{-\infty}^{\infty} s(t)\,\mathrm{d}t \tag{4.52}$$

$$V_{RMS}^2 = n\overline{Q^2} \int_{-\infty}^{\infty} s^2(t)\,dt \tag{4.53}$$

The pulse heights at the output will therefore be subject to a systematic error $\overline{V}$ and to an additional RMS fluctuation V_{RMS}. It is not possible, however, to simply relate V_{RMS} to the loss in fwhm resolution, since the distribution of the voltage is not Gaussian and depends on both the shape of the impulse response and on the distribution of pulse heights in the absence of any pile up (i.e. the distribution of Q). Several workers[21, 22, 23, 24] have published computations on the spectra distortion introduced by pile up for special cases. However, for the present purposes we will confine the discussion to a qualitative examination of the general case.

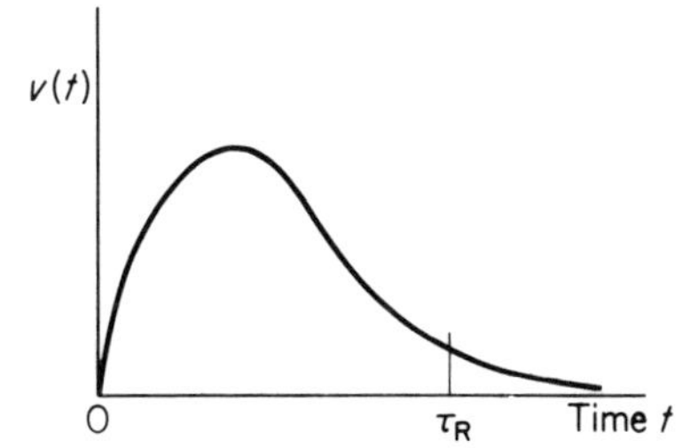

Figure 4.24 Arbitrary division of a pulse into a peak portion ($0 \leqslant t \leqslant \tau_R$) and a tail portion ($\tau_R < t \leqslant \infty$)

In determining the effect of pile up on resolution it is useful to consider two categories. We may divide the output pulse rather arbitrarily into two portions, the peak portion during the time interval $0 \leqslant t \leqslant \tau_R$, and the tail portion during the time $\tau_R < t < \infty$ as shown in Figure 4.24. The first type of pile up, peak pile up, occurs when the peak of the pulse of interest overlaps with the peak portion of another pulse as shown in Figure 4.25(a). The second type, tail pile up, occurs when the peak of the pulse of interest overlaps with the tail portion of another pulse as shown in Figure 4.25(b). Peak pile up will occur less frequently than tail pile up but will give a much larger error when it does occur.

Considering the case of peak pile up only, Figure 4.26 shows the pulse shape resulting from pile up of a typically shaped pulse. Assuming the pulse height analysis system responds truly to peak pulse height it is clear from Figure 4.26 that error occurs when the mutual pulse spacing is less than approximately $\tau_R/2$. Since all spacings occur with equal probability, the number of pulses falling in a time interval of $\pm\tau_R/2$ is $n\tau_R$ and so the rate of the pile-ups is $n^{\cdot}\tau_R$. The resultant pulse height spectrum for pulses initially of constant height is shown in Figure 4.27. The broadening of the spectrum

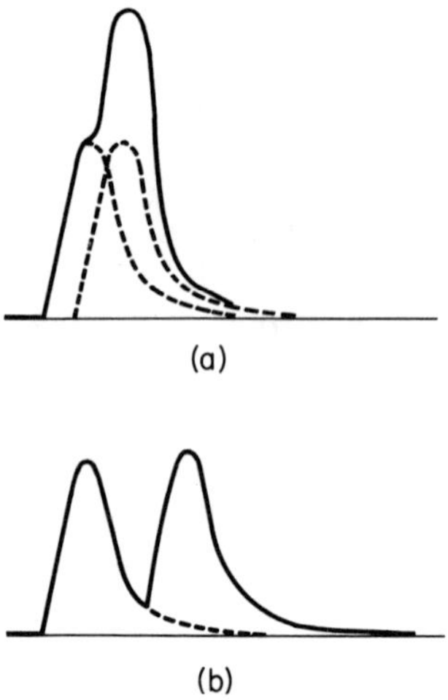

Figure 4.25 (a) Peak pile up.
(b) Tail pile up

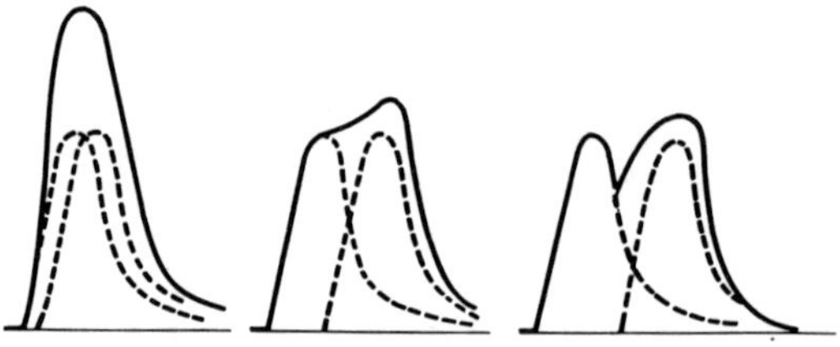

Figure 4.26 The effect of varying degrees of overlap on the pulse pile up

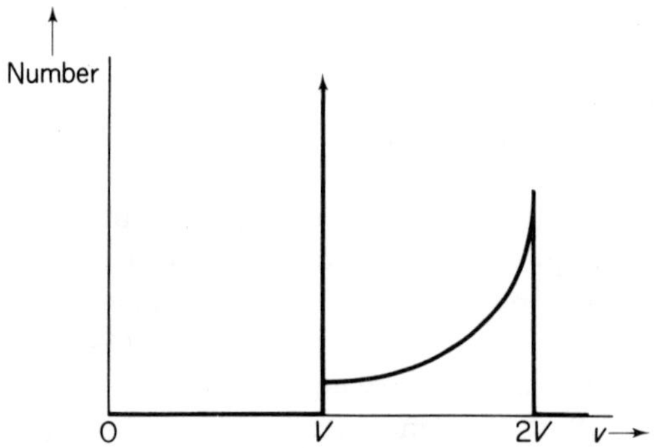

Figure 4.27 Pulse height spectrum of pulses of initially constant height V, when pile up is present

immediately to the right of the original line is dependent on the steepness of the sides of the pulses and to minimize this a square sided pulse may be regarded as ideal. When the pulses of interest are small and accompanied by large unwanted pulses an examination of the pile up process shows that the rate of pile up is better approximated by $2n^2\tau_R$ where τ_R is the mean total width of all pulses.

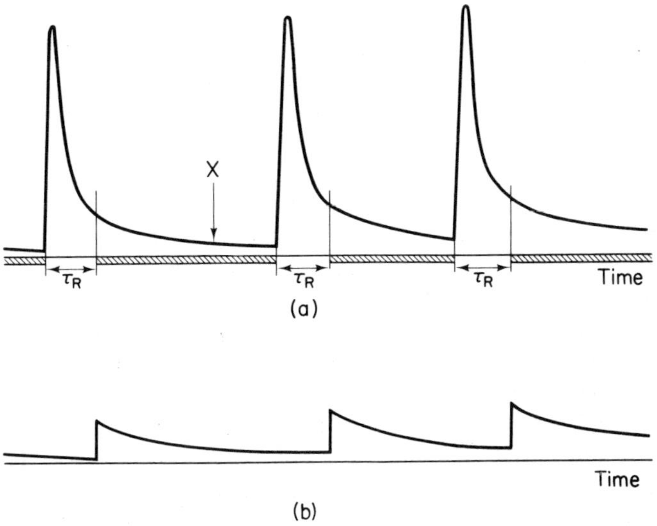

Figure 4.28 Approximation of the tail pile up voltage (a) by a continuous waveform (b)

Assuming that peak pile up has been minimized by making τ_R as small as is compatible with the requirements of a low ballistic deficit and low noise, a further reduction of peak pile up rate may be made by the use of a rejector circuit (Section 7.9). This circuit prevents measurement of the pulse heights if two pulses occur within a time $\leqslant \tau_R$ apart. Although such a circuit will fail to recognize pairs of pulses which are very close together this may not be very important as such pile-ups will usually fall outside the region of interest in the pulse height spectrum.

Although peak pile up may be effectively eliminated by the use of a pile up rejector circuit, there is still the possibility of tail pile up. We will assume that a rejector circuit is used which rejects pulses spaced closer than τ_R. The remaining tail pile up may be examined in the following way. In Figure 4.28(a) we are interested in the voltage fluctuations at point X where X is any instant during the shaded parts of the time axis i.e. wherever no pulse peak of duration τ_R is present, Since the value of the RMS fluctuations at X, V_{RMS}, is for discontinuous time periods Campbell's theorems cannot be used directly.

However, if τ_R is small compared to the mean pulse spacing the value of V_{RMS} should not be very different from that of the waveform shown in Figure 4.28(b). This is easily evaluated using Campbell's theorem and we have

$$\overline{V_B} \simeq n\overline{Q} \int_{\tau_R}^{\infty} s(t)\,dt \tag{4.54}$$

$$V_{RMS}^2 \simeq n\overline{Q^2} \int_{\tau_R}^{\infty} s^2(t)\,dt \tag{4.55}$$

What (4.55) shows is that, for a tail of given area, it is generally better to have a long shallow tail rather than a short steep one. For example, if we approximate the tail by an exponential decay $s(t) = B\,e^{(\tau_R - t)/\tau_B}$ for $t \geqslant \tau_R$, (4.55) gives

$$V_{RMS}^2 = \frac{n\overline{Q^2}B^2\tau_B}{2}$$

where for a given area $B\tau_B$ the V_{RMS} is reduced as τ_B is increased. It may be noted that where $n \gg 1/\tau_B$ the fluctuations at any point X in Figure 4.28(a) are due to the superimposed tails of many pulses and in this case the distribution will approximate to a Gaussian distribution and the equivalent fwhm resolution may be derived from V_{RMS} by (3.1) and (3.2).

As far as the relative merit of the different forms of pulse shaping with respect to tail pile up is concerned, it is difficult to examine this quantitatively since it will depend on the value of τ_R of the rejector circuit. However a comparison of the pulse shapes in Figure 4.20 shows that the unipolar methods should, from (4.55), have lower values of V_{RMS}. Since truly unipolar pulses are rarely used due to the difficulties of DC coupling all the components of the system, it is more common to employ AC coupling which imparts an undershoot to the pulse. Similarly the non-infinite value of the detector load causes the output pulse to depart from the idealized unipolar pulse and the resulting undershoot will increase the tail pile up error. These are considered in the next Subsection.

4.5.3 *Effect of AC coupling and finite detector load*

Consider the case of a truly unipolar pulse having the form shown in Figure 4.29 which is passed into a CR high pass filter of time constant τ_f. To preserve the area balance between the positive and negative lobes of the output pulse, the output pulse acquires an undershoot which, provided $\tau_f \gg \tau_w$, is given approximately by

$$\frac{S_{max}\tau_w}{\tau_f} e^{(\tau_R - t)/\tau_f} \qquad \text{for} \qquad t \geqslant \tau_R$$

From (4.54) we therefore have

$$V_{RMS}^2 = \frac{S_{max}^2\overline{Q^2}\tau_w^2 n}{2\tau_f} \tag{4.56}$$

For the cases of interest we may assume that the fluctuations are due to the superimposition of many pulse tails and their distribution would be approximately Gaussian so that (3.3) may be used. To consider a practical example, assume that the pulses are initially of constant height and of duration $\tau_w = 1\,\mu s$, and have a mean rate of $2 \times 10^4\,s^{-1}$. In order that the fluctuations given by (4.56) should give the equivalent of a fwhm resolution of 0·1 %, τ_f should be at least 0·056 seconds. In a more realistic situation the pulses would not be of initially constant height and, if the pulses of interest are accompanied by other larger pulses, $\overline{Q^2}$ would be much larger than $(\overline{Q})^2$ so the coupling time constant would have to be much larger to give a fwhm of 0·1 %. Since semiconductor detectors can give resolutions of this order it is necessary to use an amplifier for (nominally) unipolar pulses whose frequency response extends down to a few hertz if tail pile up is not to add a significant component of resolution degradation at higher counting rates.

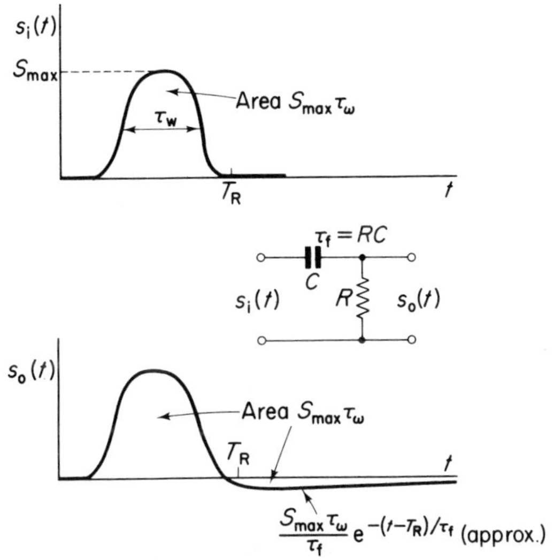

Figure 4.29 The effect of AC coupling on an idealized pulse

Similar considerations also apply to the undershoot induced through the finite value of the preamplifier input resistance R_i. As discussed in Subsection 3.2.9, even if a completely DC coupled system is used, Nowlin and Blankenship[35] showed that the effect of finite R_i is to introduce an undershoot (Figure 3.25) with a time course approximately proportional to e^{-t/τ_i} where τ_i is the time constant of the input circuit ($=R_iC_i$ in Figure 3.5). τ_i is typically a few milliseconds and taking the example of $\tau_i = 5$ ms, $n = 2 \times 10^4\,s^{-1}$, Q constant, $\tau_w = 1\,\mu s$ if we replace τ_f by τ_i in (4.56) we see that the voltage fluctuation is equivalent to a fwhm of 0·33%. As in the previous case this value will be much larger for the case of non-constant height pulses. As pointed out by Konrad[20] the removal of the effect of finite R_i is obligatory for high resolution work and the method of pole-zero cancellation discussed in Subsection 3.2.9 is the most satisfactory way. Konrad also gives an analysis of the effect of finite R_i which is more rigorous than the above in that it is for several different practical form of pulse shaping including bipolar shaping rather than for the idealized case considered here.

The tail pile up will impart a shift in the pulse height spectrum and this can be removed by the use of a baseline restorer. A baseline restorer can be used to reduce the value of V_{RMS} also since the fluctuations responsible are quite slow. In effect such a restorer removes the tail portions from the pulses, usually at the expense of some worsening of the signal to noise ratio.

4.6 Effect of baseline restoration on signal to noise ratio

Baseline restorers have been discussed in Section 3.3. The operation of the Robinson or Chase–Poulo restorers can be examined by means of the equivalent circuit of Figure 4.30(a) where the switch opens at the beginning of the pulse and closes some time after the pulse peak. In Figure 4.30(a) r represents the resistance due to finite forward resistance of the diode(s). When the switch opens at $t = 0$, the noise voltage $v_c(t)$ across the capacitor will be frozen as shown in Figure 4.30(c). At the time of the pulse peak $t = T_p$ the noise voltage will be $v_c(0) - v_1(T_p)$. To evaluate the RMS value of this noise voltage it is necessary to know the correlation function between $v_c(t)$ and $v_1(t + T_p)$. This will partly depend on the autocorrelation function of $v_1(t)$ and partly on the filtering action of C and r.

For the Gere–Miller restorer the same equivalent circuit may be used except that the switch opens at the pulse peak and closes some time after the peak so the noise voltage will be the sum of two terms similar to the case of the Robinson type restorer.

Radeka[25] has examined theoretically and experimentally the effect of baseline restorations on the signal to noise ratio for the case of CR–RC shaping. On the assumption that a fast Robinson restorer was used (i.e. with $rC \ll \tau_c$ where τ_c is the noise corner time) the signal to noise ratio of the normally optimum filter having time constants equal to τ_c was reduced by a factor of 0·89. Some of Radeka's results are plotted in Figure 4.31 and it will be seen that if the filter time constants are increased to $1{\cdot}93\tau_c$ then the use of the restorer negligibly degrades the signal to noise ratio below the optimum obtainable without the restorer. This does lengthen the pulse however and will result in increased peak pile up. Radeka also examined the case of slow restorers. In particular if rC is made comparable to τ_c then negligible degradation of the signal to noise ratio could be achieved (Figure 4.31).

Similar behaviour was observed with other types of unipolar pulse shaping. For the case of the Gere–Miller restorer the effect on signal to noise ratio will depend on the instant of switching on by the gating pulse but would be expected to give similar results for unipolar pulses. For the case of the Gere–

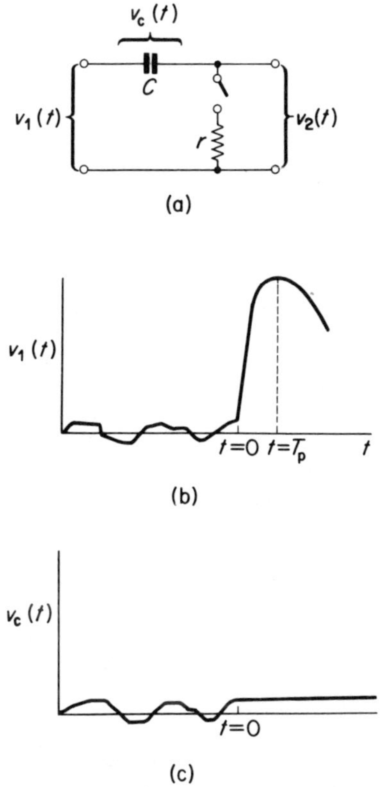

Figure 4.30 (a) Equivalent circuit of a baseline restorer where the switch opens at the beginning of the pulse. (b) Time course of the input voltage. (c) Time course of the capacitor voltage.

Miller restorer with bipolar pulses it is possible to actually improve the signal to noise ratio. This is because the Gere–Miller restorer causes a peak-to-undershoot measurement to be made and, as has been shown by Bertalaccini *et al.*,[26] this improves the signal to noise ratio attainable with a bipolar pulse over that obtainable with a single measurement of the first lobe only. This is further discussed in the next section.

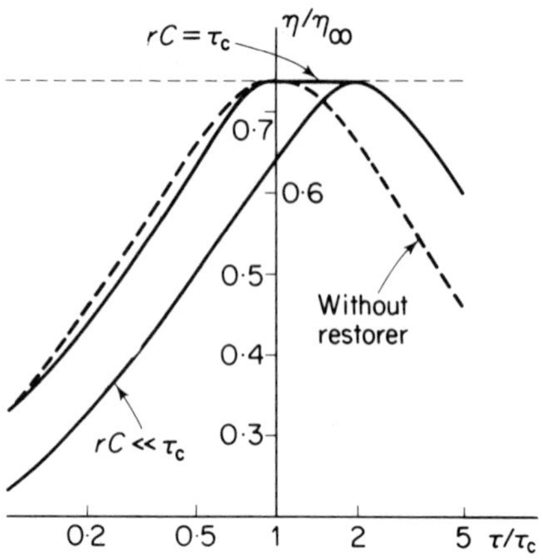

Figure 4.31 The effect of baseline restoration on the signal to noise ratio expressed as a fraction of that for the infinite cusp shaping, according to Radeka.[25] The curves are for CR–RC shaping with time constants both equal to τ, and are shown for the cases of two different restorer time constants. The dotted line shows the signal to noise ratio without the restorer

4.7 Sampling methods, time variant and non-linear pulse shaping

Various other systems of pulse shaping have been described which depart from the simpler methods already discussed. Most accept the premise that a system which is DC coupled from detector to analyser is impracticable and, from one aspect, undesirable, since parastitic slow baseline fluctuations in a DC system are difficult to supress. Hence some effort has been devoted to the optimum realization of AC coupled systems which process bipolar or tripolar pulses having equal areas in the positive and negative lobes. It may be noted again here that the maximum signal to noise ratio theoretically obtainable for any detector–preamplifier combination is that given by the infinite cusp pulse shaping (Equation (4.25)). The maximum signal to noise ratio for pulse shaping with a pulse of specific length is that given by the finite cusp (Equation (4.27)). These upper limits are valid even if time variant and non-linear shaping are considered[10, 12] and therefore give a valuable standard of comparison for the systems to be discussed.

For bipolar shaping Bertolaccini *et al.*[26] have shown that the signal to noise ratio can often be improved if the heights of both the positive and negative lobes of the pulse are sampled and summed together after suitable weighting. For a pulse approximating a symmetrical bipolar Gaussian shape, with conventional sampling of the height of one lobe only, the signal to noise ratio was 0·634 of that obtainable with finite cusp shaping having the same overall pulse duration. When the heights of both lobes were sampled this factor was increased to 0·735. On the assumption that the double sampling method is used Bertolaccini *et al.* have also derived theoretically the optimum bipolar pulse shape having symmetrical lobes and constrained to a finite duration. The signal to noise ratio for this was calculated to be 0·81 of that for the finite cusp of the same duration.

Deighton[27] has considered the case of the optimum bipolar and tripolar shapes of finite duration when a single measurement of the main peak was made. The pulse was constrained to have equal areas in the positive and negative lobes but not necessarily of symmetrical shape. His results for the signal to noise ratio are given in Figure 4.32 and as a basis of comparison the signal to noise ratios are also plotted for finite cusp shaping. In addition results for various practical shaping systems are included in Figure 4.32 where the pulse width is arbitrarily taken as between the two extreme points where the pulse had a magnitude above the baseline of 1 % of the peak height.

Bertolaccini *et al*[12, 28] have investigated theoretically and experimentally a rather different approach based on the multiple sampling technique. They have examined the case of a nominally unipolar pulse having the optimum finite cusp shape which is processed in an AC coupled system so that it has then an additional long shallow undershoot. To reduce the loss of resolution due to the tail pile up on the undershoots the output pulse was sampled at three points, just before the pulse, at the peak of the pulse and just after the main part of the pulse. It was assumed that the effect of peak pile could be made negligible through the use of a pile up rejector circuit. The weighting of the three samplings was then optimized to give the lowest overall standard deviations due to both noise and tail pile up. The overall optimum standard deviation from both these sources was dependent on the ratio of the pulse length to the mean pulse spacing and some of their theoretical results are shown in Figure 4.33. The quantity τ_R in the abcissa of Figure 4.33 is the resolving time, i.e. that time interval during which pairs of pulses must be rejected by the rejector circuit in order to avoid peak pile up, and is therefore equal to the overall pulse width when the triple sampling technique is used. The results in Figure 4.33 were obtained for a special case and the parameter β is proportional to the mean pulse rate but also depends on other factors such as the input spectrum, input noise and the shape of the pulse tail. Results for the finite cusp shaping with conventional single sampling are included and here the resolving time is equal to half the overall pulse width. It may be noted that at low counting rates the triple sampling method gives similar values of the standard deviation for the same resolving time.

Other systems that have been described make use of a switch which operates during the processing of the pulse so as to alter the filtering action for the

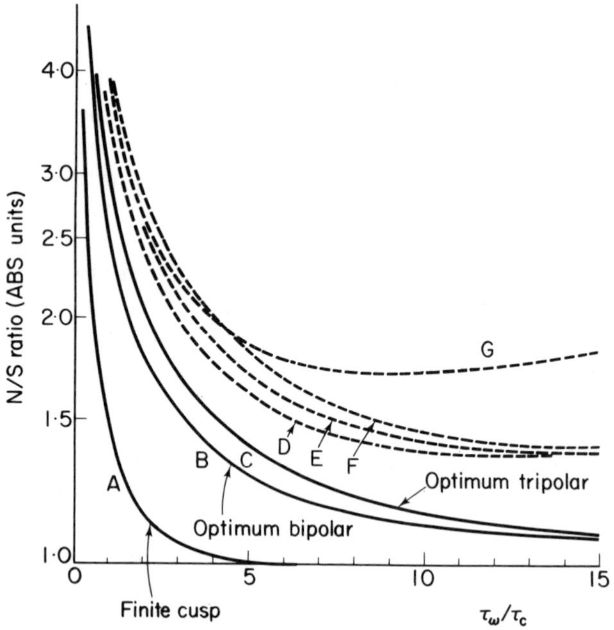

——— Theoretical minimum for:
- Unipolar finite cusp (curve A)
- Bipolar area-balanced pulse (curve B)
- Tripolar area-balanced pulse with centred peak (curve C)

- - - - Calculated performance of some practical bipolar systems:

Curve D: Double CR–DIFF^{N}, double CR–INTEG^{N} (equal time constants)

Curve E: Double delay line, single CR–INTEG^{N} ($CR = T_{D1} = T_{D2}$)

Curve F: Double-sampled single-CR shaped pulse (samples at start and at peak)

Curve G: Derivative of Gaussian pulse, i.e. $F(t) = -(t/\tau)\,e^{-t^2/2\tau^2}$

For curves D to G width is overall between 1% points

T_c = noise corner time constant

Figure 4.32 Relative signal to noise ratios for optimum unipolar (i.e. finite cusp), bipolar and tripolar shaping of finite duration, according to Deighton.[27] Also shown (dotted lines) are comparable values for various practical systems where the pulse width is arbitrarily taken as between the extreme points where the waveform has 1% of the peak amplitude pulse. The switch is operated by an auxiliary circuit from a gating pulse derived from the detector signal and since the switch must usually be operated slightly in advance of the arrival of the pulse in the main processing system it is usually necessary to delay the main pulse prior to the switch. The systems to be described differ mainly in the placement of the switch and the type of filtering applied to the pulse before and after the switch.

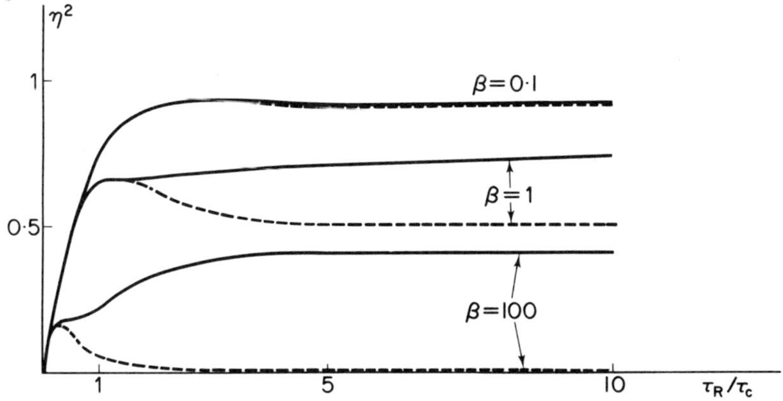

Figure 4.33 Representative results (full lines) for the triple sampling method of Bertolaccini *et al*,[28] when used with a rejector circuit of rejection time τ_R. β is a parameter proportional to the mean pulse rate. Comparable results are shown (dotted line) for finite cusp shaping with a rejector circuit

Blalock[29] has examined theoretically and experimentally several schemes for improving the signal to noise ratio of bipolar $(DL)^2$–RC shaping by means of various gating arrangements. It was possible to realize a considerable improvement in the signal to noise ratio so as to make it comparable to unipolar shaping methods. Unlike many of the other methods discussed in this section the pulse had a fairly flat top and would therefore give smaller ballistic deficit variations.

More recent work by Blalock and Nowlin[30] has extended the technique of processing bipolar pulses in AC coupled systems coupled with an active gating arrangement (Figure 4.34). By use of this arrangement they showed that it was possible to obtain a signal to noise ratio equivalent to that given by a unipolar pulse having the shape of the first lobe of the bipolar pulse. Since a triangular shape is a close practical approximation to the ideal finite cusp their shaping system used a bipolar triangular pulse as shown in Figure 4.34. The switch, a linear gate, operates on a pulse having an impulse shape. Since most detector systems use an input configuration which serves to integrate the detector signal, assumed to be a current impulse, it is first necessary to differentiate the step function from the preamplifier to regenerate the impulse form. For a bipolar pulse of length $2\tau_w$ the optimum arrangement was when the gate opened at a time $\tau_w/2$ prior to the arrival of the impulse and closed at a time $\tau_w/2$ after the impulse. Thus the impulse passed through the gate unaltered and was shaped into the final shape by conventional means, in this case by delay lines and an integration circuit. The noise to signal ratio for a (unipolar) triangular pulse is 1·075 of that of the optimum infinite cusp.

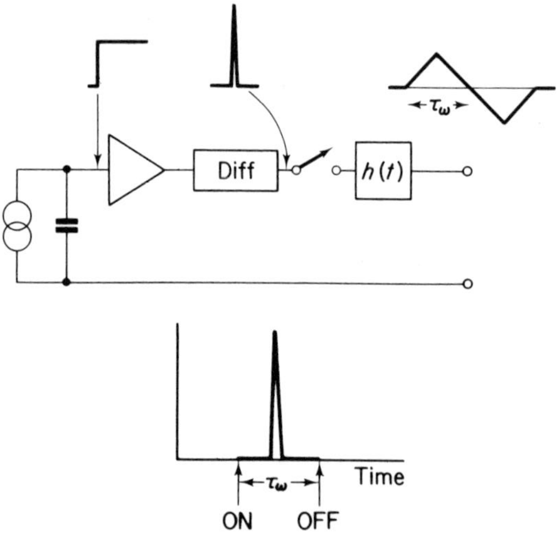

Figure 4.34 Active bipolar pulse shaping system of Blalock and Nowlin[30]

That this value was attained by the gated bipolar pulse was verified experimentally. This was obtained at the expense of the bipolar pulse having twice the duration of the unipolar triangular pulse however. By way of comparison the use of a symmetrical bipolar triangular pulse without the switch would have given a noise to signal ratio relative to the infinite cusp of 1·520 or 1·316 depending on whether the height of one or both lobes were measured.

Kandiah[31] has used several systems of the type shown in Figure 4.35 where the switch opens at the beginning of a pulse. The theory of these types of shaping has been dealt with by Deighton[32] and Table 4.2 gives his calculated noise to signal ratios (or *ENC*s) for various combinations of the time constants of the filters expressed as a fraction of that for the infinite cusp. Weise[33] has described a switched system which differs slightly from the systems considered by Kandiah and this appears to give similar results.

Radeka[38] has used a system which can give a negligible ballistic deficit. Essentially this requires the use of a flat topped pulse and although shaping by passive components was considered, in order to give greater flexibility to optimize the pulse shape for different conditions, the following scheme was

Table 4.2 Noise to signal ratios according to Deighton[32] for the pulse shaping system of Kandiah[31]. The last column gives the noise to signal ratio as a fraction of that for the theoretically optimum infinite cusp shaping (equation 4.25). τ_1 and τ_2 are the time constants of the two filters as shown in Figure 4.35 and T is the period for which the gating signal is applied.

	τ_2/τ_c	τ_1/τ_c	T/τ_c	$\frac{\eta_\infty}{\eta}$ or $\frac{ENC}{(ENC)_\infty}$
Gated integrator only	0	1·3	1·3	1·098
	0·5	0·5	1·5	1·125
Gated integrator + switched filter	0	1	1·75	1·038
	0·4	0·4	2·15	1·086
	1·05	0	2·1	1·016

adopted. The detector pulse was first shaped into a semi-Gaussian form and this was fed into a gated integrator. The period of integration included essentially all of the input pulse and so the integrated output pulse had a top which was flat for a period dependent on the period of integration. The noise to signal ratio (or the *ENC*) of the system depended on the period of integration and had a limiting value of 1·282 times that of the infinite cusp shaping.

It is difficult to summarize the relative merits of the various schemes described in this section. Some of the systems should be free from tail pile up effects since the pulse shapes are of definite duration. However, with the exception of the system used by Radeka, nearly all the systems use a pulse shape with a sharp peak and so the effect of ballistic deficit variations is likely to contribute significantly to the overall resolution. This is particularly so for the optimum bipolar shaping considered by Bertolaccini *et al.*[26] and by Deighton.[27, 32] In Table 4.2 the combinations of τ_1 and τ_2 which give the lowest signal to noise ratios are unfortunately the worst from the point of view of ballistic deficit. Particularly with Ge(Li) detectors the collection time may be of the order of a hundred nanoseconds and the variations of ballistic deficit may well be a major cause of loss of resolution. Although Deighton

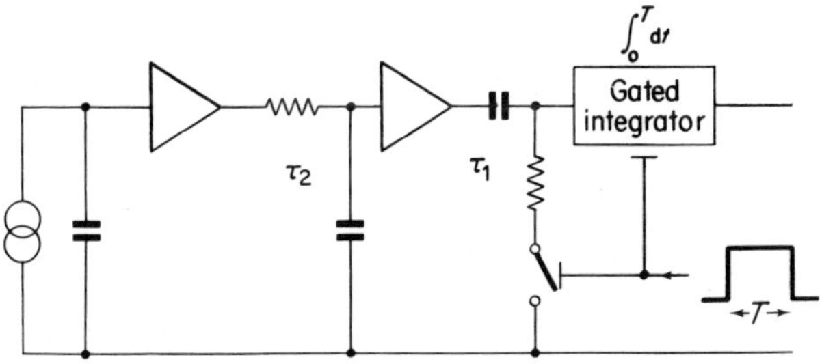

Figure 4.35 Active shaping system of Kandiah[31]

considered some aspects of this, this source of loss of resolution has not received much attention so far, due no doubt to the difficulties of adequately characterizing the variations in the detector pulse shape.

All the systems described which use a switch (i.e. a linear gate) make stringent demands on the design of the gate[34] since the effects of any gate pedestal or transients (see Section 7.8) are exacerbated by the partial integrating action of following filters. This design problem, together with the extra complication of the circuitry, offsets the improvement in resolution attainable so that the more conventional passive linear pulse shaping methods, possibly with baseline restoration, seems likely to remain the methods of choice for most applications for some time to come.

References

1. J. S. Bendat, *Principles and applications of random noise theory*, Chapman & Hall (1958).
2. S. O. Rice, *Bell Syst. Tech. J.* **23**, 282 (1944) and **24**, 46 (1945).
3. A. Van der Ziel, *Noise*, Chapman & Hall, 1955.
4. D. K. C. MacDonald, *Noise and fluctuations*, Wiley, 1962.
5. A Van Der Ziel, *Proc. I.R.E.*, **46**, 589 (1958).
6. A. Van Der Ziel, *Proc. I.R.E.*, **48**, 114 (1960).
7. F. N. H. Robinson, *Electron. Eng.*, **41**, Feb., 218, Mar., 353, April, 520, May, 77 (1969).
8. G. Dearnaley and D. C. Northrop, *Semiconductor counters for nuclear radiations*, 2nd ed., Spon, 1966.
9. E. Baldinger and W. Franzen, *Advances in electronics and electron physics*, **8**, 255 (1956).
10. M. Bertolaccini, C. Bussolati and E. Gatti, *Nucl. Instr. and Meth.*, **41**, 173 (1966).
11. F. T. Arecchi, G. Cavalleri, E. Gatti and V. Svelto, *Energia Nucleare*, **7**, 691 (1960).
12. M. Bertolaccini, C. Bussolati, S. Cova, I. De Lotto and E. Gatti, *Semiconductor nuclear particle detectors and their circuits*, p. 523, U.S. Nat. Acad. of Sciences, Nuclear Science Series, Report No. 44, Publication 1593, (1969).
13. M. Bertolaccini, C. Bussolati, S. Cova, I. De Lotto and E. Gatti, *Nucl. Inst. and Meth.*, **61**, 84 (1968).
14. A. B. Gillespie, *Signal, Noise and Resolution in Nuclear Counter Amplifiers*, Pergamon, 1953.
15. M. Tsukuda, *Nucl. Instr. and Meth.*, **14**, 241 (1961).
16. E. Fairstein and J. Hahn, *Nucleonics*, **23** No. 7, 56, No. 9, 81, No. 11, 50 (1965); **24** No. 1, 54, No. 3, 68 (1966).
17. C. H. Nowlin, J. L. Blankenship and T. V. Blalock, *Rev. Sci. Instr.* **36**, 1063 (1965).
18. J. L. Blankenship and C. H. Nowlin, *IEEE Trans. Nucl. Sci.*, NS **13**, No. 3, 495 (1966).
19. K. Hatch, *IEEE Trans, Nucl. Sci.*, NS **15**, No. 1, 303 (1968).
20. M. Konrad, *IEEE Trans. Nucl. Sci.*, NS **15**, No. 1, 268 (1968).
21. I. De Lotto and D. Dotti, *Nucl. Instr. and Meth.*, **39**, 281 (1966).
22. B. Soucek, *Rev. Sci. Instr.*, **36**, 1582 (1965).
23. R. Gold, *Rev. Sci. Instr.*, **36**, 784 (1965).
24. J. H. Williamson, *Rev. Sci. Instr.*, **37**, 736 (1966).

25. V. Radeka, *Rev. Sci. Instr.*, **38**, 1397 (1967).
26. M. Bertolaccini, C. Bussolati and E. Gatti, *Nucl. Instr. and Meth.*, **42**, 286 (1966).
27. M. O. Deighton, *IEEE Trans. Nucl. Sci.*, NS **16**, No. 5, 68 (1969)
28. M. Bertolaccini, C. Bussolati, S. Cova, I. De Lotto and E. Gatti, *Nucl. Instr. and Meth.*, **62**, 221 (1968).
29. T. V. Blalock, *Rev. Sci. Instr.*, **36**, 1448 (1965).
30. T. V. Blalock and C. H. Nowlin, *Rev. Sci. Instr.*, **40**, 1093 (1969).
31. K. Kandiah, *Semiconductor nuclear particle detectors and their circuits*, p. 544, U.S. Nat. Acad. Sciences, Nuclear Science Series, Report No. 44, Publication 1593, (1969).
32. M. O. Deighton, *Nucl. Instr. and Meth.*, **58**, 201 (1968).
33. K. Weise, *Nucl. Instr. and Meth.*, **61**, 241 (1968).
34. C. H. Nowlin, *IEEE Trans. Nucl. Sci.*, NS **17**, No. 1, 226 (1970).
35. C. H. Nowlin and J. L. Blankenship, *Rev. Sci. Instr.*, **36**, 1830 (1965).
36. C. E. Weatherburn, *Mathematical Statistics*, Cambridge University Press, 1952.
37. M. G. Strauss, R. N. Larsen and L. L. Sifter, *Nucl. Instr., and Meth.*, **46**, 45 (1967).
38. V. Radeka, *Nucl. Instr. and Meth.*, **99**, 525 (1972).

5

Amplifiers

5.1 Introduction

Scintillation counters and Geiger counters usually produce an output signal of the order of several volts; this is well above the noise level of the amplifier so the input stage often does not need to be specially designed. If the counter and the amplifier input are separated by a long cable run the signal may be distorted or attenuated. For cable lengths of under ~ 5 metres the transmission time ($\sim 2 \times 10^{-8}$ s) is short compared to the duration of the output pulse (~ 1 μs) used for pulse height measurement, so the cable can be regarded as equivalent to a lumped capacitance. Where this capacitive loading is important its effect may be reduced by driving the cable from a low impedance source, such as may be provided by an emitter follower stage. For longer lengths of cable, or for fast pulses required for timing applications, the transmission line properties of the line become dominant and reflections at the ends may become troublesome, in which case it will be necessary to terminate at least one end of the cable with its characteristic impedance (50–150 ohms for most cables).

Most other types of detector give a small enough signal that a low noise preamplifier must be used, which must be located near to the detector, since the capacitive loading introduced by a long cable would reduce the signal to noise ratio (see later). Since semiconductor detectors can give a much better resolution than other types the noise properties of the preamplifier are most critical for this case. For this reason the following discussion will be almost exclusively concerned with semiconductor detector preamplifiers but the conclusions are obviously valid for other types of detectors as well.

5.2 The field effect transistor

5.2.1 Properties of the FET

For low noise input stages the device which currently is almost universally used is the field effect transistor (FET). The principle of an N channel junction FET is shown in Figure 5.1(a). The gate is made of P type material and the PN junction so formed is reverse biased so that part of the depletion layer extends into the channel region. The resistance of the channel and hence the current flowing between source and drain are thus dependent on the extent of the depletion region in the channel. Since the gate-channel junction is reverse biased, very little current flows through the gate; the channel current is effectively controlled by the gate voltage, giving a characteristic similar to a

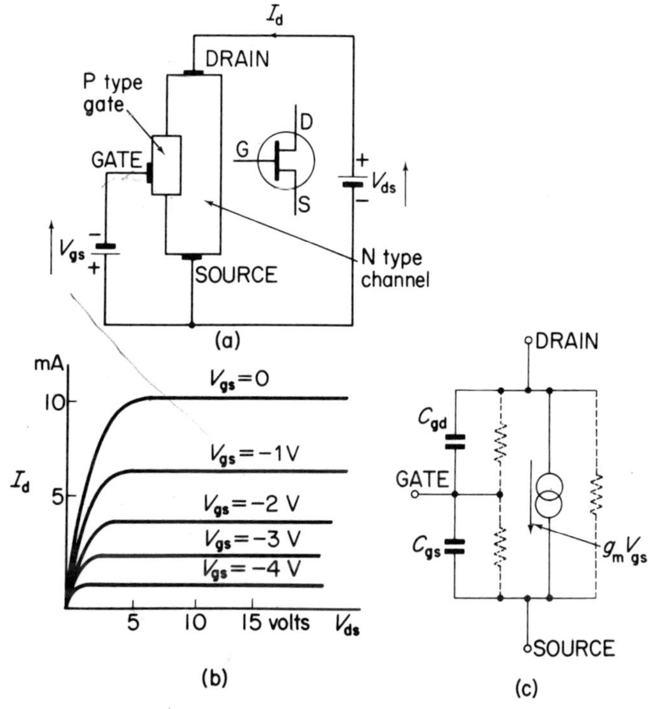

Figure 5.1 (a) Principle of the FET. (b) Characteristics of the FET. (c) Equivalent circuit of the FET for small signals

pentode thermionic valve. Figure 5.1(b) shows a typical characteristic for a FET where V_{gs} and V_{ds} are the gate and drain voltages with respect to the source, and I_d is the drain current. It will be observed that the estent of the depletion layer is not only dependent on V_{gs} but, as Shockley[1] showed, 'pinches off' the channel and the resultant current is effectively independent of the drain voltage, giving the saturated region of the characteristic. As in the case of a pentode valve, the most important parameter is the mutual conductance $g_m = dI_d/dV_{gs}$. At moderately high frequencies the behaviour of the FET in the saturated region can be approximated by the equivalent circuit of Figure 5.1(c).

5.2.2 *Noise in the FET*

The noise behaviour of FETs was first examined theoretically by Van Der Ziel.[2] One source of noise arises from the channel and, since the channel current is carried by majority carriers, this may be regarded as similar to Johnson noise. For a FET operated in the unsaturated part of the characteristic, Van Der Ziel showed that it is equivalent to a current noise generator across the source and drain of magnitude

$$d\overline{i_1^2} \simeq \tfrac{2}{3}\, 4kTg_m\, df \tag{5.1}$$

Robinson[3] has confirmed this result and has also shown that it is applicable for operation of the FET in the saturated part of the characteristic, the more commonly used part.

Another component of noise arises from the fact that the fluctuations in the channel current induce a fluctuating current through the gate-source capacitance C_{gs}. Although this is partly correlated with the drain current fluctuations, the correlation is small and a maximum error of 10% in the noise magnitude arises if they are assumed independent.[3] In this case the magnitude of this induced noise current in[2,3] the gate is

$$d\overline{i_2^2} = \frac{kT\omega^2 C_{gs}^2\, df}{g_m} \tag{5.2}$$

A third source of noise arises from the gate leakage current I_g. Since this is across the reverse biased gate-channel junction it approximates to pure shot noise and has an equivalent magnitude

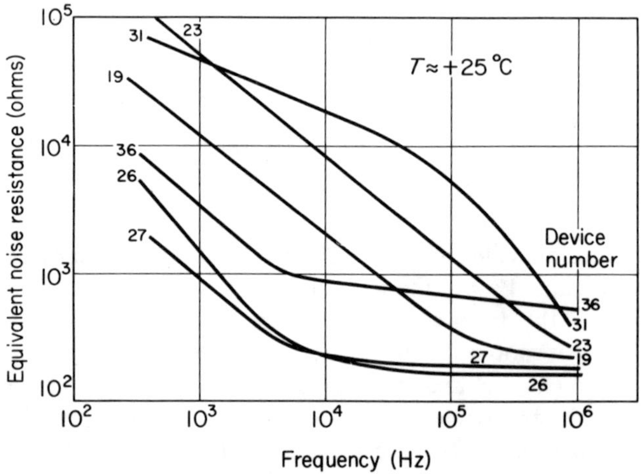

Figure 5.2 Equivalent noise resistance for a narrow frequency band, with the gate AC grounded, for various samples of the type 2N3823 FET, according to Blalock[4]

$$d\overline{i_3^2} = 2I_g e\, df \tag{5.3}$$

where e is the electronic charge. By the choice of an FET with a small enough leakage current, and operation at a low temperature, it is usually possible to make this noise source negligible.

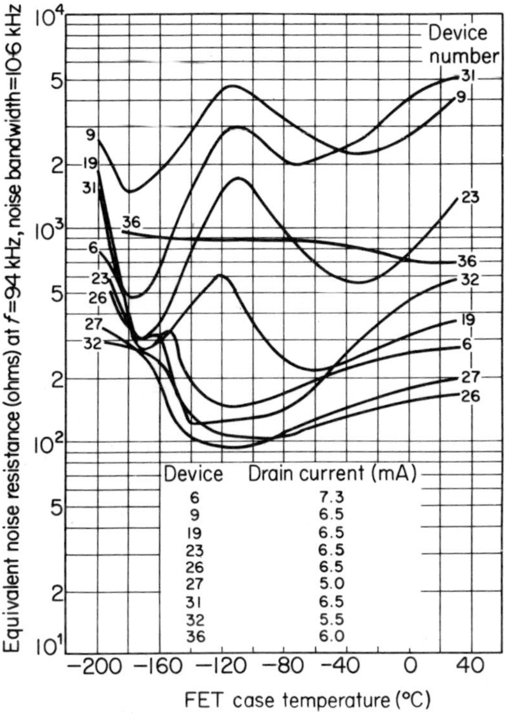

Figure 5.3 Equivalent noise resistance in a narrow frequency band centred at 94 kHz, for various samples of the type 2N3823 FET with the gate AC grounded, according to Blalock[4]

Finally, in addition to the above sources of noise, practical examples of FETs usually exhibit some other sources of noise which predominate in the low frequencies.[7] Such excess or flicker noise usually varies considerably in different examples of the same type of FET and appears to be due to generator-recombination-trapping processes both in the bulk and on the surface. Blalock[4] has examined the noise behaviour of 33 examples of the 2N3823 type FET. Figure 5.2 shows the noise behaviour, with the gate AC grounded, where the noise in a narrow band of frequency for various mean frequencies was measured. In particular numbers 26 and 27 gave noise performances close to the theoretical value (Equation (5.1)) over most of the frequency range, but it will be seen that most of the other examples tested showed a large excess noise component. Figure 5.3 shows the noise behaviour in a fixed narrow frequency band centred at 94 kHz at different temperatures, and it will be seen that the optimum temperature for operation for this type was around −110°C.

Other workers[5,6,9,55] have confirmed the critical but unpredictable dependence of the noise characteristics on the temperature and also on the quiescent biasing conditions. Goulding, Walton and Malone[5] have summarized some of the criteria for the selection of FETs and draw attention to the

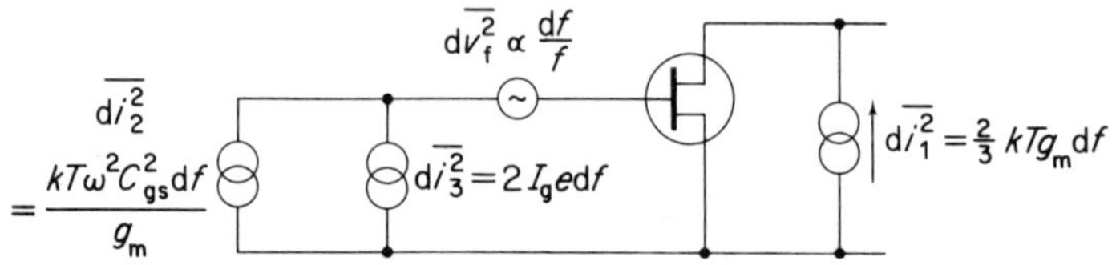

Figure 5.4 Equivalent generators for the noise sources in the FET

importance of the choice of the drain-source voltage, a quantity whose optimum value changes with temperature. Since the resolution currently attainable with semiconductor detectors is often set by the noise characteristics of the FET, there is considerable interest in reducing this noise and in developing better FETs. Junction FETs are practically the only type currently used since, as pointed out by Elad,[6] Sah[7] and Radeka,[8] insulated gate FETs (i.e. MOSFETs) have larger components of low frequency noise. Elad[6] also considered the noise performances of several different types and compared the temperature behaviour of Ge, Si and GaAs types. Howes[9] considered the noise behaviour of two types specially developed for low noise detector preamplifiers. Kern and McKenzie[48,49] have shown how the material used to encapsulate the FET can contribute a large component of dielectric noise.

In summary all the noise sources in the FET may be represented by the equivalent noise sources shown in Figure 5.4 where the excess or flicker noise is represented as having a $1/f$ frequency dependence. It appears that by suitable selection of the FET and operation at low temperature two of these noise sources, due to excess noise and to gate leakage current, may be made negligible.

5.2.3 *Equivalent circuit for detector and preamplifier*

We will initially assume that the radiation detector is coupled to the gate of the FET as shown in Figure 5.5 so that the input stage operates as a voltage sensitive amplifier. In this case two additional sources of noise appear. The first is an additional shot noise component due to the leakage current I_1 in the detector and having a magnitude

$$\overline{di_4^2} = 2I_1 e\, df \tag{5.4}$$

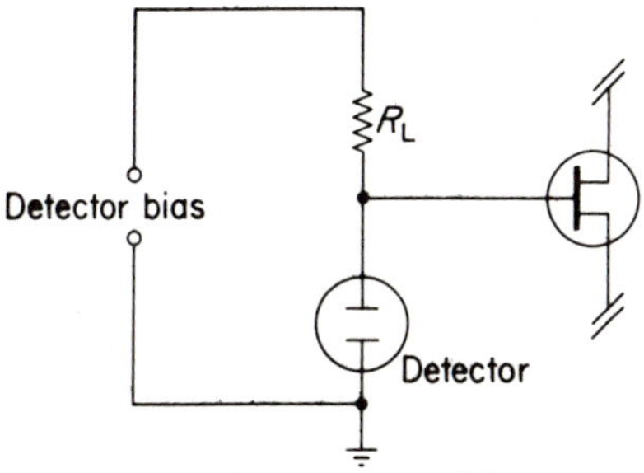

Figure 5.5 Voltage sensitive configuration

The second is due to the Johnson noise of the load resistor R_L and any other resistors connected to the input. If R_t is the total equivalent shunt resistance at the input due to these and C_t is the total shunt capacitance at the input due to the detector capacitance, input capacitance of the FET and stray capacitance, then the equivalent circuit for the signal and for the noise sources is shown in Figure 5.6 where, it will be seen, all noise sources have been reduced to their equivalent values at the input. It may here be noted that the input impedance Z_i seen by the signal which was used in previous chapters is simply seen to equal Z_t, that is the parallel combination of R_t and C_t, for this voltage sensitive amplifier.

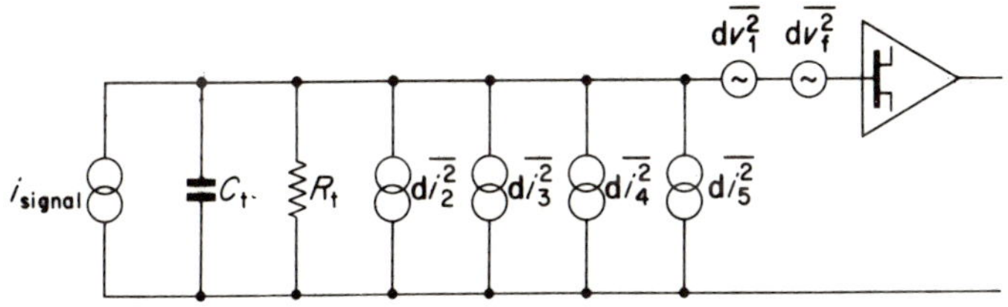

$$d\overline{v_1^2} = \frac{2}{3}\frac{4kT\,df}{g_m}$$

$$d\overline{v_f^2} = \frac{A_f\,df}{f}$$

$$d\overline{i_2^2} = \frac{kT\omega^2 \text{[illegible]}\,df}{g^{m}}$$

$$d\overline{i_3^2} = 2I_g e\,df$$

$$d\overline{i_4^2} = 2I_l e\,df$$

$$d\overline{i_5^2} = \frac{4kT\,df}{R_t}$$

Figure 5.6 Equivalent noise generators for a semiconductor detector coupled to an FET in the voltage sensitive configuration

A similar equivalent circuit may be used if the input stage utilizes a thermionic valve since the FET noise sources due to channel, gate induced, flicker and gate leakage have their counterparts in anode shot noise, grid induced, flicker and grid current noise respectively. The use of thermionic valves now appears to be of historical interest only, however, since in nuclear detector preamplifiers they cannot give as low a noise contribution as FETs at room temperature[6] and the use of a cooled FET gives a noise performance much better than any currently available device.

5.3 Preamplifier configurations

5.3.1 Conditions at the input

Certain general factors about the input configuration may be considered without regard to the type of detector or types of noise sources. Figure 5.7(a) shows a simplified equivalent circuit for the signal and noise sources where the noise sources have been grouped into two equivalent noise generators $d\overline{i_p^2}$ and $d\overline{v_s^2}$. Thus $d\overline{i_p^2}$, henceforth termed the parallel noise, contains $d\overline{i_2^2}$, $d\overline{i_3^2}$ and $d\overline{i_4^2}$ in Figure 5.6 and $d\overline{v_s^2}$ termed the series noise, represents $d\overline{v_1^2}$ and $d\overline{v_f^2}$. We will initially ignore the resistor noise contribution and this will be considered later. Z_t is the equivalent impedance at the input due to C_t and R_t.

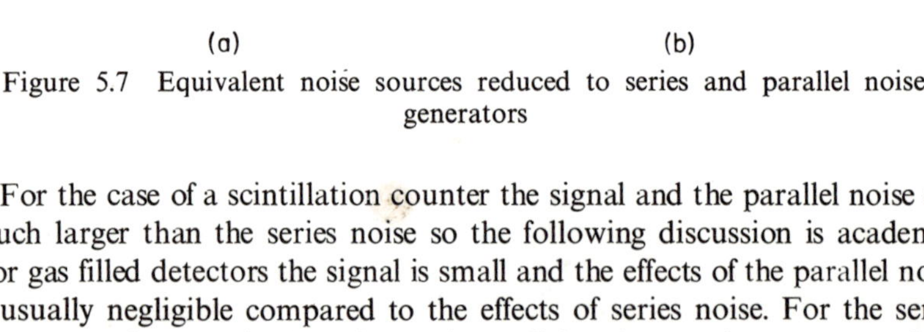

Figure 5.7 Equivalent noise sources reduced to series and parallel noise generators

For the case of a scintillation counter the signal and the parallel noise are much larger than the series noise so the following discussion is academic. For gas filled detectors the signal is small and the effects of the parallel noise is usually negligible compared to the effects of series noise. For the semiconductor detector both series and parallel noise are important and, in addition, the detector capacitance is dependent on the bias voltage and special arrangements are necessary which are dealt with later in this section.

The first point to note is that, in Figure 5.7(a), i_{signal} and $d\overline{i_p^2}$ both feed into Z_t, and so the choice of Z_t is immaterial to the signal to *parallel* noise ratio ultimately achievable, since any filtering action implicit in the choice of

Z_t could, if necessary, be cancelled out by suitable circuitry further along the amplifier. It is therefore only to the effect of *series* noise that the input configuration is critical.

The equivalent circuit may be further reduced to that of Figure 5.7(b). Confining our attention for the moment to noise in a single narrow band of frequency it is clear from Figure 5.7(b) that if $|Z_t|$ for that frequency is increased the relative contribution of series noise to the total noise can be reduced. It follows that, if $|Z_t|$ can be increased over the whole frequency spectrum, the relative contribution of series noise to the total noise in the whole spectrum will also be reduced. This reasoning is quite general for any type of signal, any pair of (independent) noise generators, and any subsequent filtering in the later pulse shaping stages. Thus, if Z_t is a parallel combination of C_t and R_t, ideally C_t should be small and R_t large. Wherever series noise is significant, that is for a semiconductor detector or gas filled detector, it is important to observe these conditions.

If we now consider the resistor noise arising from R_t this is equivalent to a current noise generator across the input of magnitude $d\overline{i_5^2} = 4kT\,df/R_t$. In order to make this negligible it is necessary to make R_t large, a condition which fortunately coincides with the optimum conditions concerning the series noise. In practice in a preamplifier for semiconductor detectors it is necessary to make R_t at least of the order of 10^9 ohms to make resistor noise negligible.

Since the depletion layer capacitance of a semiconductor is dependent on the bias voltage, unless special steps are taken the pulse height at the end of the processing system will be highly dependent on this voltage. Two alternatives appear, first to connect a large capacitor in parallel with the detector and so swamp the variations of detector capacitance or secondly to connect a low resistance in parallel with the detector so that it is effectively the detector current signal that is amplified. Neither of these alternatives can ordinarily be used directly since both serve to decrease $|Z_t|$ and the resultingly small signal would be swamped by series noise and possibly by resistor noise. The effect of a large C_t or small R_t for the signal can, however, be simulated by feedback arrangements so that there is negligible effect on the signal to noise ratio. The two possible arrangements are discussed in the next two subsections.

5.3.2 *Charge sensitive configuration*

Consider the circuit of Figure 5.8(a) where Z_t represents, as before, the impedance at the input due to the detector load, the detector capacitance, the inherent amplifier input capacitance and any stray capacitances, Whereas in the voltage sensitive version, Figure 5.7(a), the signal saw an input impedance Z_i equal to Z_t, for the feedback arrangement the equivalent circuit of Figure 5.8(b) shows that the input impedance to the signal, Z_i, is now equal to $Z_t \| (Z_f/(A+1))$. If Z_f takes the form of a capacitance C_f then the feedback arrangement provides an equivalent capacitance of $C_f(A + 1)$ shunting the input. Since this may be made much larger than any capacitance due to the detector (included in Z_t) any variations in the detector capacitance will be swamped by $C_f(A + 1)$ and the arrangement approximates to a charge sensitive amplifier.

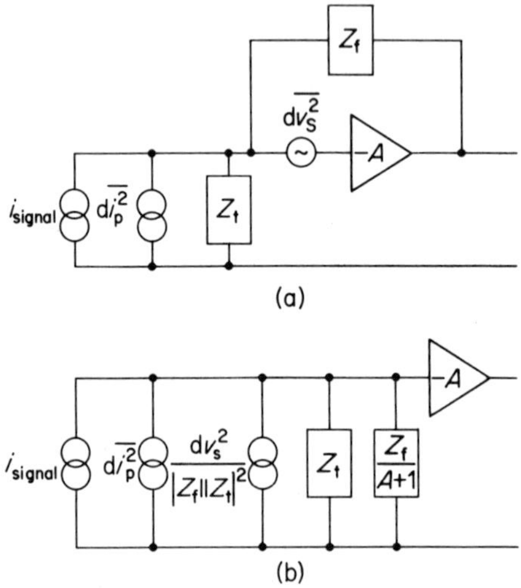

Figure 5.8 (a) Amplifier with a feedback impedance Z_f. (b) Equivalent circuit for the signal and for the series and parallel noise sources

In order to see how the inclusion of feedback via Z_f affects the noise characteristics, it is useful to reduce the series and parallel noise sources to those shown in Figure 5.8(b). It will be seen that the series noise may be represented as an equivalent noise generator across the input of magnitude $d\overline{v_s^2}/|Z_f \| Z_t|^2$. Comparing this with the equivalent circuit for the voltage sensitive configuration without feedback Figure 5.7(b), it will be seen that the contribution of the series noise will be only slightly larger for the feedback configuration provided $|Z_f| \gg |Z_t|$. For a charge sensitive amplifier where Z_f is a capacitance, C_f, this implies that $C_f \ll C_t$. The original objective of swamping the detector capacitance variations by $C_f(A + 1)$ can still be achieved provided A is large enough.

In practice the feedback impedance Z_f may consist of a parallel combination of a resistance R_f and a capacitance C_f. The inclusion of R_f may be used to stabilize the DC conditions of the preamplifer or, by means of this arrangement, it is possible to dispense with the detector load R_L as such and to apply the detector bias through R_f. This is discussed later but for the moment we may only note that R_f will also contribute a component of noise that can be included in the equivalent circuit of Figure 5.8(b) as an equivalent noise current generator across the input of magnitude $d\overline{i^2} = 4kT\,df/R_f$. It follows that to make this negligible R_f should be large and since the expression for noise current is directly comparable to that for noise from R_t which was discussed in the previous section, the conditions on R_f are identical, i.e. should have a magnitude of at least 10^9 ohms for use in a semiconductor detector preamplifier.

5.3.3 *Current sensitive configuration*

If Z_f is made a pure resistance, R_f, the input resistance of preamplifier is $R_f/(A + 1)$ and may, in principle, be madc small enough so that it is effectively the current signal from the detector that is amplified, and this will be negligibly affected by changes in the detector capacitance. The arrangement has rarely been used however since, as discussed in the previous Subsection, for the Johnson noise to be negligible, R_f must have a value of at least 10^9 ohms. To produce a low enough input impedance the gain A would have to have a very large value and, as discussed in Subsection 5.5.4, it is extremely difficult to apply resistive feedback over such a gain, as there are phase shifts at high frequencies which usually cause the system to oscillate. For low resolution work a current sensitive configuration of this type is a possibility as evidenced by the design by Millard and Blalock[10]. Using a value of R_f of 51 kΩ they obtained fwhms in the region of 10 keV.

More recently, Millard, Blalock and Hill[50] have described another type of preamplifier giving a current sensitive action, and which was developed to give a fast rise time for use in timing applications. Their configuration used the conventional feedback arrangement of the charge sensitive amplifier, i.e. a parallel capacitance C_f and resistance R_f, and so the stability problem of pure resistive feedback did not arise. In addition, the preamplifier output was provided with a load consisting of a parallel capacitance C_L and a resistance R_L, such that $R_L C_L = R_f C_f$. The current gain was then R_f/R_L. In effect the 'integrating' action of the feedback loop was cancelled by the 'differentiating' action of the load. To express it more precisely, the current transfer function had a pole at $-1/R_f C_f$ which was cancelled by a zero at $-1/R_L C_L$. The preamplifier had a rise time of 8 ns (with a detector capacitance of 10 pF) and using an uncooled FET gave a resolution of 1·4 keV (fwhm Ge) + 20 eV/pF.

5.4 Semiconductor detector preamplifiers

5.4.1 *General considerations*

Over the past few years considerable progress has been made in improving the resolution of semiconductor detector spectrometry systems. This has been partly due to the availability of detectors having better intrinsic resolutions and partly due to the decreased contribution of noise from the detector and the preamplifier.

Figure 5.9 shows the fwhm resolution to be expected solely from the detector, on the assumption of various Fano factors in Ge and Si. Assuming that the only additional source of loss of resolution is due to noise the overall $(\text{fwhm})^2$ would be expected to equal the sum of the $(\text{fwhm})^2$ of the detector

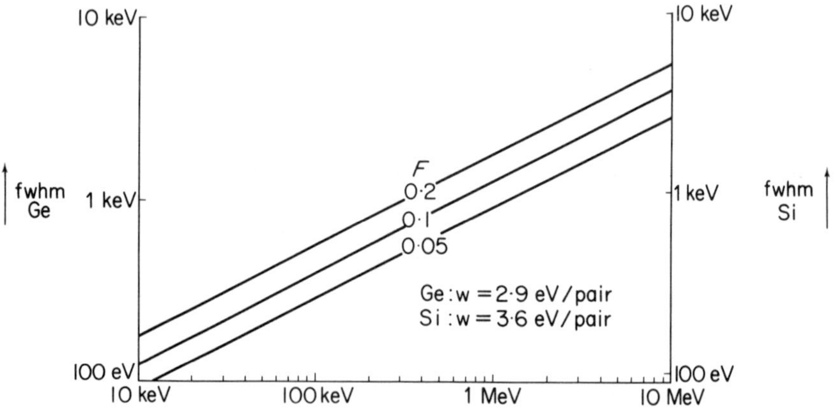

Figure 5.9 Inherent resolution of Ge (L.H. ordinate) and Si (R.H. ordinate) detectors calculated for various values of the Fano factor F

and the noise $(\text{fwhm})^2$, the latter quantity being the 'electronic' resolution which could be measured by means of a pulser at the preamplifier input. While the electronic resolution is usually good enough to be unimportant in the detection of high energy particles it is critical for the detection of low energy particles. Thus the detection of the characteristic X rays of light elements is a subject of wide interest that has become possible, by means of semiconductor detectors, only recently.

The noise sources for the semiconductor detector and FET in the voltage sensitive configuration were shown in Figure 5.6. Assuming $R_t \gg 1/\omega C_t$ in the frequency range of interest, the equivalent noise voltage at the FET input is therefore

$$\mathrm{d}\overline{v^2} = \left[\frac{kT}{g_m}\left(\frac{8}{3} + \frac{C_{gs}^2}{C_t^2}\right) + \frac{1}{\omega^2 C_t^2}\left(2e(I_g + I_l) + \frac{4kT}{R_t}\right) + \frac{A_f}{f}\right]\mathrm{d}f \qquad (5.5)$$

To conform to usage in previous chapters, this may be rewritten

$$\mathrm{d}\overline{v^2} = \left(a^2 + \frac{b^2}{\omega^2} + \frac{c^2}{\omega}\right)\mathrm{d}\omega \qquad (5.6)$$

to show the contributions of so-called 'A noise', 'B noise' and flicker noise, where

$$a^2 = \frac{kT}{2\pi g_m}\left(\frac{8}{3} + \frac{C_{gs}^2}{C_t^2}\right)$$

$$b^2 = \frac{1}{2\pi C_t^2}\left(2e(I_g + I_l) + \frac{4kT}{R_t}\right)$$

$$c^2 = A_f$$

We will confine the following discussion to the case of a system using CR–RC shaping having filter time constants τ, as this is the case for which results are most commonly quoted. It is however a simple matter to convert the results to apply to any other system of pulse shaping by using the conversion factors given in Subsection 4.4.2 (Table 4.1). We will also assume that the flicker noise component is negligible and this appears to be a reasonable assumption, as may be seen from the results given later.

For CR–RC shaping the *ENC* is given by (4.36) (Subsection 4.4.1) where $C_t = C_i$, and substituting for a and b we have

$$(ENC)^2 = (ENC)_A^2 + (ENC)_B^2 \tag{5.7}$$

to show the contribution of the A noise and B noise, where

$$(ENC)_A^2 = \frac{e^2 kT}{\tau g_m}\left(\frac{1}{3}C_t^2 + \frac{1}{8}C_{gs}^2\right) \tag{5.8}$$

$$(ENC)_B^2 = \frac{e^2 \tau}{8}\left(2e(I_g + I_l) + \frac{4kT}{R_t}\right) \tag{5.9}$$

For the charge sensitive configuration the equivalent circuit of Figure 5.8(*b*) can be used. Provided the feedback capacitance is small compared with C_t and provided the value of R_t in (5.9) is replaced by $R_t \| R_f$ the same expression for the ENC may be used.

The variation of $(ENC)^2$ with τ can best be seen by separately plotting $\log(ENC)_A^2$ and $\log(ENC)_B^2$ against $\log\tau$, giving two straight lines. $\log(ENC)^2$ should in theory be asymptotic to these two lines as shown in Figure 4.19. Figure 5.10 shows $(ENC)^2$ for a selection of designs from the data given by the authors and it will be seen that there has been a progressive improvement in the FET noise characteristics.

Representative theoretical values for the B noise are plotted in Figure 5.10 for various values of $(I_g + I_l)$ and R_t. The range of values of τ of interest is usually $\leqslant 10$ µs since larger values give a long output pulse, and pulse pile up becomes a problem. For a value of τ in the range 1–10 µs it will be observed that, so that the B noise is comparable to the A noise for the best FET in

Figure 5.10, $(I_g + I_l)$ should be $\gtrsim 10^{-11}$ amperes and $R_L \gtrsim 10^9$ ohms at 77°K. While the former condition can often be met the latter presents problems since such high valued resistors behave unpredictably at low temperatures and high frequencies and often exhibit excess noise as discussed in the next section.

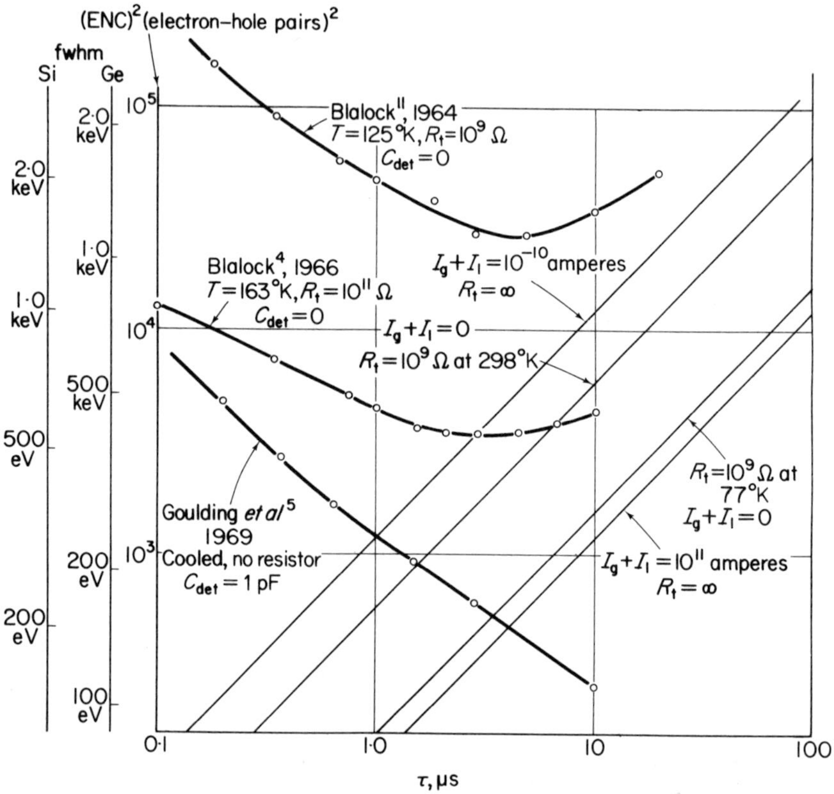

Figure 5.10 Plots of $(ENC)^2$ against the time constant of the CR–RC shaping filter for representative preamplifier designs. as estimated from the data given by the authors. Also shown are the calculated values of $(ENC)^2_B$ for various values of $(I_g + I_l)$, and for $R_t = 10^9$ ohms at two different temperatures. The equivalent fwhms for Si and Ge detectors are given on the subsidiary ordinary scales

For the A noise certain conclusions may be drawn about the choice of the FET. If the detector and stray capacitance are small the total capacitance C_t equals C_{gs} effectively. For this case, from (5.8), a FET with a low value of C^2_{gs}/g_m is desirable. If C_t is much larger than C_{gs} this conclusion is not necessarily valid. Smith and Cline[12] have shown that a reduction of $(ENC)^2_A$ may often be made by paralleling several FETs. For n FETs in parallel the expression for the A noise becomes

$$(ENC)_A^2 = \frac{e^2 kT}{\tau g_m}\left[\frac{C_t^2}{3n} + \frac{1}{8} n C_{gs}^2\right]$$

so this technique is valuable when the detector capacitance is larger than C_{gs}. Figure 5.11 shows plots of log $(ENC)_A^2$ against logτ taken from Smith and Cline's paper for various values of n, temperatures of 140°K and 298°K, and for values of C_t equal to nC_{gs} (i.e. zero detector and stray capacitance) and

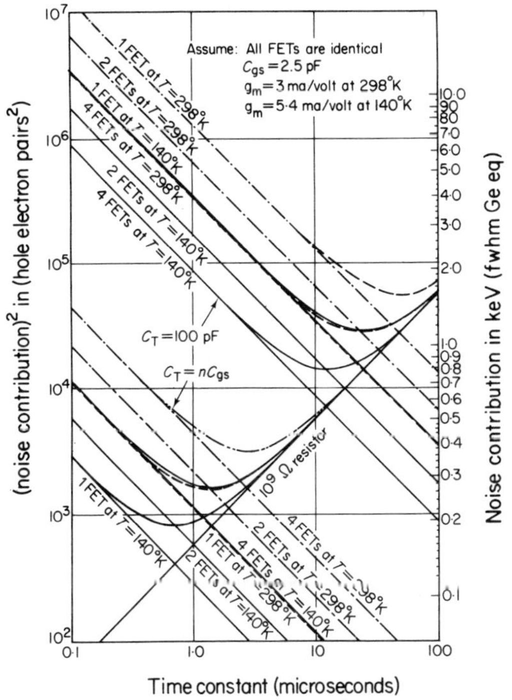

Figure 5.11 Theoretical values of $(ENC)^2$ against the time constant of CR–RC shaping for 1,2 and 4 paralleled FETs, at T = 298°K and 140°K, with C_t = 100 pF and nC_{gs}, according to Smith and Cline[12]

also for C_t = 100 pF. Although the lines are theoretical plots (assuming g_m = 3 mA/V at 298°K and 5·4 mA/V at 140°K and C_{gs} = 2·5 pF) measurements at several selected points confirmed their general validity.

5.4.2 Preamplifier circuits

Probably the simplest arrangement for coupling the detector to the FET gate is using AC coupling via a coupling capacitor as in Figure 5.12. This means the DC (quiescent) conditions on the FET can be set independently of the bias condition of the detector and so different detectors may be used with the same preamplifier without adjustment. In addition the full dynamic range of the preamplifier is available for the signal pulses alone without having to allow for the DC offset due to the detector. As Radeka[13] has pointed out, of the two arrangements (a) and (b) in Figure 5.12, (a) is better since, when high values of detector bias are used, arrangement (b) places a high voltage across the coupling capacitor and this can contribute spurious noise. One disadvantage of AC coupling is that the inclusion of the coupling capacitor causes an increase in the stray capacitance to ground and so slightly reduces the signal to noise ratio.

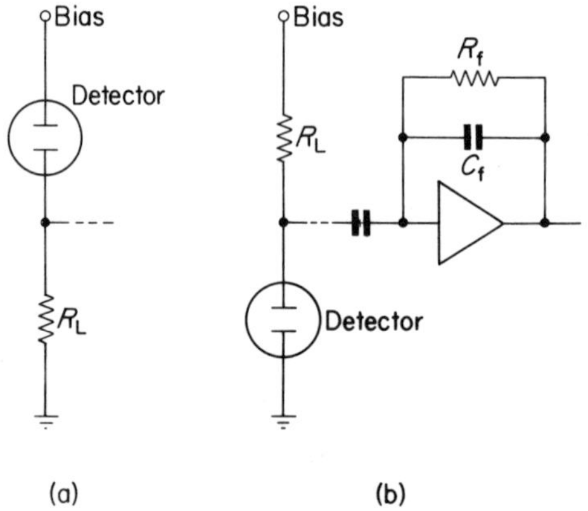

Figure 5.12 Alternative arrangements for the AC coupling of the detector to a charge sensitive preamplifier input

DC coupling is usually arranged so that the detector bias is applied through the feedback resistor, R_f, in Figure 5.13(a). For signals this is equivalent to the circuit of Figure 5.13(b). One advantage of DC coupling is that the total DC current flowing, I_t, due to detector leakage, signal current and FET gate current (although this may be of opposite sign to the previous two), may be monitored by measuring the DC voltage at the amplifier output which equals $I_t R_f(A/(A + 1)) \simeq I_t R_f$. This is usually of the order, at least, of millivolts and may be measured by a conventional meter without upsetting the circuit operation.

Although the charge sensitive amplifier would ideally have a charge sensitivity which was unaffected by the value of the detector capacity (i.e. $C_f(A + 1)$ should be much greater than the detector capacitance), there will always be a slight dependence on the detector capacitance, and it is therefore common to include a calibration system in the input circuit. This usually takes the form of a pulse generator connected to the preamplifier input via a small capacitor. For use in high resolution systems both the pulse generator,[14,51] and the capacitor,[15] should be of high stability. When the detector is connected and the bias applied, the pulse generator can inject a brief pulse of charge into the input circuit.

This pulse of charge is, to a first approximation, independent of the input capacity (including that of the detector). In this way the relative charge sensitivity of the arrangement may be found for any combination of detector and bias.

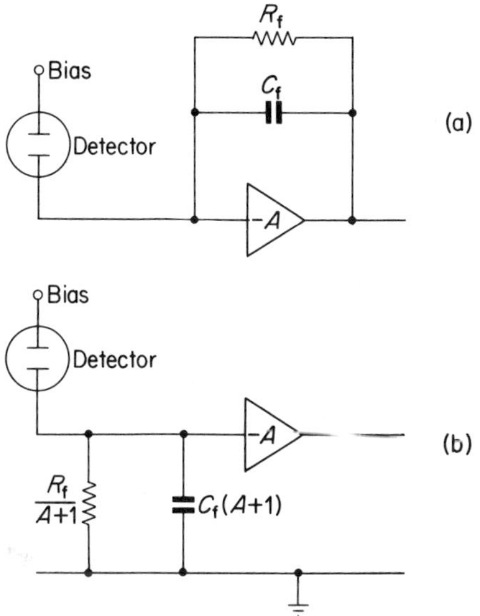

Figure 5.13 (a) Application of the detector bias via a feedback resistor R_f. (b) The equivalent circuit

As Nowlin and Blankenship[16] have pointed out, the injection of charge via a capacitor cannot be truly unipolar and is unlike the detector signal. There is therefore an associated difference in the output pulse shape and a slight dependence of the injected charge with the input capacitance. To reduce these effects they suggest the replacement of series capacitor with a parallel RC combination having the same time constant as the input circuit, so that the pulse is applied through a compensated voltage divider. The pulser should then feed an exponentially decaying pulse, having the same time constant, into the parallel RC combination.

Many designs of FET preamplifiers have been published and some examples will now be given. Figure 5.14 shows the circuit of Blalock.[4] Q_2 is a current amplifier stage interposed between the cascode pair Q_1 and Q_3. The collector load of Q_3 is bootstrapped to the output of the White emitter follower composed of Q_4 and Q_5. Feedback is applied via R_f and C_f to the FET gate and this provides a DC feedback path which stabilizes the operating points of all the transistors. The open loop gain was measured to be 3700 and had a 3 dB bandwidth extending from 100 Hz to 500 kHz. The input capacitance (including C_f) with the loop open was 8·5 pF. The rise time of the circuit was 25–80 ns depending on the detector capacitance and hence

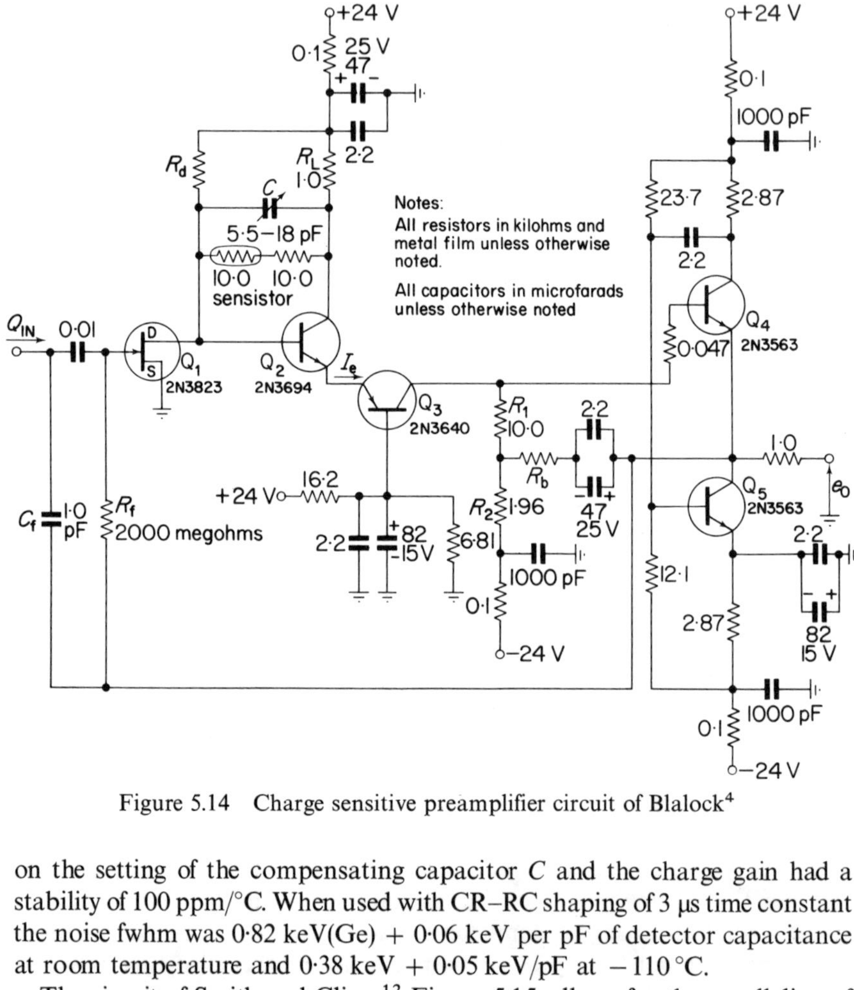

Figure 5.14 Charge sensitive preamplifier circuit of Blalock[4]

on the setting of the compensating capacitor C and the charge gain had a stability of 100 ppm/°C. When used with CR–RC shaping of 3 μs time constant the noise fwhm was 0·82 keV(Ge) + 0·06 keV per pF of detector capacitance at room temperature and 0·38 keV + 0·05 keV/pF at −110 °C.

The circuit of Smith and Cline,[12] Figure 5.15, allows for the paralleling of FETs as discussed earlier. Q_2 is bootstrapped to the output of the emitter follower composed of Q_3, Q_4 and Q_5. The open loop gain and open loop rise time were 4000 and 15 μs for a single FET at room temperature and the noise fwhm (CR–RC shaping with τ = 1–2 μs) was

$$1\ \text{FET}\ \begin{cases} 0{\cdot}53\ \text{keV(Ge)} + 0{\cdot}046\ \text{keV/pF at } 298^\circ\text{K} \\ 0{\cdot}36\ \text{keV} \quad\quad\ + 0{\cdot}03\ \ \text{keV/pF at } 140^\circ\text{K} \end{cases}$$

$$4\ \text{FETs}\ \begin{cases} 0{\cdot}91\ \text{keV} \quad\quad\ + 0{\cdot}023\ \text{keV/pF at } 298^\circ\text{K} \\ 0{\cdot}62\ \text{keV} \quad\quad\ + 0{\cdot}017\ \text{keV/pF at } 140^\circ\text{K} \end{cases}$$

so the technique of paralleling FETs is particularly valuable when the detector capacitance is large.

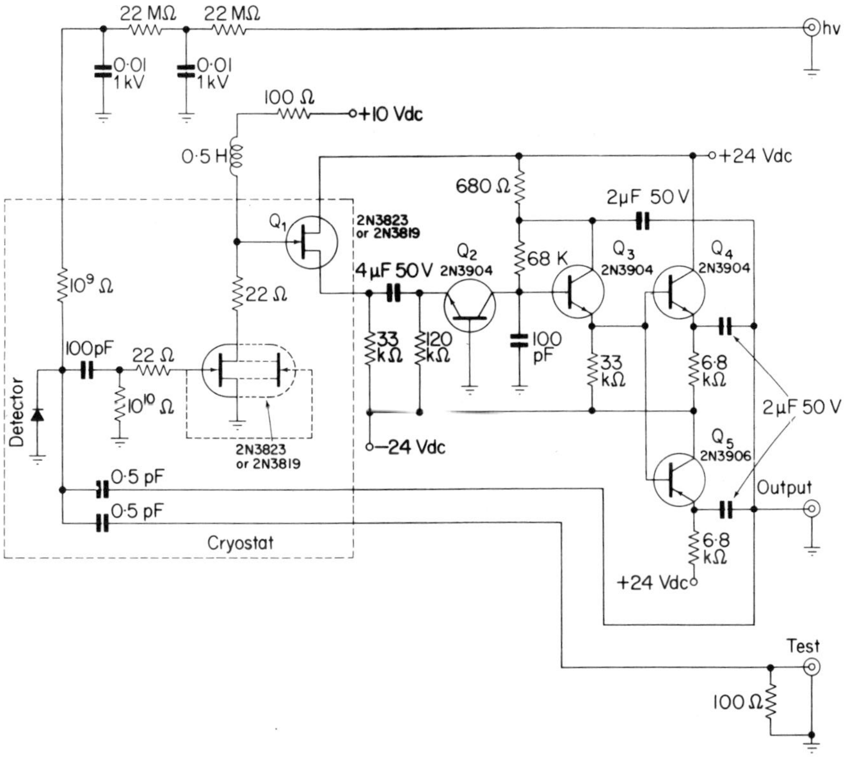

Figure 5.15 Charge sensitive preamplifier circuit of Smith and Cline,[12] allowing paralleling of FETs

For the highest resolution work several authors have called attention to other sources of noise. Radeka[13] shows how the quality of the dielectric material used in capacitors and insulators at the preamplifier input critically affects the resolution. Elad,[6] Kern and McKenzie,[48, 49] and Howes[9] also discuss this source of noise together with that due to microphonic effects produced by the bubbling of nitrogen gas. The resistors R_L and/or R_f present a major problem since such high valued resistors behave unpredictably at high frequencies and at low temperatures. Figure 5.16 shows the frequency

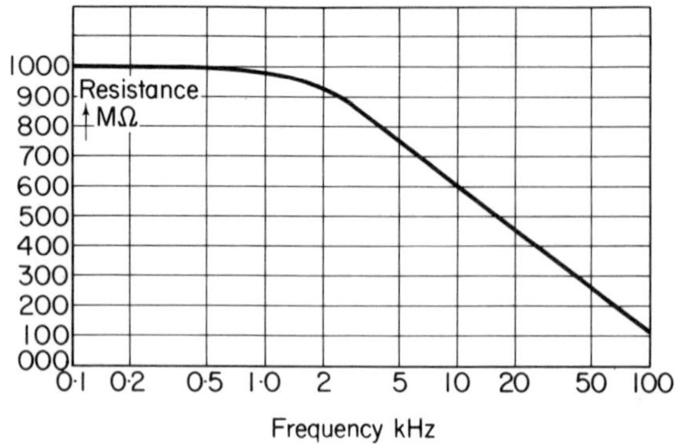

Figure 5.16 Variation of the effective resistance with frequency of a resistor of nominal value 10^9 ohms, according to Goulding, Walton and Malone[5]

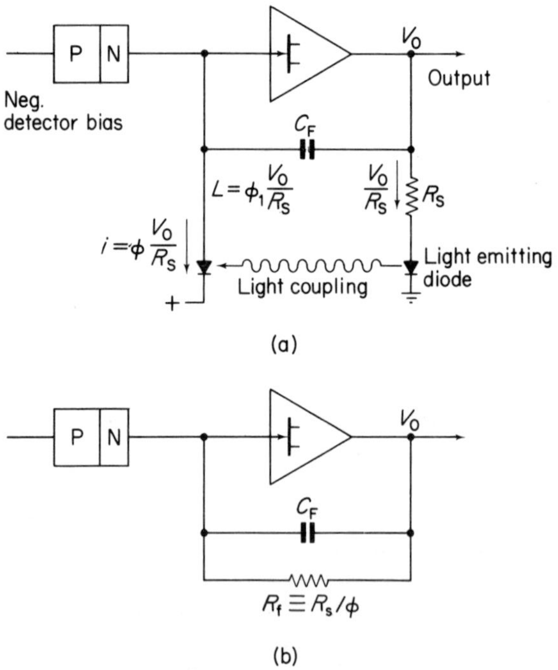

Figure 5.17 (a) Principle of the optoelectronic feedback system of Goulding, Walton and Malone.[5] (b) Equivalent circuit

behaviour of a 10^9 ohm resistor according to Goulding, Walton and Malone.[5] Similar data has been given by Radeka[13] and presumably is partly due to distributed capacitance. One result of this is that the effective resistance in the working frequency range is much less than the nominal resistance and the Johnson noise would be expected to be correspondingly larger. In any case, commercially available resistors exhibit noise in excess of the Johnson noise and, on cooling, the resistors often undergo irreversible changes.[5]

In order to avoid the necessity of using such a resistor Elad[6] advocated a voltage sensitive configuration, with the detector directly coupled to the FET gate. For use as an X-ray detector the resulting dependence of the pulse height on the detector capacitance is probably not an important factor, as the fractional resolution at such low energies is small.

In the charge sensitive configuration the function of R_f is to provide a DC path for the leakage currents and the detector signal to flow, and so Goulding *et al.*[5, 17] have circumvented the problem by replacing the resistance by a photodiode. Such a diode, being reverse biased, exhibited mainly shot noise and, at low signal rates, this will have a mean square magnitude similar to that produced by the leakage currents ($I_g \pm I_1$), and so will be very small. The photodiode was incorporated in a feedback loop by arranging for the preamplifier output to supply a light emitting diode which was optically coupled to the photodiode, as in Figure 5.17(a). The arrangement in Figure 5.17(a) may be seen to be equivalent to a feedback resistor R_f equal to R_s/ϕ, as shown in Figure 5.17(b). The full circuit is shown in Figure 5.18 where it will be seen that the gate-channel junction of the FET is used as the photodiode (any PN junction will act as a photodiode) and the circuit incorporates a second light emitting diode whose light falls on the detector junction to enable adjustments to the detector leakage current to be made. Figure 5.19 shows a plot of the noise versus τ for CR–RC shaping and is compared with that for a conventional circuit using a feedback resistor of value $5 \times 10^9\ \Omega$ and using a similar FET. The reduction of the noise from the resistor is clearly visible. With a detector capacitance of 1 pF, a noise fwhm of 115 eV (Si) was obtained with semi-Gaussian pulse shaping with a peaking time of 16 μs, and it was possible to detect the Mn K_α characteristic X-rays (5.90 keV) with a total fwhm of 180 eV (Si).

Elad[53] has developed a different method of restoring the charge at the input without using a resistor. Essentially his technique made use of the phenomenon of impact ionization in the FET. Impact ionization is the production of secondary electron-hole pairs in regions of high electric field; it ultimately leads to the phenomenon of avalanche breakdown. In an FET the region of maximum field lies between the gate and the drain and by controlling the drain-source voltage V_{ds} it was possible to control the secondary ionization. A component of the secondary ionization current flows into the gate and so, in order to control the voltage level at the input, the mean output level of the amplifier was used to control V_{ds}, thus constituting a feedback loop. Using semi-Gaussian shaping with a peaking time of 22 μs an electronic resolution of 82 eV (Si) was obtained. Used with a Si(Li) X-ray detector, the fwhm of the Mn K_α line at 5·90 keV was 152 eV.

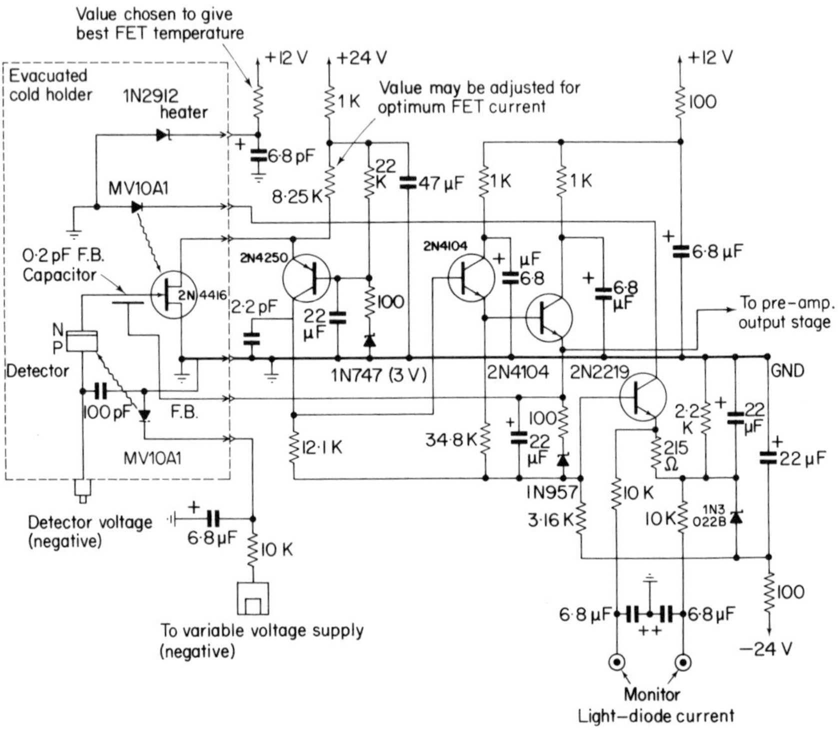

Figure 5.18 Complete circuit of the optoelectronic feedback charge sensitive preamplifier of Goulding, Walton and Malone[5]

Although the optoelectronic feedback arrangement of Goulding *et al.* and the impact ionization method of Elad both provide a DC path for the detector current, both arrangements would be expected to contribute some shot noise which will be larger at higher counting rates. Other systems have been described which, instead of providing a DC path for the detector current as in the designs discussed so far, provide an intermittent path by means of a pulsed arrangement. In most of these designs it is necessary to gate the counting system off during the restoring pulse so that spurious signals are not counted. Figure 5.20 shows the scheme of Kandiah and Stirling.[18] Two voltage level sensing circuits are placed at the output of the preamplifier.

If the voltage level moves outside these two limits the appropriate gallium arsenide light emitting diode, S_1 or S_2, emits a pulse of light and this causes a pulse of current to flow in the appropriate photodiode D1 or D2 so as to bring the input voltage back into the correct operating region. This, and other arrangements to be described, require a highly linear preamplifier if resolution is not to be degraded, since the 'quiescent' conditions will vary over the working range of input voltages. From this point of view the requirements are relaxed if the working range is restricted to a few times the expected signal magnitude.

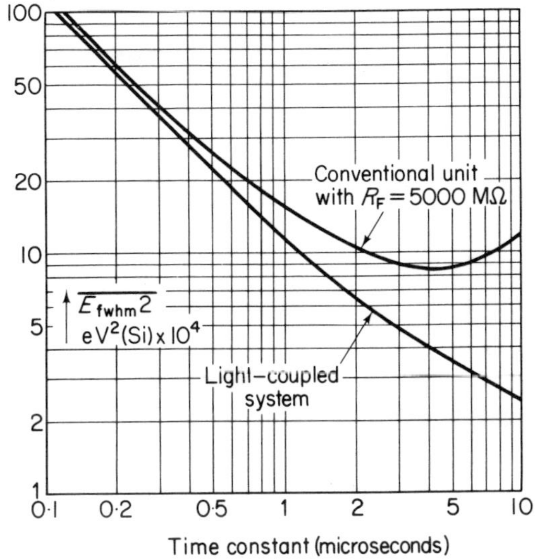

Figure 5.19 Comparison of the noise fwhm of the optoelectronic feedback system with that of a conventional system having a feedback resistor of 5×10^9 ohms and using a similar FET, according to Goulding, Walton and Malone.[5] The noise fwhm is plotted against the time constant of a CR–RC shaping system

A disadvantage of the system of Kandiah and Stirling is that the inclusion of the extra components, the two photodiodes D1 and D2, increases the input capacitance and so worsens the resolution. Radeka[19] has devised an arrangement that applies the restoring pulse to the other side of the detector (Figure 5.21), and utilizes the rectifying property of the gate-channel junction of the FET to provide a path for the restoring current pulse. In this design the restoring pulses are supplied at a constant rate but have their amplitude modulated so as to keep the input voltage in the correct region. Using filter time constants of 10 μs the electronic resolution obtained was 146 eV(Si) fwhm (210 eV (Si) total with 6·4 keV X-rays). The circuit had a dead time of 10% which corresponded to the time occupied by the restoring pulses.

Landis *et al.*[20, 52] have modified the earlier optoelectronic feedback system of Figure 5.17 to operate in the pulse feedback mode. The scheme (Figure 5.22) follows the general principle of Kandiah and Stirling in that a level sensor at the preamplifier output causes a restoring pulse of charge to be applied to the input whenever the level moves out of the working range but, as in the earlier arrangement, the photodiode took the form of the gate-channel junction of the FET and so added no extra capacitance to the input. This

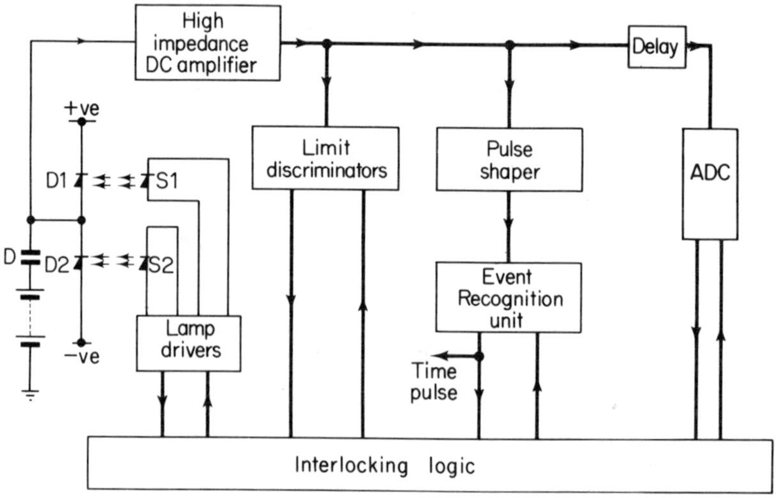

Figure 5.20 Pulsed feedback system using an optoelectronic link according to Kandiah and Stirling.[18] (*Semiconductor Nuclear-Particle Detectors and Circuits*, Publication 1593, Subcommittee on Instruments and Techniques, Committee on Nuclear Science, National Academy of Sciences–National Research Council, Washington, D.C., 1969

arrangement overcame several disadvantages of the earlier arrangement, arising from the non-linear response of the optoelectronic links which caused problems at higher counting rates and caused variation in the effective input time constant, so that exact pole-zero cancellation was not possible. Using semi-Gaussian shaping with a 45 μs peaking time and a detector capacitance of 1 pF they obtained an electronic resolution of 91 eV (Si). They also give data on the resolution with a pulser and with Mn X-rays at various counting rates and the resolution was sensibly constant for rates up to $\sim 10^4$ per second.

It has been pointed out by Kurz[54] that some of the pulse feedback systems discussed above are unsuitable for use with high energy particles at high pulse rates. This is because the rate at which charge must be restored is high and so leads to a high rate of restoring pulses which results in an unacceptable

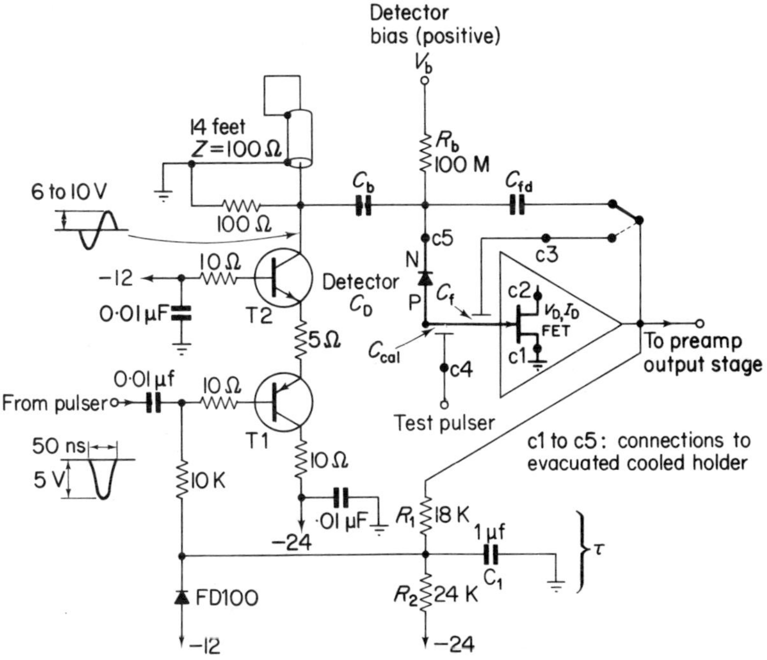

Figure 5.21 Pulsed feedback system of Radeka.[19]

amount of dead time. The system described by Kurz used a brief restoring pulse for every detector pulse. The arrival of a detector pulse at the preamplifier output caused the generation of a current pulse, of magnitude proportional to that of the detector pulse, which was fed into a GaAs light emitting diode whose light was received by the FET junction. This current pulse was generated 0·5 μs after the peak of the detector pulse. The final result of the feedback was

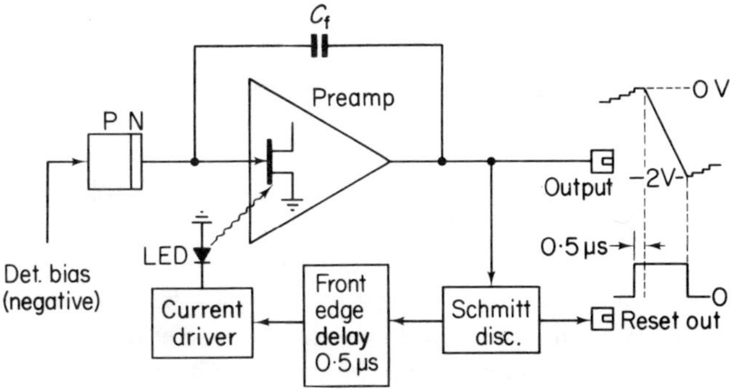

Figure 5.22 Principle of the pulsed optoelectronic feedback system of Landis *et al.*[20, 52]

that the pulse at the preamplifier output was converted into a bipolar form. In order to stabilize the baseline over a longer period, the pulse feedback was supplemented by a DC feedback via the same GaAs diode. When used with a 65 cm^3 Ge(Li) detector on the 1·33 MeV line of ^{60}Co it was possible to obtain fwhms of 2 keV at 10^3 cps, 2·2 keV at 10^4 cps and, using a baseline restorer, 3·2 keV at 5×10^4 cps. These figures were all obtained using CR–RC shaping with 2 μs time constant. Use of a 1 μs time constant, together with a baseline restorer, gave a fwhm of ~4 keV at 10^5 cps.

5.4.3 Circuits not using FETs

Devices other than FETs have been used as the amplifying element in the input stage. Thermionic valves were commonly used several years ago and gave noise fwhms in the region of 2 keV + 50 eV/pF.[6] Junction (i.e. bipolar) transistors have much larger noise fwhms of the order of 25 keV according to one of the few published designs.[21]

A design by Radeka and Chase[22] used a parametric amplifier in which the signal caused the capacitance of a varactor diode to vary and so modulated a 130 MHz frequency carrier. This system gave a noise fwhm of 1·4 keV (Si) + 20 eV/pF at 300°K and 0·64 keV + 13 eV/pF at 77°K. A design by Jonasson[23] utilized a tunnel diode in the input stage and gave a fwhm of 12 keV for an input capacity of 18 pF. Both of these types of system appear to offer some scope for improvement.

5.5 Requirements on main amplifiers for spectroscopy

5.5.1 Summary of requirements

The main amplifier which follows the preamplifier in a spectroscopy system should ideally satisfy many requirements.[24, 25, 26] It should amplify the pulse with negligible distortion and so should have sufficient linearity and a large enough bandwidth. This bandwidth should be invariant to the pulse magnitude and this means in practice that the rise of large output pulses should not be degraded due to inadequate currents being available to charge the parasitic and other capacitances in the amplifier. The amplifier should have a good recovery from overloading pulses. This may be accomplished by making sure that overloading pulses do not, by driving the transistors outside their normal working region, cause the charging of coupling or other capacitors which can only discharge slowly. This may in part be achieved by placing limiting diodes at a suitable place in the circuit, preferably where the pulse is of short duration. (See Subsection 3.2.9 dealing with pole-zero cancellation methods for shifting the pulse duration.) The input stage should have a reasonably low noise and the output stage should be capable of driving the full output voltage swing into an impedance low enough (50–150 ohms) to provide matching for a coaxial cable.

As with most other electronic components, the use of semiconductor detectors with their high resolution has caused a considerable design problem in the realization of satisfactory main amplifiers. It should be remarked right away that there have been few designs published which examine their design and performance in the light of all the above criteria.

5.5.2 *Bandwidth requirements*

Once the pulse shaping system has been selected the ideal amplifier would have a frequency response extending sufficiently far that negligible distortion of the chosen pulse shape would occur. Since, in a linear system, the order of pulse shaping networks and amplifying stages is of no importance in determining the output pulse shape it is convenient to imagine the pulse shaping to be performed first and to be followed by the whole of the amplification. It is therefore sufficient that the amplifier should be capable of amplifying with negligible distortion a pulse of the chosen output shape.

The effect of the amplifier's limited high frequency response on pulse shape distortion has been discussed in Subsection 2.1.3 and the main consequence is a lengthening of the pulse rise and decay times. A useful index of performance of an amplifier in this respect is its rise time t_r which is the rise time of the output for a step function input. If the amplifier has an upper 3 dB frequency f_h then for most amplifiers the rise time is quite accurately predicted by the relationship $f_h t_r \simeq 0{\cdot}35$. For negligible distortion of the signal pulse the rise time of the amplifier should be considerably smaller than that of the signal pulse and a rise time of the order of 100 ns is usually considered the largest acceptable. This corresponds to an upper 3 dB frequency of at least 4 MHz. If the amplifier consists of a cascade of n identical stages, the overall rise time t_r is approximately related to that of each state t'_r by the relationship $t_r = \sqrt{n} t'_r$, so that typically the stage rise time would have to be $\sim$40 ns, and the upper 3 dB frequency for each stage would be at least $\sim$10 MHz.

For unipolar signal pulses the low frequency response should ideally extended down to DC. Although DC coupled amplifiers have been used, the design problems are more complicated and some AC coupling is often used. Such an AC coupled system will not, of course, deliver truly unipolar output pulses. If a baseline restorer is not used with such a system there will be a baseline shift at the output and a baseline fluctuation due to the random nature of the signals. To make these fluctuations, and the associated loss of resolution in the pulse height spectrum, negligible it is necessary (Subsection 4.5.3) to make the coupling time constants of the order of $\sim$0·1 s. Any other distortion of the signal pulse due to the limited low frequency response is negligible, even when the coupling time constants are much smaller than this.

5.5.3 Gain stability

Most of the parameters determining the gain of a transistor stage are dependent on fluctuating quantities, such as ambient temperature and quiescent collector current, and a large amount of negative feedback is necessary to stabilize the gain. A typical nuclear pulse amplifier might have an overall gain of 10^3 and since the resolution in the pulse height spectrum may be 0·1 %, the gain should have a temperature stability in the region of 0·01 % per °C. Assuming the temperature stability of individual stages to be about[25] 1 % per °C, the overall gain stability must be about 100 times that of a single stage, and this factor will here be termed the stability factor, *S*.

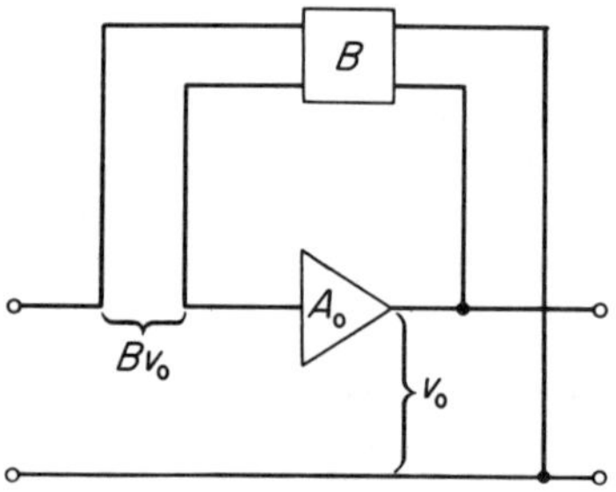

Figure 5.23 Principle of negative feedback

We will assume that an amplifier of natural gain (the 'open loop gain'), A_0, at some particular frequency of signal, has negative feedback applied as shown in Figure 5.23 so that a fraction *B* of the output voltage is subtracted from the input voltage. It will be seen that the gain now, (the 'closed loop gain') °A_c, is given by $A_c = A_0/(1 + BA_0)$. If the open loop gain A_0 changes by a small amount δA_0, the corresponding change in A_c is given by:

$$\frac{\delta A_c}{A_c} = \frac{1}{1 + BA_0} \frac{\partial A_0}{A_0} \qquad (5.10)$$

so that by making $(1 + BA_0)$ large enough the fractional variations in the closed loop gain may be made as small as required. In practice it is difficult to apply negative feedback over many stages since the cumulative phase shift through the amplifier may become 180° at high frequencies and the resulting feedback would become positive with consequent oscillation. The overall amplifier must therefore be made up of several feedback sections each with its own feedback loop.

If we consider an amplifier consisting of *m* identical feedback sections, each with a feedback factor *B*, and each containing *n* identical stages of gain *A*, then the open loop gain A_0 of each section is A^n. The overall gain of the amplifier A_t is then given by

$$A_t = \left(\frac{A^n}{1 + BA^n}\right)^m \tag{5.11}$$

and the stability factor S is given by

$$\frac{dA_t}{A_t} = \frac{1}{S}\frac{dA}{A} \tag{5.12}$$

so that

$$S = \frac{A^n}{A_t^{1/m} nm} \tag{5.13}$$

Assuming the overall gain A_t is fixed, it is clear from (5.13) that S increases rapidly with n, the number of stages per section, and so this would be chosen to be the maximum number that can be accommodated within a feedback loop without oscillation. Assuming the value of n is fixed the stability factor first increases with m and then decreases. From (5.13), for the cases of $A_t = 10^2$, 10^3 or 10^4, the optimum values of m are 5, 7 or 10 respectively.

Although this is a highly simplified treatment in that it assumes all the stages are identical, both in regard to stage gain and stage gain variation, it is nevertheless of interest to pursue the example and see how the value of the stability factor mentioned earlier might be obtained. For an overall gain, A_t, of 10^3 Table 5.1 gives the minimum values of open loop gain $A_0(=A^n)$ for each feedback section which will give a stability factor of 100. It will be seen from Table 5.1 that the optimum with respect to m is a fairly broad one and a value of $m = 5$ might be taken as a resonable choice. For this case the closed loop gain of each section would have to be 3.98 (i.e. $\sim$12 dB) which yields a feedback factor $B \simeq 1/3{\cdot}98$.

The expression for the gain stability strictly applies to sinusoidal signals only. Since the open loop gain of any section is reduced at high frequencies, it follows that the stability is reduced for such signals. A pulse signal may be regarded as containing an infinite spectrum of frequencies and so it is important to know how this affects the stability with respect to the pulse height. Viewed in another way, if a rapidly changing pulse signal is applied to the input, the feedback takes a finite time to develop the voltage fed back to the input and so stabilization is not instantaneous. Fränz and Paucksh[27,28] have examined this problem and defined a build up time for stabilization as the time necessary for the stability to build up to half its full value obtained with slow signals. They examined feedback systems of various types designed to give a non-overshooting response to a step function input signal, and showed that the stabilization time was approximately four times the rise time of the closed loop when large amounts of feedback ($BA_0 = 1000$ at low frequencies)

Table 5.1 Minimum values of open loop gain, A_o $(=A^n)$, of each feedback section in order to give a stability factor of 100 in an amplifier of overall gain of 10^3

n, number of stages per section	m, number of feedback sections: 1	2	4	5	7	10	100
1	10^5	$6{\cdot}32\times10^3$	$2{\cdot}24\times10^3$	$1{\cdot}99\times10^3$	$1{\cdot}87\times10^3$	$2{\cdot}14\times10^3$	$1{\cdot}08\times10^4$
2	2×10^5	$12{\cdot}6\times10^3$	$4{\cdot}48\times10^3$	$3{\cdot}98\times10^3$	$3{\cdot}74\times10^3$	$4{\cdot}28\times10^3$	$2{\cdot}16\times10^4$
4	4×10^5	$25{\cdot}3\times10^3$	$8{\cdot}96\times10^3$	$7{\cdot}96\times10^3$	$7{\cdot}48\times10^3$	$8{\cdot}56\times10^3$	$4{\cdot}32\times10^4$

were used. Unfortunately, no data has been presented on the effect on the stabilization of the heights of typical nuclear pulses, but on the basis of the above it would appear that, to obtain adequate pulse height stabilization, the rise time of the closed loop sections should be less than that of the pulse by a factor of at least ten.

Assuming the gains of the feedback sections are well enough stabilized, an additional source of instability arises from the coupling factor between

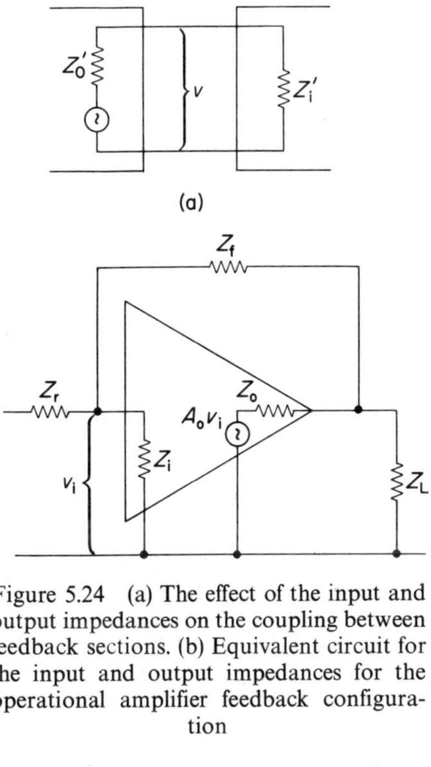

Figure 5.24 (a) The effect of the input and output impedances on the coupling between feedback sections. (b) Equivalent circuit for the input and output impedances for the operational amplifier feedback configuration

sections[25]. If Z_o' and Z_i' are the output and input impedances of two feedback sections coupled together as in Figure 5.24(a), then the variations in these cause an uncertainty of voltage transmission dV of a signal V of

$$\frac{dV}{V} = \left(\frac{Z_o'}{Z_o' + Z_i'}\right)\left(\frac{dZ_i'}{Z_i'} - \frac{dZ_o'}{Z_o'}\right) \tag{5.14}$$

Z_i' and Z_o' are variable quantities since their magnitude is dependent on the open loop gain A_o of the sections. For example in the operational amplifier feedback configuration, the input and output impedances are approximately:

$$Z'_i \simeq Z_r + \left(\frac{Z_f}{A_o + 1}\right) \| Z_i \qquad \text{assuming } Z_o \ll Z_L \| Z_f$$

$$Z'_o \simeq Z_o \left(\frac{Z_r + Z_f}{A_o Z_r}\right) \qquad \text{assuming } Z_i \gg Z_f \| Z_r \text{ and } Z_o \ll Z_f$$

where Z_i and Z_o are the input and output impedances of the amplifier section without feedback (Figure 5.24(b)). For other feedback configurations similar considerations apply, and to make coupling variations negligible (5.14) shows that the ratio Z'_o/Z'_i must be sufficiently small. Equation (5.14) is for the case of voltage transmission. Where current is the parameter carrying the signal information similar considerations apply except to obtain negligible variation in the current coupling factor Z'_o/Z'_i must be made sufficiently large.

5.5.4 *Stability to oscillation*

The essential problem in applying large amounts of feedback to a section is that of realizing a design which does not oscillate. The parasitic capacitances, associated with the individual stages within the section, will cause a cumulative phase shift at high frequencies, which may turn the feedback into positive feedback and cause oscillation. To take the example of common emitter configurations in cascade, the gain-frequency characteristic of a single stage, on a double logarithmic scale, is shown in Figure 5.25 where, assuming $R_c \ll \beta r_e$, the break in the characteristic is at $f \simeq 1/2\pi C_{b'e} R_c$; this arises from a single pole due to the action of the equivalent low pass RC filter formed by R_c and the base-emitter diffusion capacitance $C_{b'e}$ of the simplified hybrid-pi common emitter equivalent circuit (see Figure 5.25). It may be noted that the cut off frequency f_T of the transistor is usually defined as that frequency at which the short circuit current gain is reduced to unity and so is equal to $f_T \simeq 1/2\pi C_{b'e} r_e$. Reduction of R_c, or the introduction of local degeneration by means of a resistor in the emitter lead, increases the 3 dB bandwidth at the expense of gain, the gain-bandwidth product remaining approximately constant, and equal to f_T. The presence of $C_{b'c}$ has so far been ignored but may be important at high frequencies, since the Miller effect (Figure 5.26) causes it to be equivalent to a capacitance $(A + 1)C_{b'c}$ supplementing $C_{b'e}$ across the input, where $-A$ is the voltage gain of the stage. This effect may be minimized by using the cascode arrangement, where the common emitter stage is followed by a common base stage (Figure 5.26). The overall voltage gain is practically unaltered but the voltage gain, $-A$, across the common emitter stage alone is then small.

An amplifier section containing several stages will contain several poles in the transfer function for an open loop gain, $A_o(j\omega)$, and on the double logarithmic plot might typically have a characteristic asymptotic to the lines shown in the 'Bode plot' of Figure 5.27. With feedback, the closed loop gain equals $|A_0(j\omega)|/|1 + BA_0(j\omega)|$ and at low frequencies this is asymptotic to $|1/B|$ and at high frequencies it is symptotic to $|A_0(j\omega)|$. For resistive feedback B is independent of frequency and the case for $B = 1/100$ is plotted in Figure

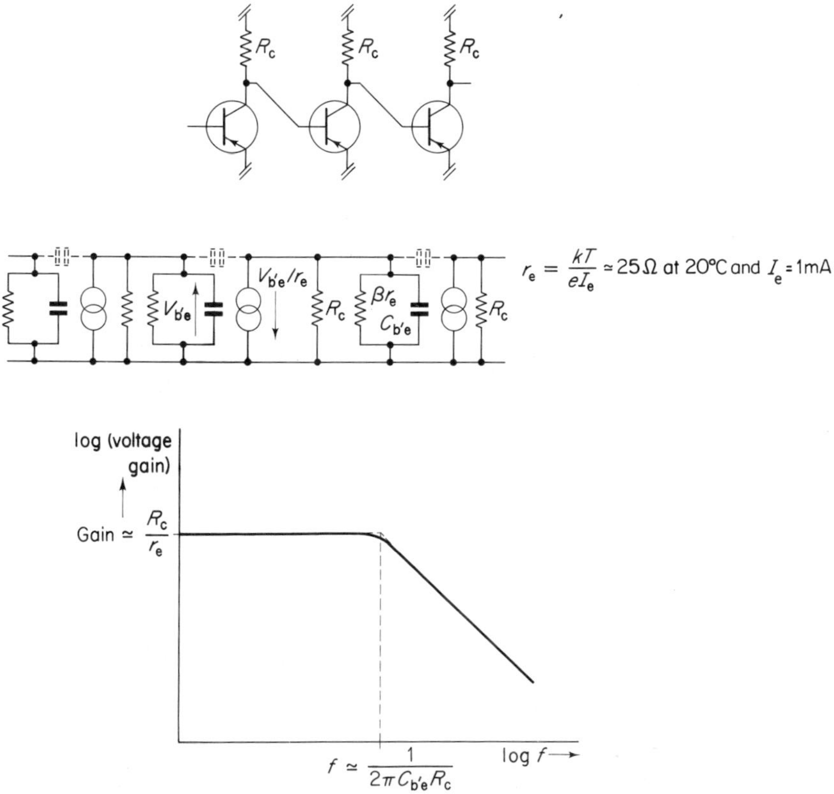

Figure 5.25 Cascaded identical common emitter stages, the hybrid pi equivalent circuits, and a gain/frequency plot for a single stage assuming $C_{b'c}$ to be negligible and $R_c \ll \beta r_e$

5.27. The intercept of the lines of $\log|1/B|$ with the open loop gain characteristic corresponds to the condition $|BA_0(j\omega)| = 1$, and a sufficient condition for stability is therefore that the phase of $BA_0(j\omega)$ should not have reached 180° at this point. For the case $B = 1/100$, this is not the case but, if the amount of feedback is increased to give $B = 1/10$, the arrangement is unstable. To apply the large amounts of feedback necessary in a nuclear pulse amplifier it is necessary to take special steps to obtain a stable arrangement. It is only

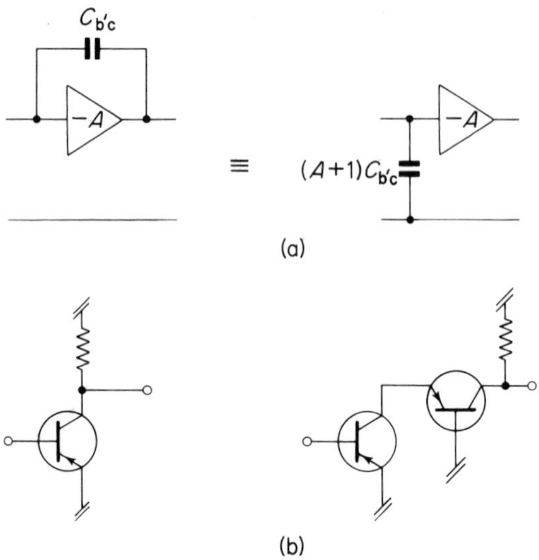

Figure 5.26 (a) The Miller effect of $C_{b'c}$ and the equivalent circuit. (b) The reduction of the Miller effect by the cascode connection

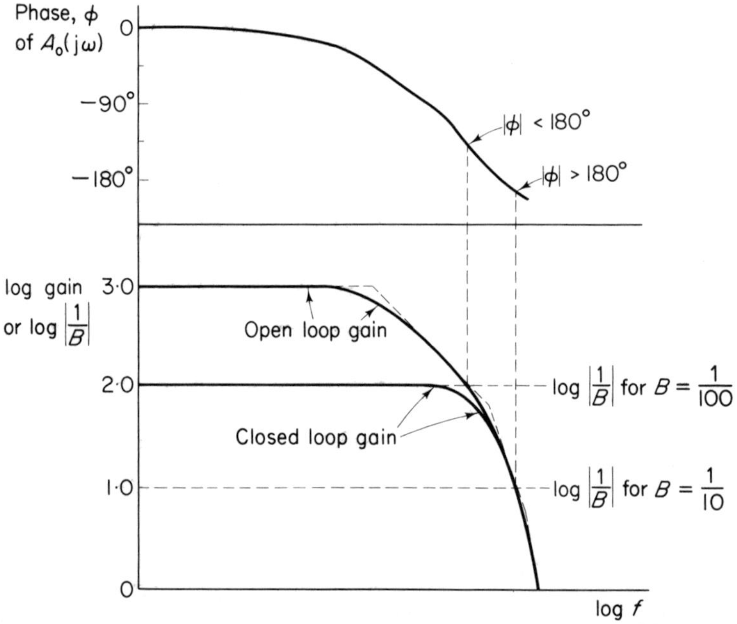

Figure 5.27 Bode plot showing the magnitude and phase of the open loop gain and the magnitude of the closed loop gain

possible to give the briefest outline here of these techniques and for further information the reader is referred to the standard texts.[29, 30, 31]

For the above case, $B = 1/10$, two approaches may be used to give a stable arrangement, modification of the open loop gain and modification of the feedback. For resistive feedback (i.e. B constant with frequency), a useful 'rule of thumb' for the stability, with some safety margin, is that the line of $\log|1/B|$ should intersect the open loop characteristic where it has a slope of less than 45° (i.e. 6 dB per octave). Hence the problem resolves into moving the poles in the open loop response so that the break frequencies are so situated that the response has the requisite slope at the intercept with $\log|1/B|$. This may be done for example as shown in Figure 5.28. The first break frequency, f_1, is shifted to a lower value by, for instance, adding capacitance

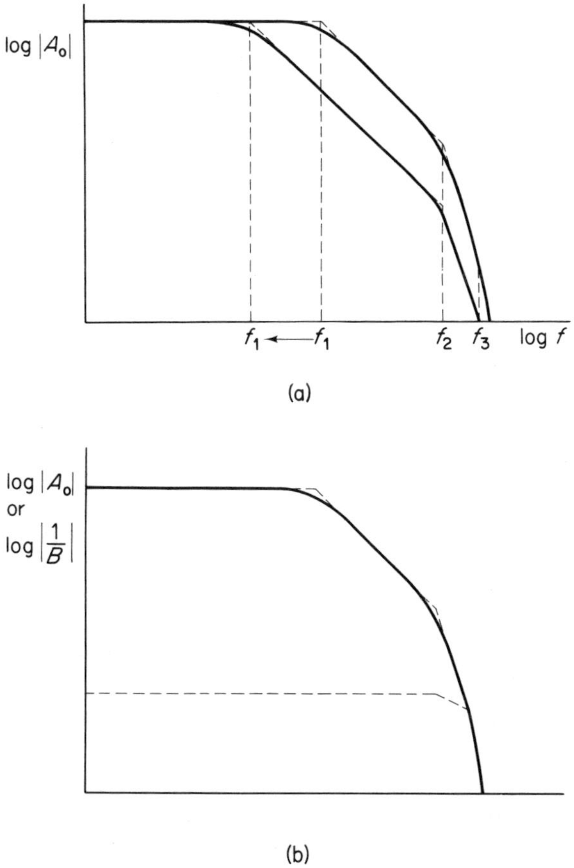

Figure 5.28 (a) Modification of the open loop gain.
(b) Modification of the feedback characteristic

at the appropriate point. Additionally, or alternatively, the second break frequency f_2 could be shifted to a higher value by using pole-zero cancellation.

If the feedback network contains a shunting capacitor to form a high pass network then the phase shift produced by this will counteract the phase shift in the open loop gain. The form of $\log|1/B|$ for this case is plotted in Figure 5.28 and the rule of thumb for stability may then be restated that the angle of intersection of the $\log|1/B|$ characteristic with the open loop gain should be more than 45°.

In general the design of feedback section is aimed at producing an open loop characteristic with a single dominant pole which gives a 6 dB per octave fall off in response over most of the range down to unity gain. In addition it is necessary that the closed loop gain, assuming a stable configuration has been arrived at, should give a response to a step function input that is non-overshooting. This condition is not easily amenable to theoretical analysis although Fränz[28] has discussed some sufficient conditions for this. In practice most designs are made to give a non-overshooting response by a small trial and error adjustment of one of the components controlling the break frequencies of the open loop gain or the feedback characteristic.

Finally, if coupling capacitors are used between amplifier stages, there is a possibility that the cumulative phase shift at very low frequencies may become 180° and the resulting positive feedback may produce low frequency oscillations. This can be circumvented by procedures similar to those used at the high frequency end; however, to avoid the problem it is more common to employ DC coupling between stages within a feedback section.

5.6 Practical examples of high stability amplifiers

Hatch[25] has considered in detail the design of a feedback section for a high stability amplifier. The section was developed from the basic configuration of Figure 5.29(a) to give that of Figure 5.29(b). The addition of the common base input stage and the emitter follower output stage increases the open loop gain and reduces the dependence of the input and output impedances on the transistor parameters. The open loop gain essentially had a high frequency response characteristic of a single pole, due to the Miller effect of Q_2 and of a small additional capacitance between the base and collector of this transistor. With R_I and R_F chosen to give a closed loop gain of 50, the rise time of the section was 15 ns and the gain stability was 10 ppm per °C.

Hatch[32] has also described a complete nuclear amplifier based on the use of six of the above sections. The overall gain was 1024 and the amplifier incorporated a modified first section to accept a differential input, the common mode rejection of which excluded hum and noise picked up in the screened twin input cable. The output stage was modified to enable it to drive 5 volts into up to three 50 ohm cables. The overal gain stability to pulses was ~50 ppm per °C, the rise time was 70–90 ns, and the non-linearity was less than ±0·01 %. A 500 × overloading pulse caused a recovery time (for recovery of the output to 0·1 volts above the baseline) of 5–7 μs. Further work on the basic arrangement of Hatch has also been done by Bradley.[33]

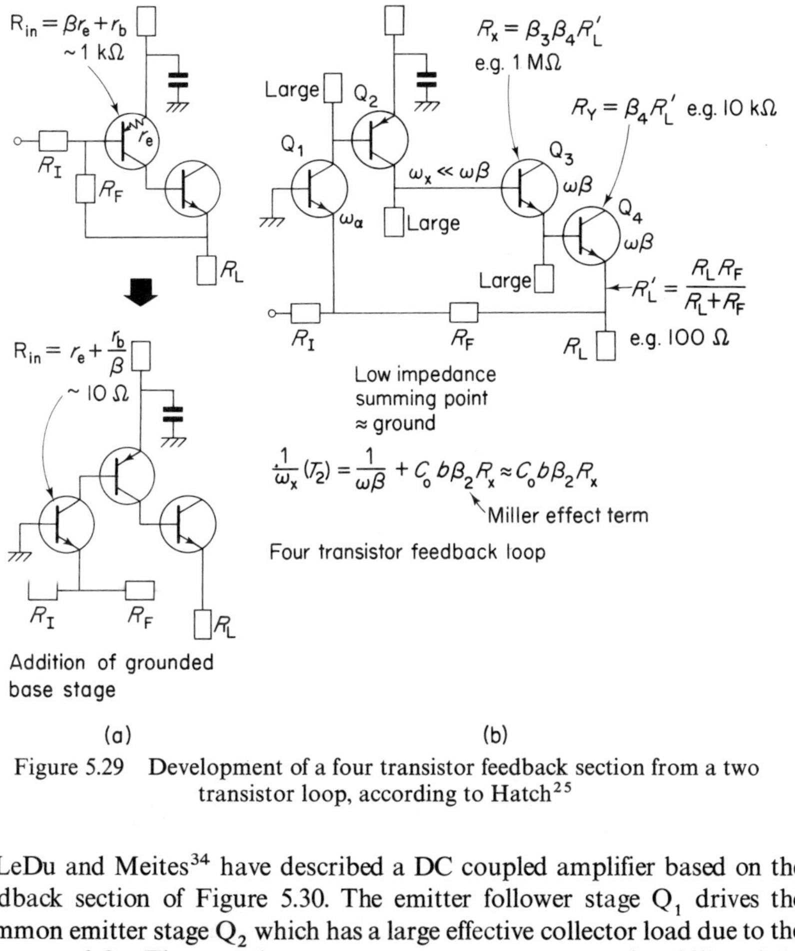

Figure 5.29 Development of a four transistor feedback section from a two transistor loop, according to Hatch[25]

LeDu and Meites[34] have described a DC coupled amplifier based on the feedback section of Figure 5.30. The emitter follower stage Q_1 drives the common emitter stage Q_2 which has a large effective collector load due to the presence of Q_3. The complementary output stage gave good stability of the baseline and allowed up to ±10 volts to be driven into a 100 ohm load. When the section was adjusted to give a closed loop gain of 5·04 the rise time was <10 ns and the stability of the gain was 20 ppm per °C.

The complete amplifier used nine feedback sections, to give gains up to 4096, with a rise time of 25 ns, a non-linearity of <1 %, and a gain stability of 50 ppm per °C. The final output baseline was stabilized by a special circuit to give a stability of <40 μV per °C, and this also incorporated a baseline restorer of the Robinson type for use with pulses having long undershoots.

Following earlier thermionic valve practice, several authors have used a long tailed pair configuration as giving good overloading performance, since the maximum current either of the pair of transistors can conduct is limited to

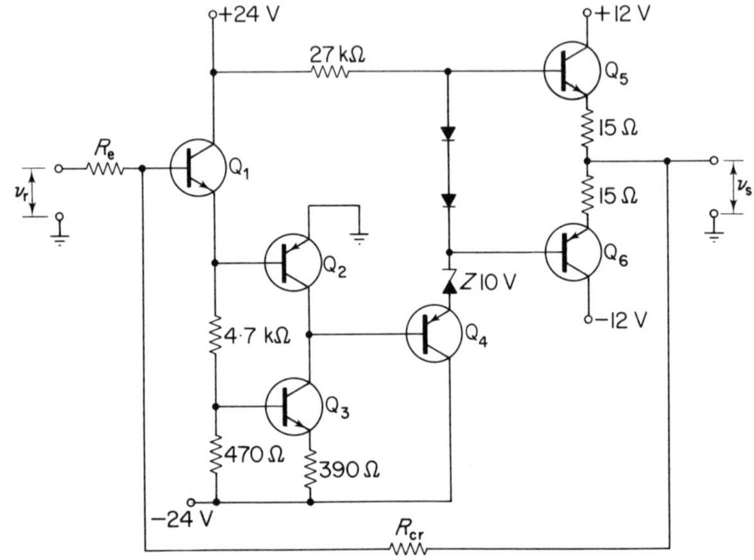

Figure 5.30 Simplified feedback section as used by LeDu and Meites[34]

twice its quiescent current. Collinge, West and Lloyd[35] used the basic configuration of Figure 5.31. Three of these sections gave a gain of up to 1000 with a stability of 0·015% per °C, a rise time of 80 ns, and could drive up to ±10 volts into 100 ohms with a non-linearity of 0·3%. Positive overloads caused the common emitter stage to bottom and negative overloads were limited by diodes placed across the output of each section. However, recovery times were not given.

Millard and Blalock[26] considered several long tailed pair configurations in the light of the criteria of the previous discussions. The circuit finally evolved, Figure 5.32, could be DC coupled and applies the feedback to the base of Q_2. Q_4 and Q_5 form two cascode arrangements to reduce the Miller effects of Q_1 and Q_2. The phase shifts through the path Q_1, Q_4 and Q_6 ideally should be similar to that through Q_2 and Q_5 but this is not so in Figure 5.32 because of the Miller effect of Q_6, so a further cascode stage was interposed between Q_6 and the point P (not shown in Figure 5.32). The high frequency response of the forward path was then essentially that of a single pole due to the equivalent capacitance C_T and resistance R_T. To give a high open loop gain R_T was kept large by using a bipolar driver with FETs in the first stage. The rest of the bipolar driver consisted of a complementary pair of cascaded emitter followers (Figure 5.32) which was capable of driving up to 100 mA at 10 volts into the load.

An alternative low noise version used FETs to replace the input transistors Q_1 and Q_2. The closed loop gain (including the driver) in either arrangement

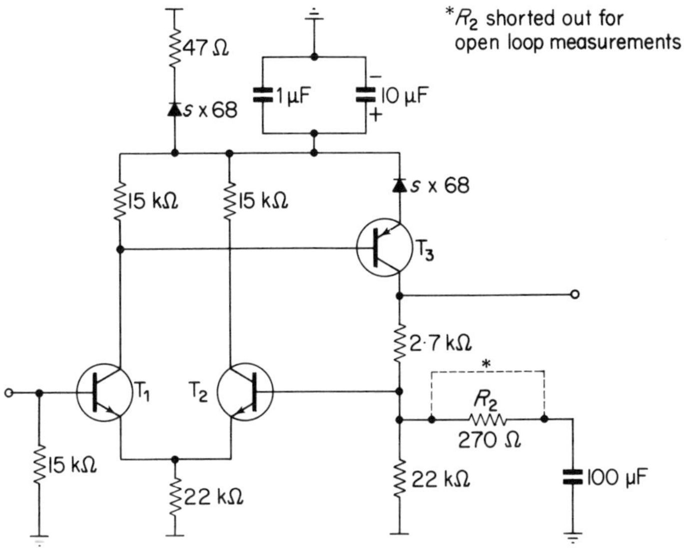

Figure 5.31 Feedback section of Collinge, West and Lloyd[35]

was 16 and had a stability of ~50 ppm per °C; the rise time was of the order of 10 ns for both small and large signals. The amplifier had excellent overloading properties with a recovery time of 20 ns from a 5 volt overloading input pulse. This was achieved partly through the use of DC coupling throughout, with the consequent elimination of the non-linear charging of coupling capacitances. In addition, to prevent the transistors being driven into saturation, with a consequently large charge storage time, a special circuit whose action approximated to that of a pair of catcher diodes, was placed at point P.

The commercial availability of integrated circuit amplifiers has simplified some of the design problems and these appear to offer considerable scope for future developments. Some preliminary work has been reported,[36,37] but it remains to be seen whether a complete amplifier can be realized which gives a good performance with respect to all the necessary criteria, and in particular with respect to the response to large and overloading pulses.

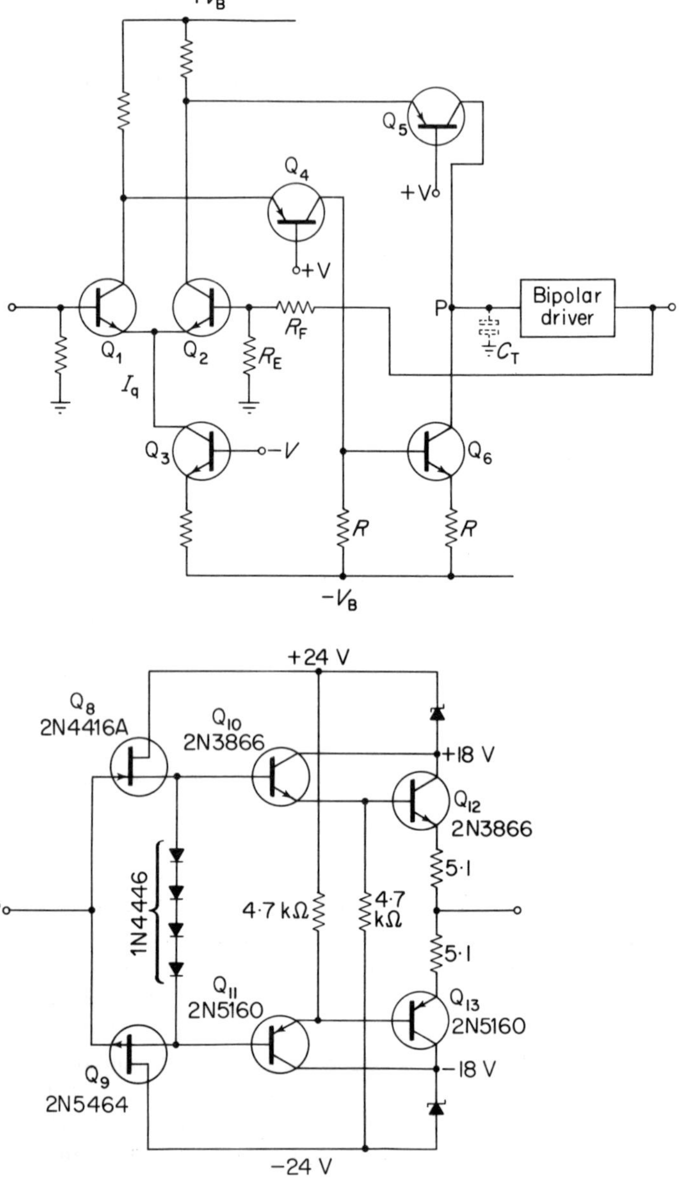

Figure 5.32 Basic feedback section of Millard and Blalock[26] and the circuit of their bipolar driver section

5.7 Fast amplifiers

For some applications, such as in high counting rate systems or high speed timing systems, it is necessary to use amplifiers with a short rise time. Although the distinction between such fast amplifiers and the amplifiers for spectroscopy already discussed is not a clear one, certain general differences may be noted. Typically, fast amplifiers have a rise time an order of magnitude less than that of spectroscopy amplifiers and this is usually obtained at the expense of some gain stability. For some applications a suitable design may be realized as a cascade of single stages, with or without local feedback for each stage. Most commonly feedback is applied over two stages however.

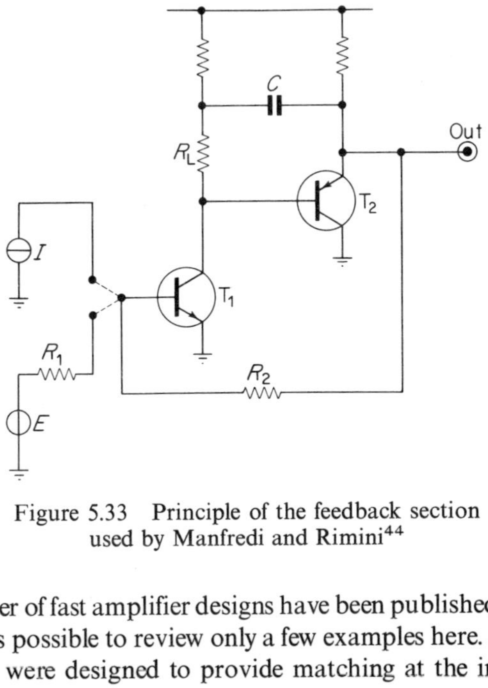

Figure 5.33 Principle of the feedback section used by Manfredi and Rimini[44]

A large number of fast amplifier designs have been published (e.g. References 38 to 47) and it is possible to review only a few examples here. All the examples to be described were designed to provide matching at the input and output for a 50 ohm cable.

In the design by Manfredi and Rimini[44] the basic feedback section is shown in Figure 5.33. The emitter follower T_2 is driven by the common emitter stage T_1 which has a bootstrapped collector load, and feedback is provided via R_2. Two of these sections, plus an additional emitter follower output stage gave a gain of 100 with a rise time of 2·5 ns. The gain stability was 0·1 % per °C and the linearity was 0·2 %.

Goyot, Samueli and Sarazin[45] described a DC coupled amplifier whose circuit is shown in Figure 5.34. The basic section may be seen to be that of a common base stage driving a common emitter stage. Feedback over both

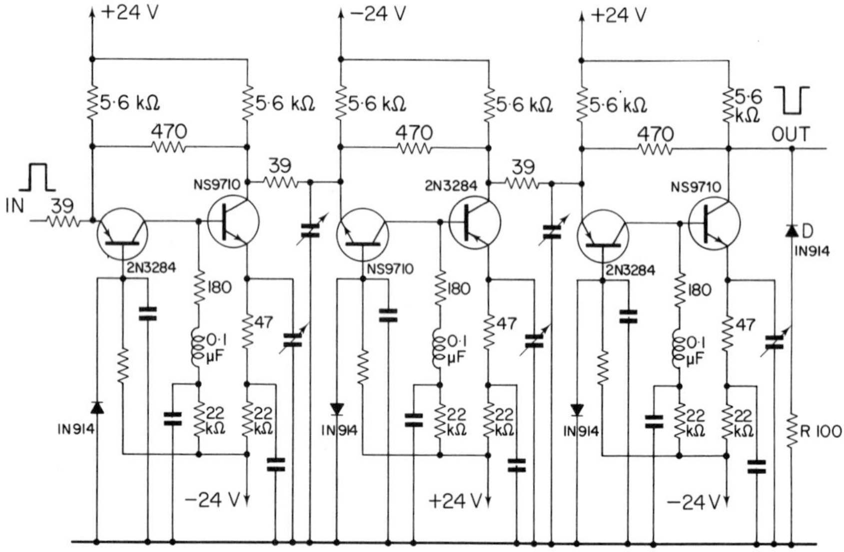

Figure 5.34 Amplifier section of Goyot, Samueli and Sarazin[45]

stages was provided by the 470 ohm resistor and, in addition, the 47 ohm resistor in the emitter lead provided local feedback for the common emitter stage. The overall amplifier had a gain of 10, with a stability of 0·1% per °C, and frequency compensation was provided by the inductors to give a rise time of 1·0 ns. The amplifier had a recovery time (for return to baseline) of 30 ns when a ×10 overloading pulse of 5 ns duration was applied.

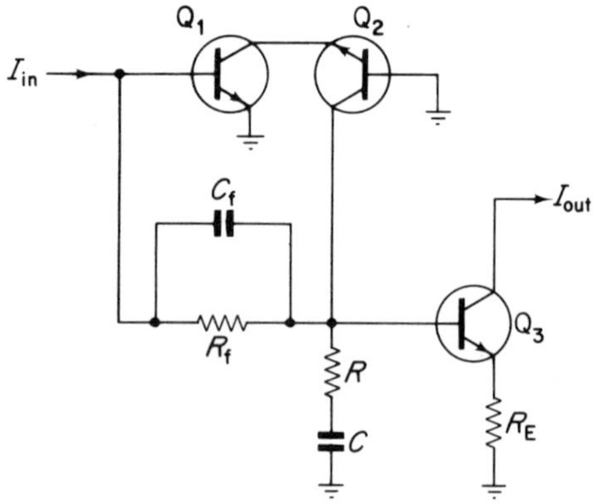

Figure 5.35 Principle of the feedback section used by Millard[46]

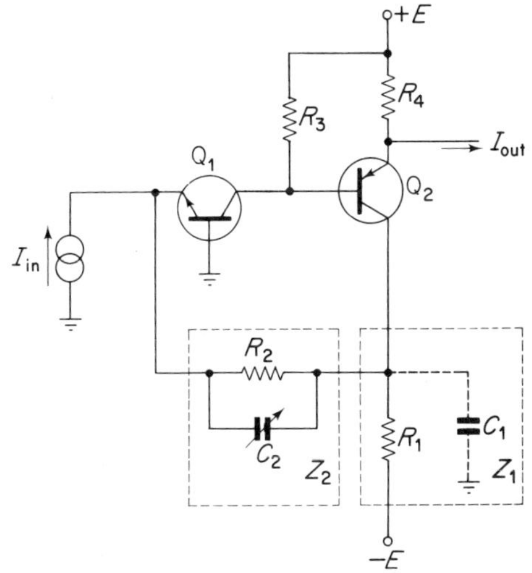

Figure 5.36 Basic current amplifier section of Rush[47]

Millard[46] used a basic current amplifier section shown in Figure 5.35 where feedback is provided by R_f and also locally by R_E. Three such sections gave a gain of 270 with a 1·0 ns rise time. Rush[47] used the current amplifying section of Figure 5.36. Five such stages, separated by common base stages, gave a gain of 1580 with a rise time of 3·05 ns. The non-linearity was 1·2% and a × 64 overload caused a recovery time to baseline of 300 ns.

References

1. W. Shockley, *Proc. I.R.E.*, **40**, 1365 (1952).
2. A. Van Der Ziel, *Proc. I.R.E.*, **50**, 1808 (1962) and **51**, 461 (1963).
3. F. N. H. Robinson, *Electron. Eng.*, **41**, Mar., 353 (1969).
4. T. V. Blalock, *IEEE Trans. Nucl. Sci.*, NS **13**, No. 3, 457 (1966).
5. F. S. Goulding, J. Walton and D. F. Malone, *Nucl. Instr. and Meth.* **71**, 273 (1969).
6. E. Elad, ISPRA Nuclear Electronics Symposium, Euratom Report EUR 4289, p. 21 (1969).
7. C. T. Sah, *Semiconductor Nuclear Particle Detectors and their Circuits*, p. 402, U.S. Nat. Acad. Sciences, Nuclear Science Series, Report No. 44, Publication 1593 (1969).
8. V. Radeka, *Semiconductor nuclear particle detectors and their circuits*, p. 393, U.S. Nat. Acad. Sciences, Nuclear Science Series, Report No. 44, Publication 1593 (1969).
9. J. H. Howes, ISPRA Nuclear Electronics Symposium, Euratom Report EUR 4289, p. 35 (1969).
10. J. K. Millard and T. V. Blalock, *Semiconductor nuclear particle detectors and their circuits*, p. 476, U.S. Nat. Acad. Sciences, Nuclear Science Series, Report No. 44, Publication 1593 (1969).

11. T. V. Blalock, *IEEE Trans. Nucl. Sci.*, NS **11**, No. 3, 365 (1964).
12. K. F. Smith and J. E. Cline, *IEEE Trans. Nucl. Sci.*, NS **13**, No. 3, 468 (1966).
13. V. Radeka, Int. Symp. on Nuclear Electronics, Versailles, Vol. 1, p. 46 (1968).
14. M. G. Strauss, L. L. Sifter, F. R. Lenkszus and R. B. Brenner, *IEEE Trans. Nucl. Sci.*, NS **15**, No. 3, 518 (1968).
15. R. Brenner, *Rev. Sci. Instr.*, **40**, 1011 (1969).
16. C. H. Nowlin and J. L. Blankenship, *Rev. Sci. Instr.*, **36**, 1830 (1965).
17. F. S. Goulding, J. T. Walton and R. H. Pehl, *IEEE Trans. Nucl. Sci.*, NS **17**, No. 1, 218 (1970).
18. K. Kandiah and A. Stirling, *Semiconductor nuclear particle detectors and their circuits*, p. 495, U.S. Nat. Acad. Sciences, Nuclear Science Series, Report No. 44, Publication 1593 (1969).
19. V. Radeka, *IEEE Trans. Nucl. Sci.*, NS **17**, No. 3, 433 (1970).
20. D. A. Landis, F. S. Goulding and J. M. Jaklevic, *Nucl. Instr. and Meth.*, **87**, 211 (1970).
21. T. L. Emmer, *IEEE Trans. Nucl. Sci.*, NS **9**, No. 3, 305 (1962).
22. V. Radeka and R. L. Chase, *IEEE Trans. Nucl. Sci.*, NS **13**, No. 3, 477 (1966).
23. L. G. Jonasson, *Nucl. Inst. and Meth.*, **26**, 104 (1964).
24. E. Fairstein and J. Hahn, *Nucleonics*, **23**, No. 7, 56, No. 9, 81, No. 11, 50 (1965); **24**, No. 1, 54, No. 3, 68 (1966).
25. K. F. Hatch, *IEEE Trans. Nucl. Sci.* NS **12**, No. 1, 314 (1965).
26. J. K. Millard and T. V. Blalock, *IEEE Trans. Nucl. Sci.*, NS **17**, No. 1, 307 (1970).
27. K. Fränz and H. Paucksh, *Nucl. Instr. and Meth.*, **27**, 125 (1964).
28. K. Fränz, *Nucl. Instr. and Meth.*, **47**, 217 (1967).
29. R. F. Shea, (Ed.), *Amplifier Handbook*, McGraw–Hill, New York (1966).
30. M. S. Ghausi, *Principles and design of linear active circuits*, McGraw–Hill, New York, (1965).
31. R. R. Benedict, *Electronics for scientists and engineers.* Prentice Hall, New Jersey, (1967).
32. K. F. Hatch, *IEEE Trans. Nucl. Sci.*, NS **13**, No. 1, 351 (1966).
33. A. E. Bradley, *IEEE Trans. Nucl. Sci.*, NS **13**, No. 1, 611 (1966).
34. R. LeDu and J. D. Meites, Int. Symp. on Nuclear Electronics, Versailles, Vol. 1, p. 42 (1968).
35. B. Collinge, C. West and G. H. Lloyd, *Nucl. Instr. and Meth.*, **35**, 313 (1965).
36. N. P. Lero and P. B. Frantlović, ISPRA Nuclear Electronics Symposium, Euratom Report EUR 4289, p. 85 (1969).
37. R. Benoit and V. Mandl, Int. Symp. on Nuclear Electronics, Versailles, Vol. 1, p. 39 (1969).
38. H. Verweij, *Nucl. Instr. and Meth.*, **24**, 39 (1963).
39. R. J. Epstein, *Nucl. Instr. and Meth.*, **24**, 333 (1963).
40. L. M. Welter, *Rev. Sci. Instr.*, **36**, 487 (1965).
41. A. Tojo, *Nucl. Instr. and Meth.*, **50**, 45 (1967).
42. K. Wille, *Nucl. Instr. and Meth.*, **72**, 314 (1969).
43. L. Papadopoulos, *Nucl. Instr. and Meth.*, **73**, 186 (1969).
44. P. F. Manfredi and A. Rimini, *Nucl. Instr. and Meth.*, **49**, 71 (1967).
45. M. Goyot, J. J. Samueli and A. Sarazin, *Nucl. Instr. and Meth.*, **46**, 149 (1967).
46. J. K. Millard, *Rev. Sci. Instr.*, **38**, 169 (1967).
47. C. J. Rush, *Rev. Sci. Instr.*, **35**, 149 (1964).
48. H. E. Kern and J. M. McKenzie, *IEEE Trans. Nucl. Sci.*, NS **17**, No. 1, 260 (1970).
49. H. E. Kern and J. M. McKenzie, *IEEE Trans. Nucl. Sci.*, NS **17**, No. 3, 425 (1970).
50. J. K. Millard, T. V. Blalock and N. W. Hill, *IEEE Trans. Nucl. Sci.*, NS **19**, No. 1, 388 (1972).

51. A. E. Bradley, *IEEE Trans. Nucl. Sci.*, NS **19**, No. 1, 429 (1972).
52. D. A. Landis, F. S. Goulding, R. H. Pehl and J. T. Walton, *IEEE Trans. Nucl. Sci.*, NS **18**, No. 1, 115 (1971).
53. E. Elad, *IEEE Trans. Nucl. Sci.*, NS **19**, No. 1, 403 (1972).
54. R. Kurz, *IEEE Trans. Nucl. Sci.*, NS **19**, No. 3, 418 (1972).
55. D. C. Young and D. Hardie, *Nucl. Instr. and Meth.*, **100**, 553 (1972).

6

Pulse Height and Shape Discriminators

6.1 Types of pulse height discriminator or analyser

Previous chapters have dealt mainly with 'linear' or 'analogue' pulses, that is, pulses whose amplitudes are linearly related to the detector pulse amplitude and hence to the energy deposited by the particle in the detector. These linear pulses now require further processing such as the selection by pulse height, counting or the operation of coincidence circuits.

Once the linear pulses have been selected according to pulse height, their amplitude is of no further interest and they are fed into a trigger circuit which, for each pulse, gives an output 'logic' pulse of standardized height and duration. The term 'logic' is used somewhat loosely in this context and derives from the computer terminology. The logic pulses carry time information about the original nuclear events and so they are still randomly distributed in time. It is worth noting, that such a random pulse train could not be processed directly by a conventional digital computer, as such computers are arranged to process pulses occurring at predetermined intervals controlled by a 'clock'.

In order that various circuit modules using logic pulses should be compatible with one another when interconnected, the NIM committee[1] has recommended certain standard logic levels which module outputs should deliver and to which module inputs should respond. These standards have gained wide recognition and are given in Table 6.1.

For linear pulses the simplest operation likely to be performed is that effected by a 'biased amplifier' whose action is shown in Figure 6.1(a). This serves to pass pulses having an amplitude greater than the 'threshold bias' and it will be seen that the output pulses are still linear. An 'integral discriminator' has a similar action but delivers the output as a logic pulse (Figure 6.1(b)) and is thus equivalent to a biased amplifier plus a trigger circuit. A 'differential discriminator' pulse height analyser differs in that it yields an output pulse only if the linear input pulse falls between prescribed

Table 6.1 Preferred practice standards laid down by NIM committee[1] for logic signals to ensure compatability of interconnected circuit modules.

Medium speed positive logic signals

	Module outputs (must deliver)	Module inputs (must respond to)
Logical '0'	−2 to +1 volts	−2 to +1·5 volts
Logical '1'	+4 to +12 volts	+3 to +12 volts

Fast negative logic signals

The standards here relate to the case of an output working into a 50 ohms impedance, and an input having an input impedance of 50 ohms to provide matching for an interconnecting cable of this characteristic impedance.

	Module outputs (must deliver)	Module inputs (must respond to)
Logical '0'	+1 to −1 mA	+20 to −4 mA
Logical '1'	−14 to −18 mA	−12 to −36 mA

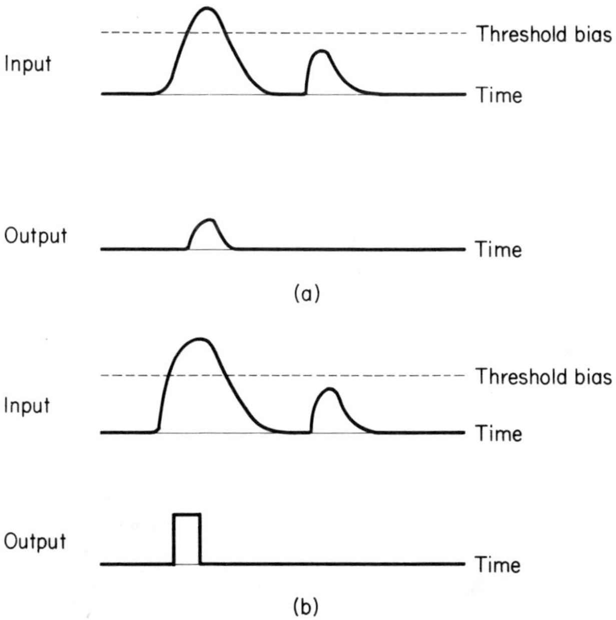

Figure 6.1 (a) Action of a biased amplifier. (b) Action of an integral discriminator

upper and lower levels, the so-called 'window'. The function of a differential discriminator is usually synthesized by the arrangement in Figure 6.2. The linear input pulse is fed simultaneously to two integral discriminators whose thresholds are set to the upper and lower levels. The two logic pulse outputs from these are fed to an anticoincidence gate which only delivers an output when the input from the lower threshold discriminator alone is present.

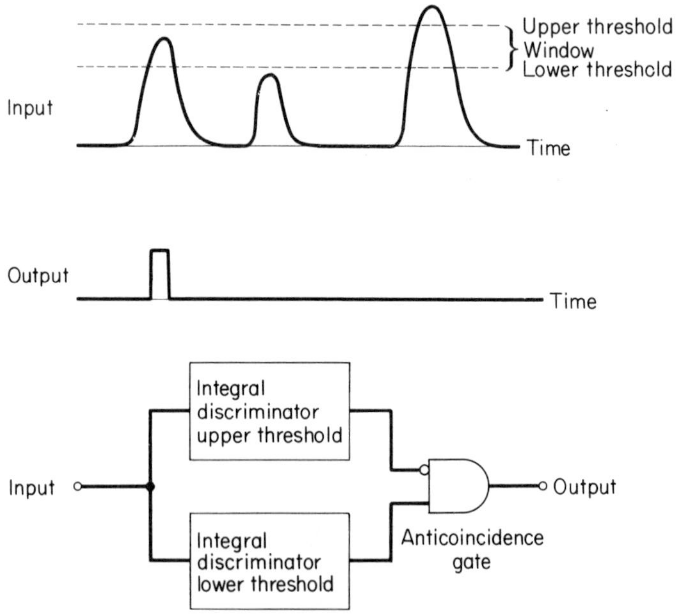

Figure 6.2 Action of a differential discriminator or pulse height analyser and the principle of operation

A multichannel analyser is equivalent in principle to many single channel pulse height analysers in parallel. The number of channels may be between 100 and ~10000 and they are arranged so that their windows are contiguous. Thus if 100 channels are set to analyse input pulses of heights in the range 2 to 12 volts, each channel would cover a range of 0·1 volt and so channel 37, for example, would select pulses in the height range 5·7 to 5·8 volts. Since a multichannel analyser is much slower than a single channel analyser, it is rarely practical to take the logic pulse outputs of the individual channels for use in other circuits. For this reason the output of each channel is usually fed directly to a scaler, or register, and a memory unit.

6.2 Preservation of timing information

If all that is required of an experimental system is the counting of pulses whose heights fall in a certain range, where the mean pulse rate is low, then a simple pulse height analyser and scaler, or alternatively a multichannel analyser, is all that is required. Other applications require the use of coincidence or anticoincidence circuits, and for these cases the derivation of the logic pulse from the linear pulse must accurately preserve the time information of the original nuclear event. Discriminators (integral and differential) usually show some dependence of the relative timing of the output pulse on the amplitude of the input pulse; this phenomenon is known as time 'slewing' or time 'walk', as illustrated in Figure 6.3. The way in which this may be of importance is best illustrated by considering some representative systems.

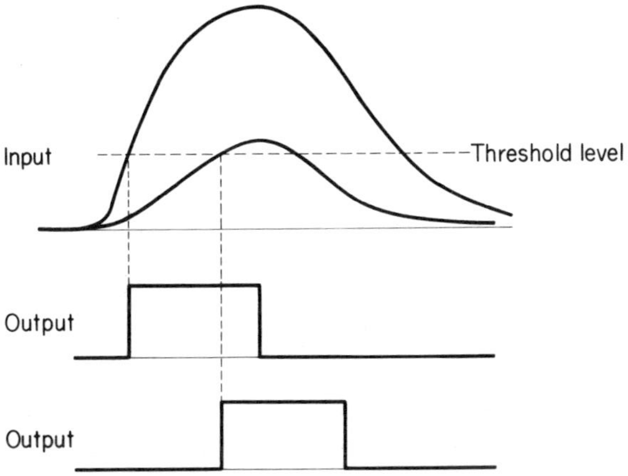

Figure 6.3 The effect of input pulses of differing amplitude on the timing of the output pulse from an integral discriminator

Figure 6.4 shows a simple arrangement for the detection of a pair of particles emitted in coincidence from the same nuclear event. This may be of interest in the detection of, for example, two products of a nuclear reaction, or two particles emitted rapidly in cascade during radioactive decay, or the two gamma quanta produced by the annihilation of a positron. Such events will be termed true coincidences. However, a certain number of accidental coincidences will also be passed by the coincidence gate in Figure 6.4. We will assume the coincidence gate has a resolving time of $\pm\tau$, that is, it delivers an output pulse if its two input pulses are within a time τ of one another.

The characterization of the time jitter is discussed more fully in the next chapter. We may only note here that in the system of Figure 6.4 the effects of time walk, illustrated in Figure 6.3, would make the time jitter very large, so that, to collect most of the true coincidences, the resolving time of the coincidence circuit would have to be correspondingly large. Since the accidental rate is proportional to the resolving time of the coincidence circuit, it follows that this rate would be correspondingly large. In many practical cases it would exceed the true rate since a true coincidence is the comparatively unlikely outcome of the detection of both particles from a single nuclear event. In these circumstances, while it is possible to correct the measured rate for the inclusion of accidental coincidences, this procedure severely limits the statistical accuracy in the computation of the true rate (see Appendix B.6).

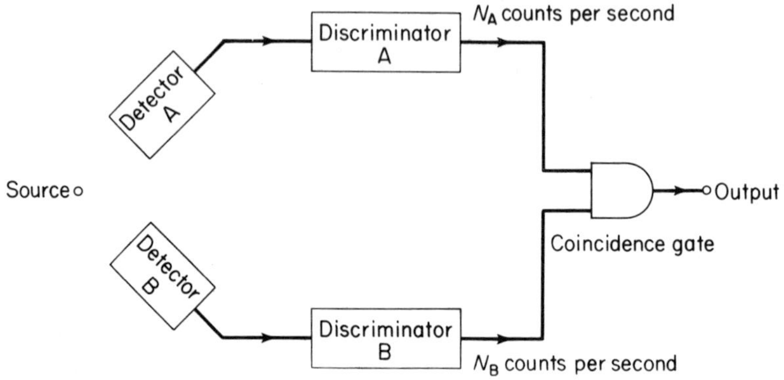

Figure 6.4 Simple system for detecting coincidences from two detectors

In order to avoid this it is necessary to use a 'fast' coincidence circuit, i.e. with a small resolving time, and this in turn means the input pulses to the coincidence circuit must have a small component of time walk. While it is possible to reduce the time walk of the discriminators by the incorporation of compensating devices, another solution is possible. This consists of having a separate circuit, a time pick-off circuit, whose function is primarily to generate an accurate (i.e. fast) timing pulse from the linear pulse and with a small time walk. Each discriminator can now be arranged, as in Figure 6.5, so that its output opens a gate to allow the timing pulse to pass on to the

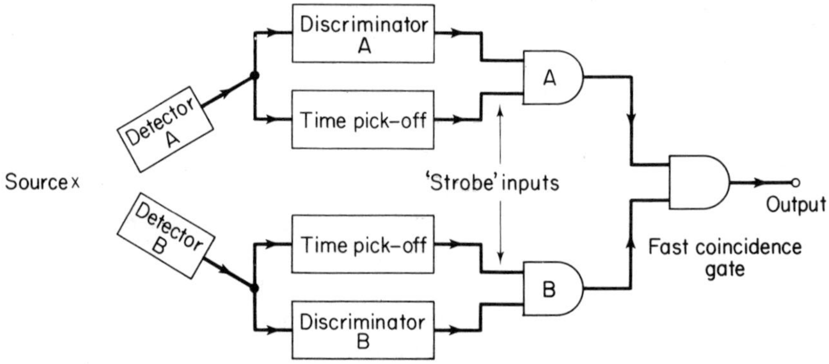

Figure 6.5 Slow–fast coincidence arrangement

main fast coincidence gate. This arrangement has the advantage of relegating the two functions, that of the pulse height discrimination and that of precise timing, to two separate units whose designs can then be separately optimized for the two functions.

It should be noted that the gates A and B in Figure 6.5 should deliver an output pulse with a timing (and possibly a duration) derived from the time pick-off input if the object of a low time jitter at the input of the main fast coincidence circuit is to be realized. In other words, the input from the time pick off should 'strobe' the input from the discriminator (from the analogy of a stroboscope). This will usually be ensured in an AND gate if the strobing input is of shorter duration and falls entirely within the time period of the other input.

For the sake of completeness it should be pointed out that fast coincidence selection can be achieved by other arrangements and the various methods for performing both coincidence and anticoincidence selection are discussed more fully in Chapter 7.

The timing of the logic pulses to the anticoincidence gate of a differential discriminator shown in Figure 6.2 needs some attention. Due to the time walk effect, for an input pulse above the upper threshold, the lower discriminator will deliver its logic pulse to the gate before that of the upper discriminator, as shown in Figure 6.6. There is therefore a possibility that

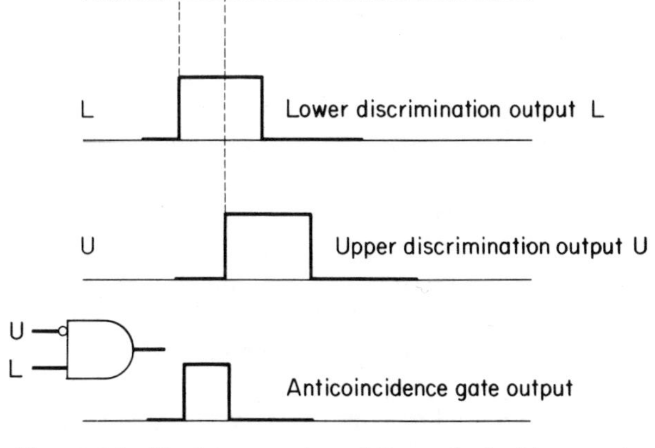

Figure 6.6 Faulty operation of the anticoincidence gate of a differential discriminator due to the effect of time walk

the gate may incorrectly deliver an output in the time interval between the arrival of the leading edges of the two pulses. Various solutions may be found to avoid this, for example, by delaying and shortening the logic pulse from the lower discriminator so that in all cases it falls within the duration of the logic pulse from the upper discriminator. Other solutions that are possible depart from the use of the simple gate, and make use of bistable circuits which are set by the outputs of the discriminators and whose states are interrogated after a suitable time delay. Church[50] has discussed the relative merits of some arrangements and proposed a new method based on a sequential logic technique.

6.3 General requirements of pulse height discriminators

Pulse height discriminators (integral or differential) have a critical role as the interface between the linear pulse and the logic pulse. Ideally a discriminator should deliver an output pulse when the input pulse falls within a certain height range which is precisely defined regardless of the shape of the input pulse. The discriminator should therefore be able to respond quickly and, in addition, it should recover rapidly and completely so that it can process a closely following pulse with the same precision. The dead time, that time during which it remains unable to accept a following pulse, should be constant regardless of input pulse shape, amplitude and rate, so that the counting losses may be calculated accurately. The timing and duration of the output pulse should be independent of the shape and amplitude of the input pulse.

Inevitably all these virtues cannot be combined in a single instrument. Furthermore, the shaping of the linear pulse enters partly into the considerations. For good energy resolution it is necessary to have a small contribution from noise and pile up to the overall fwhm. The noise contribution is usually a minimum for linear pulses of length of $\sim 1\ \mu s$ (see Chapter 4) and, to keep pile up small, the pulse rate must be low. A discriminator used in such a system need not have a fast recovery but should have good precision and stability with respect to pulse height. On the other hand, if the pulse rate is high, loss of resolution due to pile up would become the dominant factor and the linear pulses must be shorter, so the discriminator need not have as good pulse height discrimination precision, but should have a rapid response and recovery. In the limit this type of discriminator becomes effectively a time pick-off circuit where the discrimination is minimal (against noise for example) but the circuit responds rapidly and with precise timing.

In the case of a differential discriminator synthesized by an arrangement of two integral discriminators in parallel with their outputs placed in anti-coincidence (as in Figure 6.2), the stability of the window width in such an arrangement is dependent on the stability of both threshold levels. If a narrow window is being used, the percentage stability of the window width is highly dependent on the threshold stabilities. Thus for analysis of pulses in the range 0 to 5 volts height with a window width of 50 mV, if the two thresholds have a stability of ± 5 mV, the window width will therefore have a 'differential' stability of up to $\pm$ 10 mV, that is $\pm$ 20 %, with a corresponding variation in the counting rate (assuming a uniform pulse height spectrum). In fact this would be a pessimistic estimate since most sources of instability (changes in temperature for example) would tend to move both thresholds in the same direction. Nevertheless the stabilities of the component integral discriminators in such an arrangement need to be of a high order to give acceptable stability of the window width. Similar considerations apply to the linearity of the integral discriminators. The actual threshold of an integral discriminator will deviate slightly from a linear relationship with the threshold setting voltage (the discrepancy being termed the 'integral non-linearity'), and this leads to a variation in window width with the window position (sometimes termed the 'differential non-linearity').

The dead time of the discriminator acts to some extent as an anti-pile up device if it is set to a value equal to or greater than the input pulse width. Many of the types of trigger circuit which define the threshold give an output pulse whose length τ_t is dependent on the length of the linear input pulse. It is usual to follow the threshold defining trigger circuit by a clipping circuit (for example a CR high pass filter and a diode) and use the clipped pulse to trigger a monostable circuit in order to give an output pulse of fixed duration τ_m. If τ_m is made shorter than τ_t, the dead time of the circuit is dependent on τ_t and its therefore variable, so it is usually preferable to make $\tau_m > \tau_t$ when the dead time is then τ_m and is fixed. However, this is still not entirely satisfactory as there is a possibility that the threshold trigger circuit may deliver a second pulse to the monostable before it has reset.

A related problem concerns the recovery time of the discriminator, since trigger circuits require an extra time period after resetting in order to attain the truly quiescent state. A closely following pulse would probably see an apparent threshold level higher than the nominal level during this recovery period. These uncertainties can usually only be resolved by additional circuitry which prevents a following input pulse from being accepted until all the essential circuit elements have fully recovered. Several designs with this facility are discussed in the following sections.

Even when such measures are taken, two closely following input pulses will pile up on top of one another and be seen by the discriminator as a single pulse. This possibility can be reduced by the use of a pile up rejector circuit which 'inspects' the input and blocks the discriminator action where necessary. Pile up rejector circuits are discussed in Section 7.9.

Meiling and Stary[43] have comprehensively reviewed many types of discriminator, with particular emphasis on high speed circuits.

6.4 Practical circuits

6.4.1 Schmitt trigger circuits

The Schmitt trigger circuit is widely known and, probably for this reason, few designs using transistors have been published. The switching time of transistor trigger circuits is greater than that of devices such as tunnel diodes and this is a source of error, since the effective threshold shows a dependence on the rate of rise of the input pulse. For a short switching time in a Schmitt trigger circuit, the open loop gain should be large (see Subsection 2.3.3), but this gives a larger hysteresis effect; the threshold to an input pulse must be large enough for the Schmitt trigger to be able to reset after the input pulse. In the data given by Mullard[2] for instance, when switching transistors having $f_T \simeq 500$ MHz are used in a conventional emitter coupled Schmitt trigger without limiter diodes, a rise time of the output pulse of 25 ns is obtainable with a hysteresis of 0·7 volts at the input. Where it is necessary to analyse small input pulses, Nizan and Elad[3] have described an arrangement of two coupled Schmitt trigger circuits, one of which initiates the switching action and the other of which terminates it. The hysteresis can then be adjusted to near-zero.

6.4.2 Diode type discriminators

This class of discriminator uses a biased diode as the discrimination element as shown in Figure 6.7. If the input pulse height exceeds the diode bias, the diode will conduct and the output will deliver that part of the input pulse exceeding the threshold bias. This truncated pulse needs shaping into a logic pulse by a trigger circuit which also provides a means of rejecting parasitic pulses transmitted through the diode capacitance. Since the trigger circuit generating the logic pulse has its own threshold which will add to the diode bias to give the overall threshold, the trigger circuit threshold should be small compared to the diode bias. This can be arranged if an amplifier is interposed between the diode circuit and the trigger circuit. Alternatively a tunnel diode trigger circuit may be used since this can have a small trigger threshold. Such an arrangement was described by Ward and York[4] and is shown in Figure 6.8. The tunnel diode trigger is a monostable and the output pulse is partly dependent on the magnitude of the input pulse. A second tunnel diode monostable was therefore placed after the first to give an output pulse of effectively standardized height and duration. The arrangement was capable of accepting input pulses up to 10 volts height with rise times 0·5 ns to 20 ns and durations 5 ns to 100 ns.

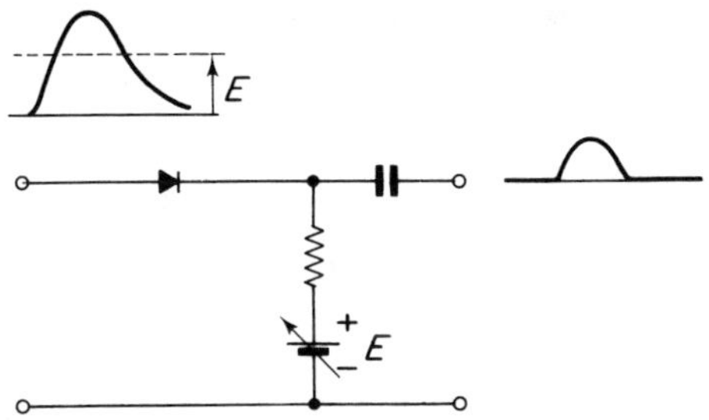

Figure 6.7 Principle of a diode type of discriminator

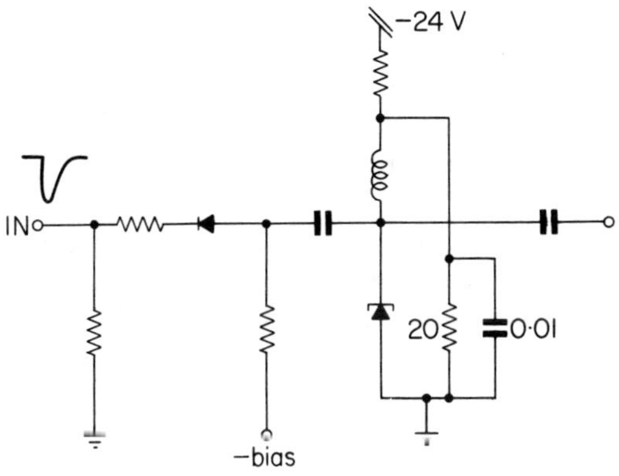

Figure 6.8 Principle of the circuit of Ward and York[4]

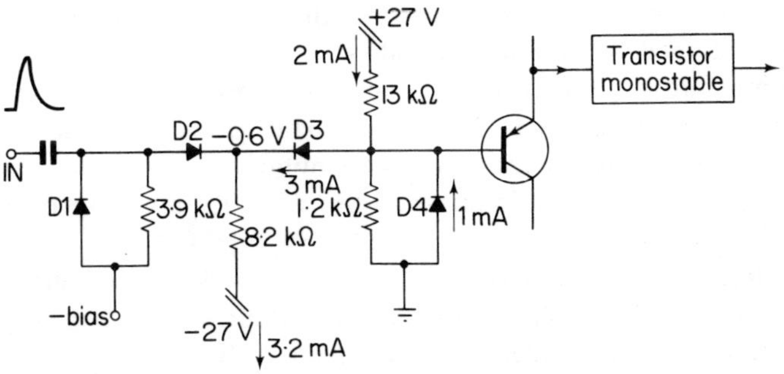

Figure 6.9 Principle of the circuit of Barna, Marshall and Sands[5]

Barna, Marshall and Sands[5] have used two diodes connected back to back (Figure 6.9) and this has the advantage that temperature variations in diode thresholds are to some extent compensated. D1 is a baseline restorer diode, D3 and D4 are conducting in the quiescent state and the reverse voltage across D2 sets the threshold level. A sufficiently large input pulse causes D2 to conduct 1 mA and causes D4 to cut off. The voltage appearing at the base of the transistor amplifying stage is that part of the input pulse exceeding the bias across D2 (less the forward voltage across D3), and this is passed on to a transistor trigger circuit. Diodes D3 and D4 serve to protect the circuit from overload pulses, since large input pulses cause a large current to flow into the 8·2 KΩ resistor and D3 cuts off. The discriminator accepts input pulses up to 10 volts, has a threshold stability of <5 mV/°C, and can handle input pulses of widths 20 to 200 ns with a shift in effective threshold of <10 mV. For a change in input pulse width from 20 to 10 ns the change in threshold was <0·15 V. The recovery of the circuit was such that a second pulse following within 150 ns saw a change in effective threshold of <0·1 V over that seen by the first pulse. A similar arrangement was described by Rozen and Altman[6] and had a threshold stability of 0·02 %/°C. For 2 V input pulses having rise times 2 μs and 50 ns, the effective threshold changed from 2·00 V to 2·25 V.

The diode discriminator has been employed in many other configurations and the following are a representative sample. LeComte and Allemand[7] described an arrangement where the diode was placed across the collectors of a transistor pair differential amplifier. The bias in this case is applied to one of the transistor bases, and the input pulse is applied to the other base. The discriminator had a threshold stability of 1 mV/°C and accepted input pulses up to 5 V with widths down to 20 ns with a shift in level of <1 mV. The recovery was such that a second pulse, spaced 200 ns after a first, saw a threshold shift of <1 mV. Klein, Hustman and Blok[8] incorporated a diode in the feedback loop of a transistor monostable so that an input pulse applied to the diode blocked it and caused the feedback to become operative so the monostable triggered. Other configurations have used the base-emitter diode junction of a transistor as the discriminating element. Thus Verweij[9] obtained a threshold stability of 0·2 mV/°C. The circuit could handle input pulses up to 2 V height and down to 20 ns width, with a shift in effective threshold of 10 mV, and had a shift of 2 mV for pulse spacings of 80 ns. Coli[10] using the same principle obtained a threshold stability of <0·5 %/°C and the threshold changed by <5 % for an input pulse rate of 100 MHz.

For a differential discriminator it is possible to use two integral discriminators in parallel, set to the upper and lower thresholds, and to place the two outputs in anticoincidence as discussed earlier. With diode discriminators another solution is possible and this is shown in Figure 6.10. The bias on the first diode sets the lower threshold and so the second diode placed in series receives only the truncated pulses passed by the first; the window width is then set effectively by the bias on the second diode only. If an amplifier, a so called window amplifier, is interposed between the diode circuits as in Figure 6.10 then, provided the gain of the window amplifier is stable, the window width can be made insensitive to changes in the discrimination level of the second diode; the arrangement can give a very good differential stability. Such a system was described by Welter[11] and gave integral and differential stabilities of 1% over 20–45°C and could accept pulses at a rate of 500 kHz with negligible change in threshold level or window width. With a 2% window the differential non-linearity was ±3·5%.

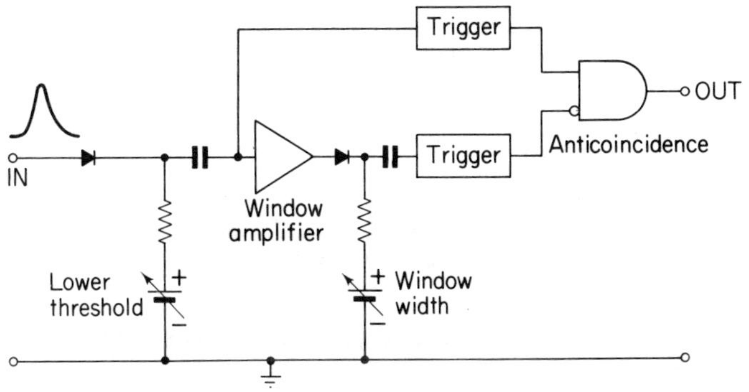

Figure 6.10 Principle of a differential discriminator using a window amplifier

The availability of integrated circuit DC differential amplifiers has led to the development of a variant of the diode discriminator. These amplifiers have a fast rise time (typically ~40 ns), high gain (~60 dB) and due to their monolithic construction have very good common mode rejection and high stability of the input offset voltage. (The input offset voltage is the differential DC voltage that must be applied to the input to give an output of zero volts.) Figure 6.11 shows the principle of their use in a discriminator circuit. One of the differential inputs is biased and the input pulse applied to the other input so that, as the input pulse exceeds the bias, the amplifier output is swung rapidly from zero to maximum output. This output can now be fed into a trigger circuit in the usual way. Due to the high gain of the differential amplifier, changes in the threshold level of the trigger circuit are unimportant—the amplifier effectively acts as a comparator or null detection method. It may be noted however that it is not practical to perform differential discrimination by adding a second circuit as a window amplifier in series, as in Figure 6.10, as the gain of the integrated circuit amplifiers is too high.

Designs using the configuration of Figure 6.11 have been published by Lauch and Nachbar[12], by Rudnick, English and Howard[13] by Cole[51] and by Bertolaccini and Cova[54]. These designs are particularly interesting because they make extensive use of commercially available logic integrated circuits, a technique that seems likely to become common, and have special provision for avoiding undesirable effects at high count rates. Two of these designs will be discussed in detail since they serve to illuminate the problems encountered generally when high pulse rates have to be dealt with. Before doing this it is necessary to examine the sources of possible malfunctioning at high pulse rates.

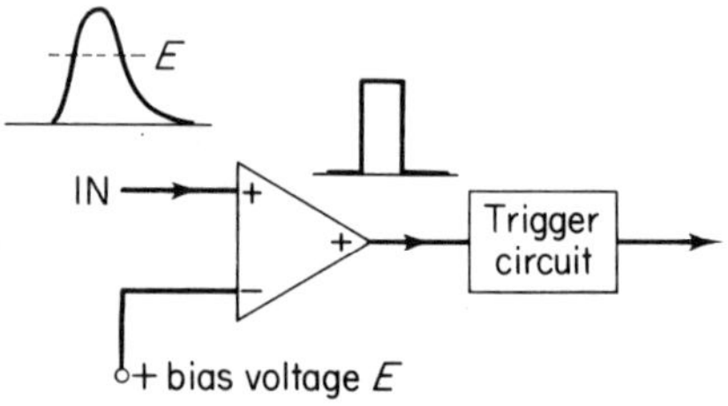

Figure 6.11 The use of a differential DC amplifier as a discriminator element

When a discriminator accepts and processes a pulse, the trigger circuits are first triggered to an 'excited' state and then, after a certain relaxation time, reset to their original state. However they do not necessarily all reset at the same instant. Thus in the arrangement of a Schmitt trigger feeding a monostable, the Schmitt trigger will perhaps reset before the monostable and there is then a possibility that a subsequent input pulse may again trigger the Schmitt circuit before the monostable has reset. This may cause the monostable to prolong its relaxation time and hence the output pulse length, to the detriment of any coincidence gating further along the system. It will also extend the dead time of the circuit, although this would only be responsible for a second order effect in the calculation of counting losses. A more important source of malfunctioning concerns the recovery, as opposed to the resetting, of the trigger circuits. Most trigger circuits, after they have just reset, require a further period to fully recover, that is, to re-attain the truly quiescent state; during this period they do not have the same characteristics, such as triggering sensitivity, as they do in the quiescent state. Thus a Schmitt trigger during recovery requires a larger input pulse to retrigger it than it would in the truly quiescent state.

In the circuit of Rudnick *et al*[13], the use of the differential amplifier gave an integral stability of $<0{\cdot}12\,\mathrm{mV}$ for pulses in the height range 0–4 V. Figure 6.12 shows a simplified version of the logic scheme used, where all the signal lines are shown in their quiescent state. When the linear input pulse exceeds the bias, the output from the differential amplifier changes from the quiescent value of 3 volts (logical '1' in Figure 6.12) to 0 volts (logical '0'). The input to the bistable changes to '1' and the bistable triggers, which in turn triggers the first monostable, which then changes line A to '0'. This changes the output of the AND gate to '0', which blocks the input gate G1 and at the same time resets the bistable. This has all occurred rapidly and some time later when the relaxation time of the first monostable has expired it resets and in the process triggers the second monostable whose output B now continues the blocking of the gate G1. Eventually the second

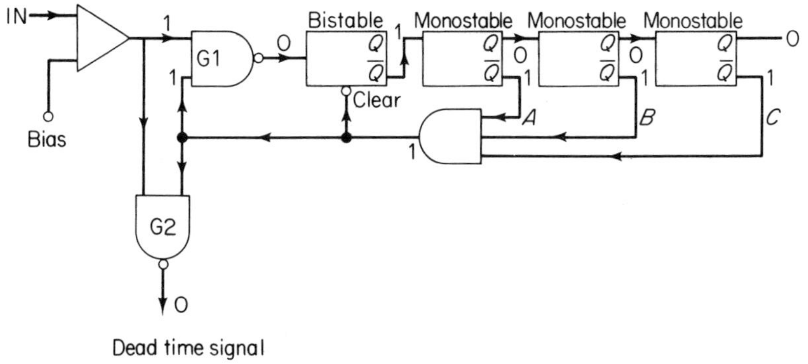

Figure 6.12 The logic scheme of Rudnick, English and Howard,[13] simplified and slightly modified

monostable resets and triggers the third monostable whose output C continues the blocking of the gate G1. Only when the third monostable resets do the lines A B and C become simultaneously '1', and the gate G1 becomes enabled to permit another signal into the bistable. The dead time is therefore that time when gate G1 is blocked, i.e. the sum of the relaxation times of the monostables. The reason for using the chain of three monostables is that, if they all have the same relaxation time τ_m, only one is in the excited state at any time and so all the monostables have a time of at least $2\tau_m$ to recover fully to their quiescent states before they can be retriggered.

It may well happen that at the end of the timing cycle, when the gate G1 is enabled, the amplifier output is in the '0' state due to the presence of another pulse at the input (or the tail of the original pulse for that matter if it is very long). It is important that the bistable should not be triggered in this eventually, since the signal to it does not correspond to the leading edge of the input pulse. This does not occur, in fact, since the bistable can only trigger when its input goes from '0' to '1' and the '0' state can only be generated by the NAND gate G1 when both its inputs are '1' i.e. when the amplifier output is in quiescent condition. The eventuality does however extend the dead time until the amplifier output becomes quiescent '1' and this will make a second order correction to the counting loss. Where this is important it may be assessed by means of the gate G2. The total dead time may be seen to be the time that the output of G2 is in the "1" state. A practical way of measuring the total or average dead time during an experiment would be to use this dead time signal to gate a high frequency oscillator and count the number of cycles passed by the gate during the experiment.

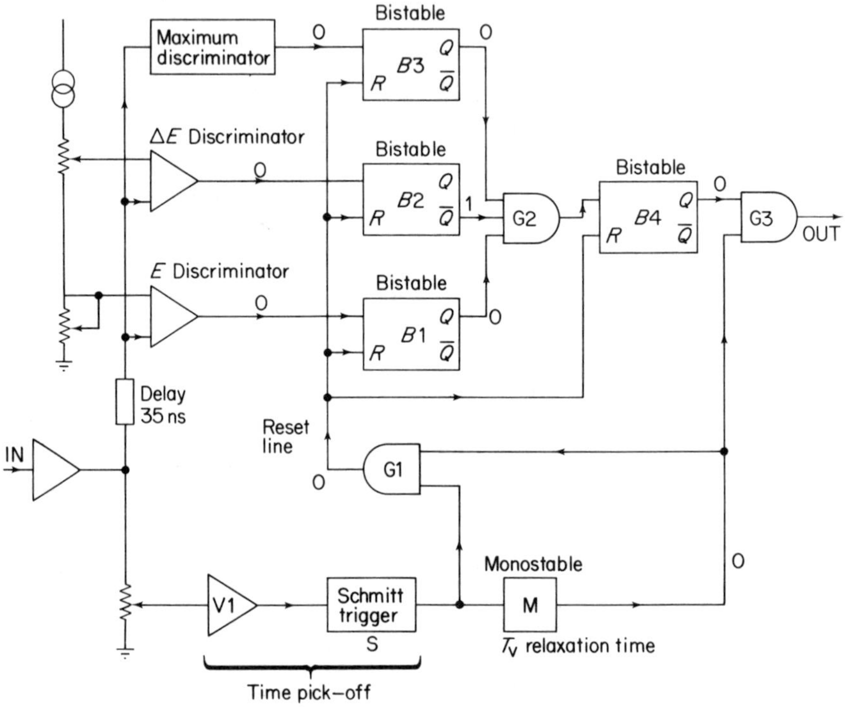

Figure 6.13 Simplified block diagram of the differential discriminator of Lauch and Nachbar[12]

The arrangement described by Lauch and Nachbar[12] was for a differential discriminator and also utilized integrated circuit differential amplifiers as the pulse height sensing elements. The full circuit incorporated a time pick-off which gated the output of the discriminator proper, similar in principle to that shown in Figure 6.5, a pile up rejector circuit and other gating facilities. Figure 6.13 shows a simplified version of the block diagram of the circuit. The time pick-off could derive its timing either from the leading edge of the input pulse or from the zero crossover point of a double differentiated pulse (see Chapter 7), but only the former mode will be considered here. In this case the early part of the input pulse caused the output of the amplifier V1 to swing rapidly to maximum output, so triggering the Schmitt trigger S. The output of the Schmitt trigger, through the AND gate G1, resets the bistables B1, B2, B3, B4 in readiness for the analysis; it also triggers the monostable M, which has an adjustable relaxation time T_v, so that its output becomes logical '0'. The logical states of all the lines at this instant are shown in Figure 6.13. Some time later, the input pulse will have risen to the lower threshold E and will have caused the E amplifier output to swing to maximum, hence triggering the bistable B1 so that its output goes from '0' to '1'. Assuming the ultimate pulse height does not exceed $E + \Delta E$ (i.e. it lies within the window ΔE) the ΔE amplifier output will not swing and the $\bar{Q}$ output of the bistable B2 remains at '1'. The function of the maximum discriminator MD is to sense the instant at which the input pulse reaches its peak and to produce a '1' output signal from its bistable B3 slightly after this occurs. At this stage all three inputs to the AND gate G2 are '1' and the output from this gate triggers the bistable B4 so its output is now '1'. The purpose in including the input to G2 from the maximum discriminator was to ensure that both the E and $E + \Delta E$ discriminators had been exposed to the full height of the input pulse before their outputs were sampled through the G2 gate. This is necessary since, with an input pulse larger than $E + \Delta E$, due to the finite rise time there will be a time interval between the pulse exceeding the E and the $E + \Delta E$ thresholds during which both discriminators would give '1' outputs and so would give a spurious output.

The final output is only obtained from gate G3 some time later when it is enabled by the input from the monostable M relaxing to '1'. In this way the ultimate timing of the output pulse derives from the time pick-off provided the enablement of G3 by the monostable is after that from the input from B4. In order that T_v may be adjusted to a minimum and yet accomplish this safely for any type of input pulse a special monitor was incorporated (not shown in Figure 6.13).

The output of the monostable M was also taken to gate G1 which was therefore disenabled during the period T_v. This prevented accidental re-setting of the bistables during an analysis due to the arrival of a second rapidly following pulse. The monostable M was designed to have a negligible recovery time so that a rapidly following pulse could be processed. To achieve this a conventional monostable could not be used. Instead a circuit was used in which the input signal triggered the rapid discharge of a capacitor. The capacitor then recharged linearly from a constant current source, the magnitude of which controlled the relaxation time T_v, until the original voltage was attained. The output signal was derived, via a differential amplifier, from the capacitor voltage. The gain of the amplifier was high enough for its output to be swung to maximum for almost all the period during which the capacitor voltage was below its original value.

In addition, the circuit incorporated a device for rejecting overloading input pulses and a pile up rejector which operated on the principle of sensing the time between leading edge and the zero crossover of the double differentiated pulse. The circuit would accept input pulses in the range 0 to ±2 V, of duration 40 ns to 5 μs (DL and $(DL)^2$ shaped) or of time constant up to 8 μs (RC shaped). The stability of both the E and $E + \Delta E$ thresholds was $(0{\cdot}25\text{–}0{\cdot}05\ E_{max}/E) \times 10^{-3}/°C$.

6.4.3 *Tunnel diode discriminators*

A large number of discriminator circuits have been described using tunnel diodes and it is only possible to review a small selection. The attraction of the tunnel diode is its speed of operation and it is nearly always used in circuits designed to process the highest pulse rates. In practice the tunnel diode may be used in several circuit configurations to give the trigger characteristic and in order to use it as a pulse height discriminating element the input pulse is usually added to a bias voltage or current. Thus in the simple circuit of Figure 6.14(a) the input signal is applied as a current to the

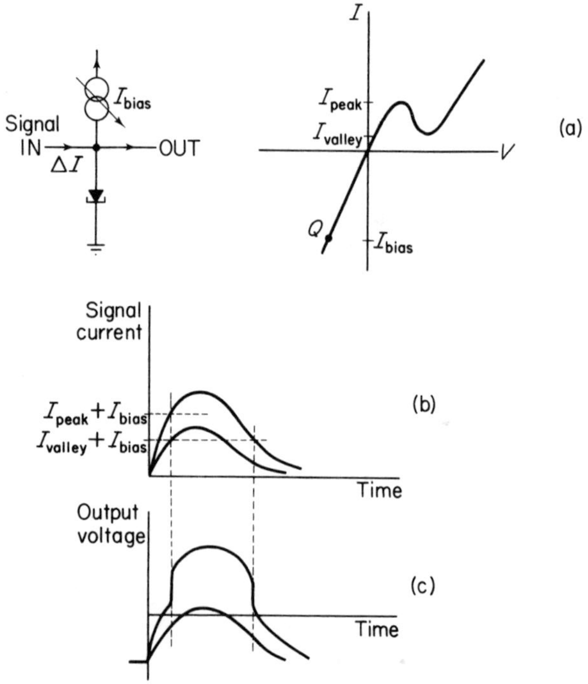

Figure 6.14 Simple tunnel diode discriminator approximating the Schmitt trigger action, together with the input and output signals

diode. Figure 6.14(a) shows the quiescent operating point Q, which is set by the bias current I_{bias}, and a signal of height ΔI, larger than the threshold $I_{peak} + I_{bias}$, lifts the diode into the high voltage state. It is necessary that the biasing point Q is lower than I_{valley} or the circuit will not reset at the end of the pulse. This means that the configuration is not suitable for use with small pulses. Figure 6.14(c) shows the output pulse shapes for input pulses above and below the threshold. It will be seen that subthreshold pulses gave a significant feed-through compared to pulses above the threshold; for this reason it is necessary to include a second discrimination circuit to exclude the subthreshold feed-through pulses. This is especially important since Q is usually placed well down the left hand side of the characteristic so that temperature variations in the value of I_{peak} have only a minor influence of the overall threshold. It will be observed that the reset time of the circuit is determined by the input pulse and in this respect it is similar to the Schmitt trigger. The standardization of the output pulse may be performed by a monostable circuit, usually using a second tunnel diode, which can be also used to perform the secondary discrimination referred to above.

Figure 6.15(a) shows the circuit of Gorodetzky *et al.*[14] using this arrangement. The function of the capacitor C is to differentiate the pulse so the relaxation time of the monostable is independent of the reset time of the first tunnel diode. In addition it serves to isolate the quiescent DC voltages in the two tunnel diode circuits. Other workers have used DC coupling via a diode, as in the circuit of Figure 6.15(b) of Van Zurk.[15] The diode is normally conducting a small forward current in the quiescent state. The input pulse above the threshold triggers the first tunnel diode which in turn triggers the second tunnel diode. With a short input pulse TD1 will soon reset but since the coupling diode is then blocked TD2 will not be affected and will only reset when the relaxation time, determined by the delay line, expires. The two delay lines are of the same length and each has a characteristic impedance of 50 ohms so that together they match the source impedance. The circuit had a recovery time of 12 ns for pulses in the range 1 V to 4 V. The biasing of a tunnel diode with a constant current source must be below I_{valley} and so gives a low trigger sensitivity, and the minimum threshold for the circuit in Figure 6.15(b) was 1 V. Van Zurk described an alternative arrangement where a series inductance and resistance were connected across TD1. In this form the minimum threshold was 12 mV and the recovery time was 20 ns.

In the circuit of Herbst[16] (Figure 6.15(c)) TD1 is biased at P between A and the peak on the characteristic, depending on the required threshold level. If P is below I_v TD1 is monostable but if it is above I_v, the tunnel diode is in a nominally bistable arrangement and the resetting of both TD1 and TD2 is by the auxiliary circuit in Figure 6.15(c). The sequence of events is that, when TD1 triggers to the high voltage state, the back diode BD

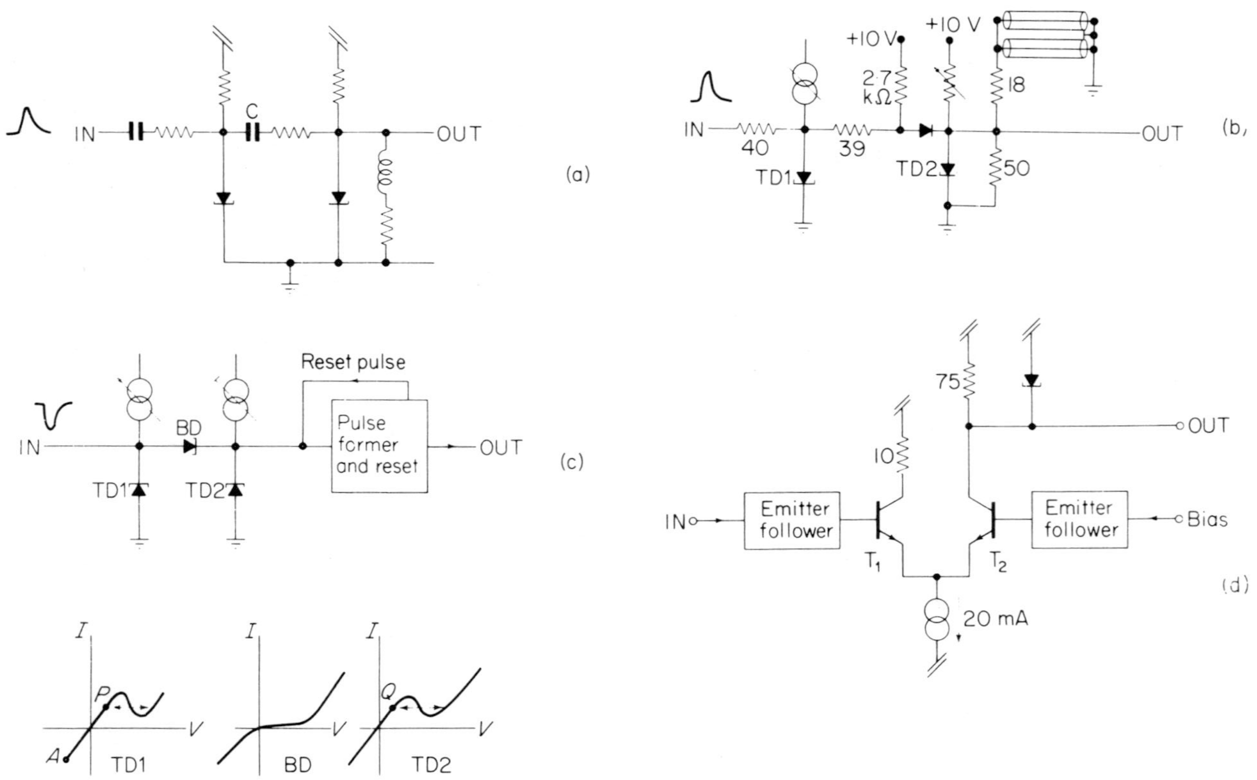

Figure 6.15 Principles of the discriminator elements used by various authors. (a) Circuit of Gorodetzky *et al.*[14] (b) Circuit of Van Zurk.[15] (c) Circuit of Herbst.[16] (d) Circuit of Barna and Cisneros[53]

conducts and triggers TD2, which was originally biased at Q in the bistable region, to its high voltage state. After a time the reset circuit generates a pulse that resets TD2 and, via BD, resets TD1 if it has not already reset. The back diode BD is used in preference to an ordinary diode since it exhibits a low conductivity to small voltages such as are generated by TD1. The advantage of biasing TD1 at points near the peak is that the circuit has a small minimum threshold. Herbst gave details of two reset circuits, the first being a transistor monostable giving reset times of 100 ns to 20 μs, and the second a transistor delay cable arrangement giving reset times of <40 ns. The circuit accepted input pulses from 1 to 8 mA into a 50 or 100 ohm input load and had a threshold stability of 5 μA/°C.

Another arrangement which gives a quiescent point near the peak of the characteristic, and hence gives a low threshold, was described by Harms[52]. The tunnel diode was provided with a non-linear load due to the inclusion of a back diode. The effect of this was to provide a high impedance load when the tunnel diode was in the low voltage state (the quiescent state), and a low impedance when in the high voltage state. This results in a Schmitt trigger action with little hysteresis. The circuit accepted input pulses in the range 50–500 mV height (into a 50 ohm input impedance). All stages were DC coupled and the operating points of critical stages were specially stabilized to give a threshold stability of 0·1 mV/°C.

Modern ECL integrated circuit logic elements can operate at high speeds and Barna and Cisneros[53] extensively used such units in a tunnel diode discriminator. Figure 6.15(d) shows the principle of the threshold determining stage. The long tailed pair T_1 and T_2, part of an integrated circuit, are quiescently biased so that most of the 20 mA flows in T_1. The arrival of a sufficiently large input pulse causes some of the current to be diverted into T_2 and so causes the tunnel diode to exhibit a Schmitt trigger action. The output from the tunnel diode was shaped by a monostable which was made up from ECL AND/NAND gate elements. The discriminator was DC coupled throughout and could accept input pulses in the range 150 mV to 1·15 V with a pulse pair resolution of ~10 ns.

Many designs have been published using a single tunnel diode arranged in a monostable configuration. The resetting is accomplished by means of an LR or CR circuit or by a delay line. Ideally the input pulse should be short compared to the reset time. Two arrangements used by Grieder[17] are shown in Figure 6.16(a). In the first arrangement the reset time is controlled by the LR circuit, and the secondary discrimination against subthreshold feedthrough is performed by suitable biasing of the transistor base. In the second arrangement the reset time is controlled by the delay line when the clipped input voltage drops to zero. It should be noted however that it is not the inverted peak which resets the diode since this will be small with slowly decaying input pulses and may cause failure to reset.

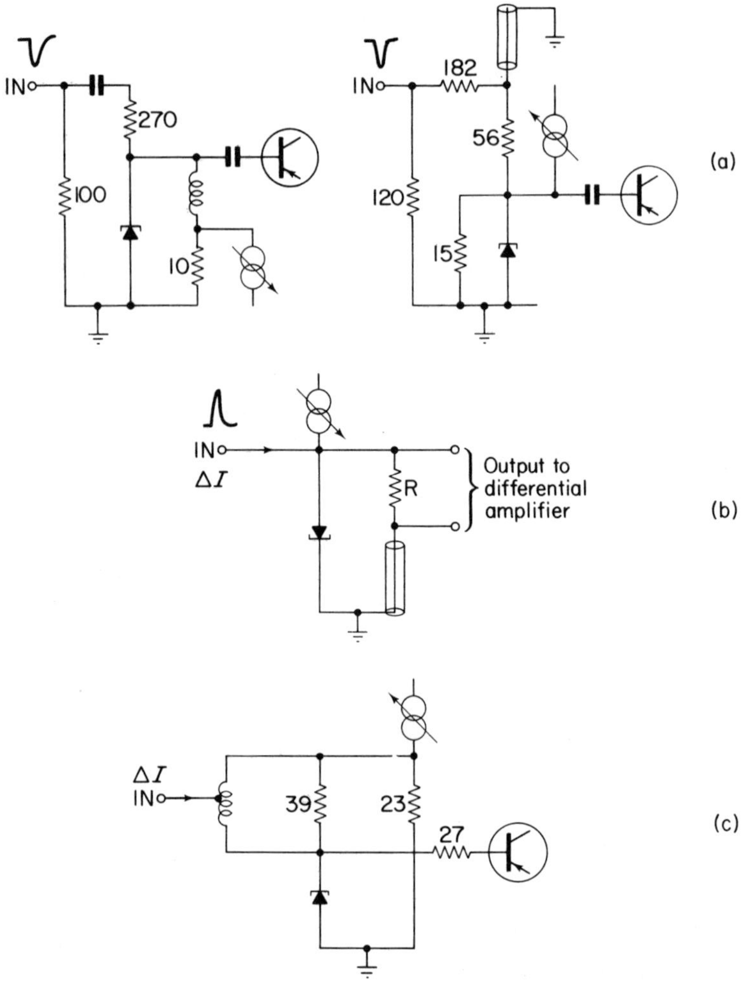

Figure 6.16 Principles of the discriminator elements used by various authors. (a) Circuit of Grieder.[17] (b) Circuit of Van Zurk.[18] (c) Circuit of Van Zurk[19]

Van Zurk[18] has used the arrangement shown in Figure 6.16(b) where the output signal is taken from across the resistance R by means of a differential amplifier. This circuit had a dead time of 10 ns for short input pulses and had a negligible recovery time. In another configuration Van Zurk[19] applied the input pulse to the mid-point of the inductance in Figure 6.16(c). This circuit has a Schmitt trigger action with a very short recovery time. It was followed by a tunnel diode monostable to standardize the output pulse and can handle input pulses of 2 ns rise time with a dead time of 10–14 ns. The temperature variations of the threshold were reduced by incorporating a positive temperature coefficient resistor in the bias supply and this gave a stability of 0·2 mV/°C (maximum input pulse height 500 mV).

Bernard *et al.*[20] have incorporated the above discriminator element in a differential discriminator arrangement (Figure 6.17(a)) which does not use the more common AND NOT anticoincidence gate. The outputs from the upper and lower discriminators are differentiated and applied to a tunnel diode bistable and gates P1 and P2. An input pulse above the window causes the front edge of the upper discriminator output to trigger the bistable output $\bar{Q}$ from the quiescent '1' to '0', thus blocking P1; at the same time the Q output, in going from quiescent '0' to '1', enables gate P2. Eventually a pulse derived from the trailing edge of the lower discriminator output, through the enabled gate P2, causes the bistable to reset, no output pulse having been delivered through the blocked P1. An input pulse lying in the window produces only the input from the lower discriminator which in this case finds P1 enabled and so gives an output pulse. As it stands the arrangement has the disadvantage that the timing of the output pulse is derived from the instant the input pulse decays past the lower threshold. However, it is possible to arrange the timing of the output pulse to derive from the instant that the input pulse rises past the lower threshold provided a delay is inserted in the lower threshold channel. By interchanging the connections A and B the circuit may alternatively be used as a pile up rejector. For input pulses in the range 0–600 mV the stability was 0·25 mV/°C and, for input pulses of duration 5 ns, the dead time was 12 ns.

Another design using delay cable reset has been described by Sugarman, Higinbotham and Yonda[21] and the circuit, shown in Figure 6.17(b), was capable of processing input pulses of 1·8 ns width with a recovery time of 14 ns. The diode at the input prevents the tunnel diode being turned off by overshoots on the input pulse. In the design by De Blust, Mandl and De Lotto,[22] Figure 6.17(c), the reset pulse is derived from a stage along the amplifier. This circuit could accept pulses down to 50 ns width with a change in threshold of less than 0·1 mV and had a recovery time of 400 ns for 5 mV pulses to within 1 mV of the original threshold.

A more elaborate form of discriminator has been described by De Volpi *et al.*[23] suitable for use in fast–slow and slow–fast coincidence arrangements where the dead time needs to be constant but adjustable. The discriminator consisted of a tunnel diode monostable which was incorporated in the simplified logic circuit of Figure 6.18. When an input pulse exceeds the threshold, the discriminator gives out a logic pulse which, via the enabled gate, sets the bistable A. The Q output of this immediately goes to '1' and after a time set by the delay cable A sets bistable B. Its Q output, in going to '1', resets bistable A and blocks the gate. After a further time set by delay

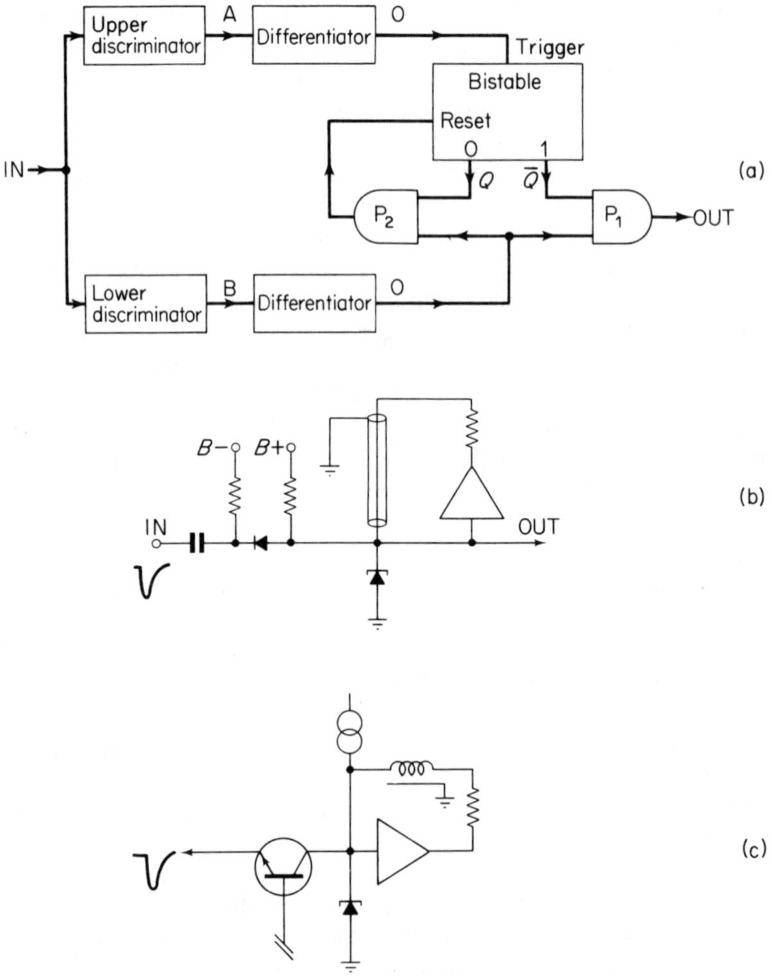

Figure 6.17 (a) Differential discriminator of Bernard *et al.*[20] (b) Discriminator element of Sugarman, Higinbotham and Yonda.[21] (c) Discriminator element of De Blust, Mandl and De Lotto[22]

cable B the bistable B also resets and the gate becomes enabled again. The advantage of the arrangement is that bistable A recovers while bistable B is set, the input gate being blocked in the meantime. The circuit therefore has a well defined dead time (adjustable from 100 nm to 10 ms) although the arrangement does not prevent certain malfunctions at high pulse rates in the discriminator itself. In this respect the later logic arrangement from the same laboratory (Figure 6.12) which has already been described is superior.

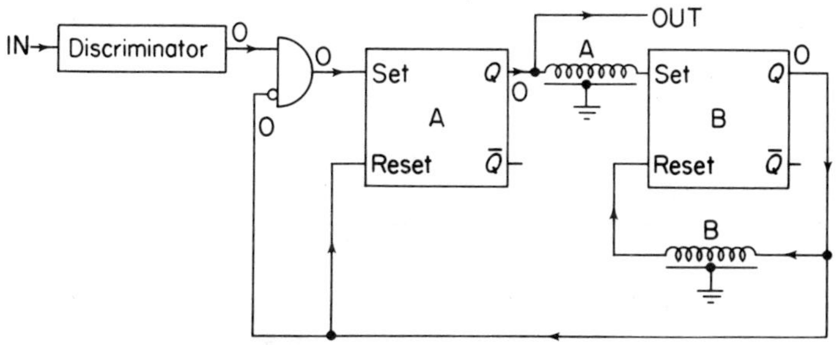

Figure 6.18 Simplified logic scheme as used in the discriminator of De Volpi *et al.*[23] to give a constant dead time

6.5 Pulse shape discrimination (PSD)

6.5.1 Introduction

Pulse shape discriminators have been used in several applications but the most common use is to discriminate between events caused by gamma rays (i.e. by electrons) and neutrons (i.e. via recoil protons) in organic scintillators. The behaviour of organic scintillators to nuclear particles of different specific energy loss has been discussed in Subsection 1.5.5. The basis of particle discrimination in organic scintillators is well illustrated by Figure 6.19, taken from Kuchnir and Lynch,[24] showing the time course of the light intensity from various scintillators when excited by gamma rays and neutrons. Figure 6.20, from the same authors, shows the total light output for stilbene for gamma rays and neutrons and illustrates the effect of ionization quenching.

Gatti and De Martini[25] have derived the theoretically optimum linear processing system for pulse shape discrimination (PSD) in scintillators on the assumption of simple statistics. They assumed that the number of electrons N_t collected in any small constant time interval Δt had a variance proportional to N_t. For α and γ induced events giving the same amount of collected charge, and where the time courses of the collector currents in the two cases were assumed to be $\alpha(t)$ and $\gamma(t)$ respectively, these were weighted by a function $P(t)$. Assuming the weighted signals were sampled after a time T to give S_α and S_γ where

$$S_\alpha = \int_0^T \alpha(t)\, P(t)\, \mathrm{d}t$$

$$S_\gamma = \int_0^T \gamma(t)\, P(t)\, \mathrm{d}t$$

$P(t)$ was then optimized to give a maximum value for the ratio of $(\bar{S}_\alpha - \bar{S}_\gamma)$ to the standard deviation of $(S_\alpha - S_\gamma)$. The optimum form of $P(t)$ was then given by

$$P(t) = \frac{\overline{\alpha(t)} - \overline{\gamma(t)}}{\overline{\alpha(t)} + \overline{\gamma(t)}}$$

In other words those parts of the collector pulse where the difference between the α induced and the γ induced response is largest should be weighted by the largest factor. Gatti and De Martini derived the optimum form of $P(t)$ for anthracene and described a circuit which approximately realised the function. Figure 6.21 shows their experimental results for ^{210}Po α particles and ^{137}Cs γ rays where the axis numbered 0 to 18 represents the

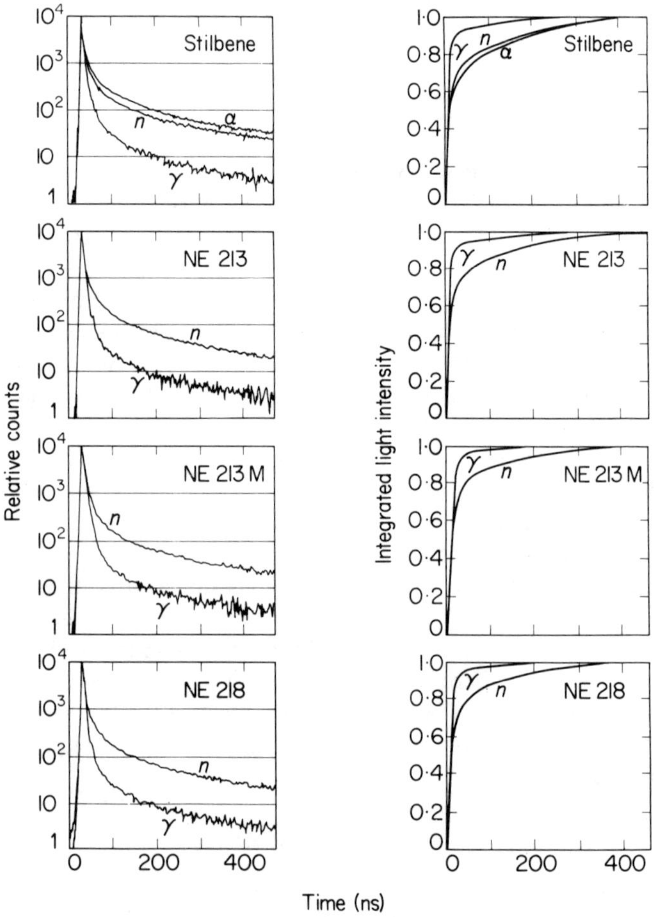

Figure 6.19 The time course of the light intensity (left) and of the integrated light (right) from various scintillators excited by γ rays and neutrons, according to Kuchnir and Lynch.[24] In all cases the curves are normalized to constant maximum height

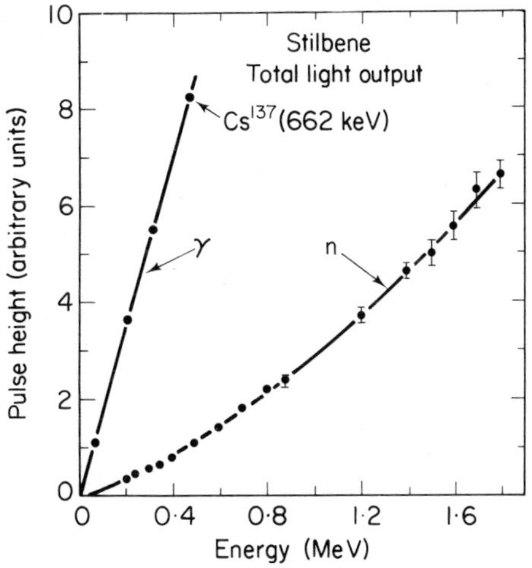

Figure 6.20 Total light output as a function of energy for Compton electrons and for recoil protons, according to Kuchnir and Lynch[24]

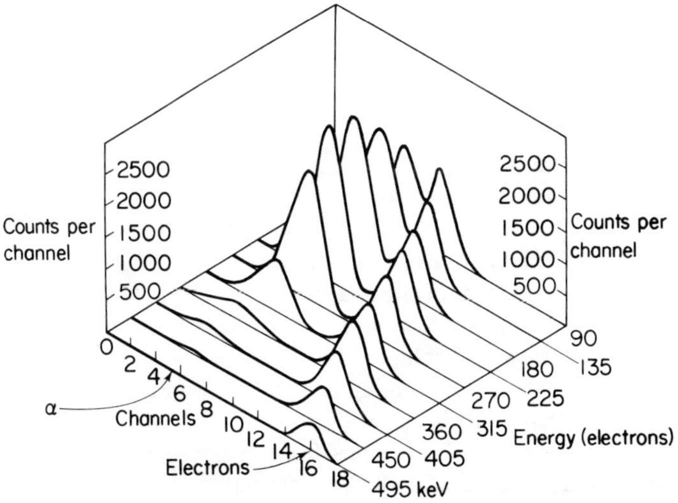

Figure 6.21 Experimental results of Gatti and De Martini[25] obtained with ^{210}Po α particles and ^{137}Cs γ rays in an anthracene crystal. The axis numbered 0 to 18 represents the parameter on which the discrimination method is based

parameter S. According to the theory the minimum electron energy at which discrimination is just possible (as defined in their paper) is 13 keV but the experimental arrangement gave a value of 39 keV. The discrepancy is presumably partly due to the imperfect practical realization of $P(t)$ and to the assumption of the simple statistics in the theory.

More recently Gatti *et al.*[55] have extended the theory to include both the effects of thermionic noise in the photomultiplier tube and the effect arising from the non-instantaneous response of the tube to the light signal. They calculated the performance of the optimum processing system with NE 213 liquid scintillator and compared it to that calculated for the zero crossing system (see later).

Techniques for PSD in scintillators described before 1962 have been reviewed by Owen[26]. More recently progress has been reviewed by McBeth, Winyard and Lutkin[27]. Ideally a PSD system should be capable of completely discriminating between both types of particles with 100% efficiency over all particles energies. For n/γ discrimination in scintillators most systems can approximate to this for particle energies above the equivalent of 0·5 MeV (electron energy) and over an energy range of 100:1. A large number of publications have described PSD systems and it is usually difficult to compare their performances as different authors report their results in different ways. In this context the figure of merit M, as defined by Reid and Hummel[28], is a useful parameter. If S is the parameter generated by the PSD circuit on which the discrimination is based, the distribution of S will show two peaks corresponding to the two particle types. M is defined as

$$M = \frac{\bar{S}_1 - \bar{S}_2}{\Delta S_1 + \Delta S_2}$$

where $\bar{S}_1 - \bar{S}_2$ is the difference between the mean peak positions, and ΔS_1 and ΔS_2 are the fwhms of the peaks for particle types 1 and 2. For PSD with a zero crossing type system McBeth, Winyard and Lutkin[27,56,57] have given values of M for many different scintillators at different energies and it is interesting to note that, of the organic types tested, xylene based liquid scintillators give the largest values of M for n/γ discrimination, heptane based scintillators showing negligible PSD properties.

The properties of the light coupling arrangement and the photomultiplier also enter into consideration as evidenced by the work of Kalyna and Taylor[29], who considered the reasons for the poor performance of large scintillator systems. Figure 6.22 shows their values of M for two different sizes of scintillator used in a zero crossing type system. The variations of photoelectric efficiency and transit times over the tube cross-section were shown to be important factors and ways of minimizing these effects were considered.

PSD has also been used in gas filled and semiconductor detectors to discriminate between particles giving different ionization densities. These will produce different pulse shapes provided the orientation of the ionization track of at least one particle type is reasonably well defined, as in the case of a collimated beam. In semiconductor detectors it is possible to use PSD to discriminate against, for instance, γ events in which part of the ionization occurs outside the depletion layer since such charge will be collected more slowly. Examples of these applications are given in the following sections.

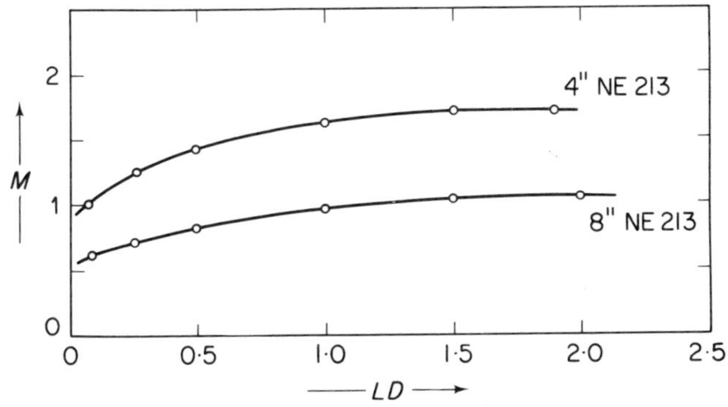

Figure 6.22 Figure of merit plotted against the discriminator bias (1·0 is equivalent to the energy of the Compton edge of ^{60}Co) for two diameters of NE 213 scintillator, according to Kalyna and Taylor[29]

6.5.2 Zero crossing method

This method was originated by Alexander and Goulding[30] and there have been a large number of publications describing circuits using the same principle. The basis of the method is to take the detector pulse and doubly differentiate it, either by CR filters or by delay line techniques. The time interval T in Figure 6.23(a) between the beginning of the pulse and the zero crossing point is then measured by means of one or two time pick-off circuits (see Section 7.2 for examples of time pick-offs). This time interval is in principle independent of the energy of the event, but depends on the detector pulse shape, and so it is the parameter on whose value the PSD is based. The instant of the beginning of the pulse is sensed by a leading edge time pick-off and since this must have a finite threshold for triggering, there will be a slight dependency of the triggering instant on the amplitude of the pulse (i.e. a 'time walk'). To compensate for this many systems define the second time instant measurement at a point slightly below the true zero crossing as shown in Figure 6.23(b). The setting of the threshold for the zero crossing time pick-off is usually adjusted experimentally to give a minimum variation of the measured time interval with pulse amplitude. The time interval, carrying the information about the particle type, is often converted into a pulse of proportional height by means of a time to amplitude converter (see Section 7.7). If this pulse is now fed into a pulse height discriminator with a suitably adjusted threshold it is possible to use the logic pulse output from this to gate the counting system.

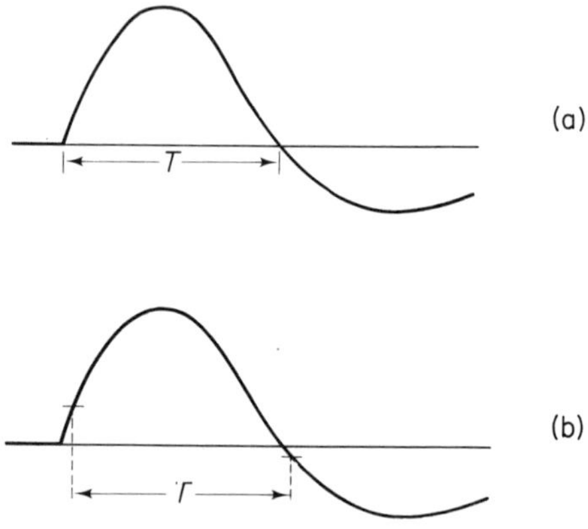

Figure 6.23 (a) Principle of the zero crossing method. (b) Compensation of the time walk by the choice of a near-zero crossing point

The characteristics of zero crossing timing are examined more fully in Subsection 7.2.3 and it is shown that if the pulse is double delay line shaped, the instant of zero crossing is derived from the time of the instant at which 50% of the total detector charge has arrived (see Figure 7.13). For pulses 'differentiated' by CR filters the instant of zero crossing is not simply related to the instant of collection of any fraction of charge. For n/γ discrimination in various scintillators Roush, Wilson and Hornyak[31] have considered the arrangement where the detector signal is first integrated by a RC low pass filter of time constant τ_1 and then differentiated by a CR high pass filter of time constant τ_2. They defined a factor of merit equal to the product of the slope of the signal at the zero crossing instant and the difference in the zero crossing times for the two particle types. For most organic scintillators their analysis showed that the optimum values of τ_1 and τ_2 should be in the region of 500 ns and they conclude that delay line differentiation should give a slightly better figure of merit. It should be noted, however, that this figure of merit does not take into account the statistical variations of the pulse shape as such.

Gatti *et al.*[55] have considered the performance of the same arrangement used with the liquid scintillator NE 213 and on the assumption of Poisson statistics. They predicted that the performance of the arrangement could be made to be nearly as good as that of their theoretically optimum system, provided the values of τ_1 and τ_2 were in the region of a few μs. The values of τ_1 and τ_2 were not very critical in this respect however. Experimental

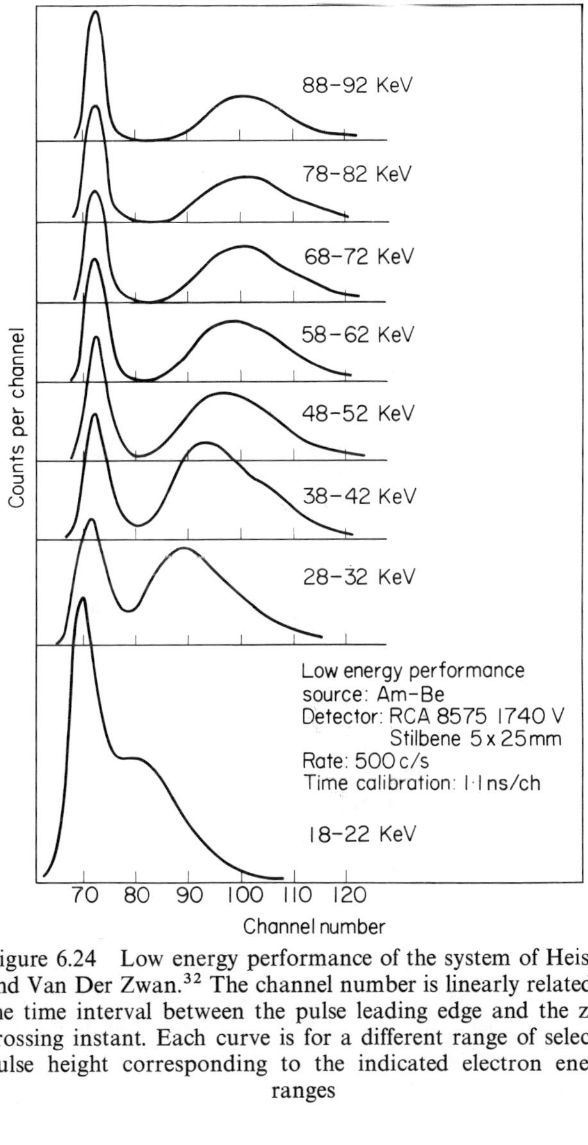

Figure 6.24 Low energy performance of the system of Heistek and Van Der Zwan.[32] The channel number is linearly related to the time interval between the pulse leading edge and the zero crossing instant. Each curve is for a different range of selected pulse height corresponding to the indicated electron energy ranges

measurements with the zero crossing system yielded performances which agreed reasonably well with those predicted.

Some recent examples will indicate the type of results obtainable for n/γ discrimination in scintillators using the zero crossing method. Heistek and Van Der Zwan[32] used delay line shaping and used an integrated circuit differential amplifier to detect the zero crossing point. The arrangement could accommodate a pulse amplitude range of 500:1 and Figure 6.24 shows the low energy performance for neutron and gamma events in stilbene.

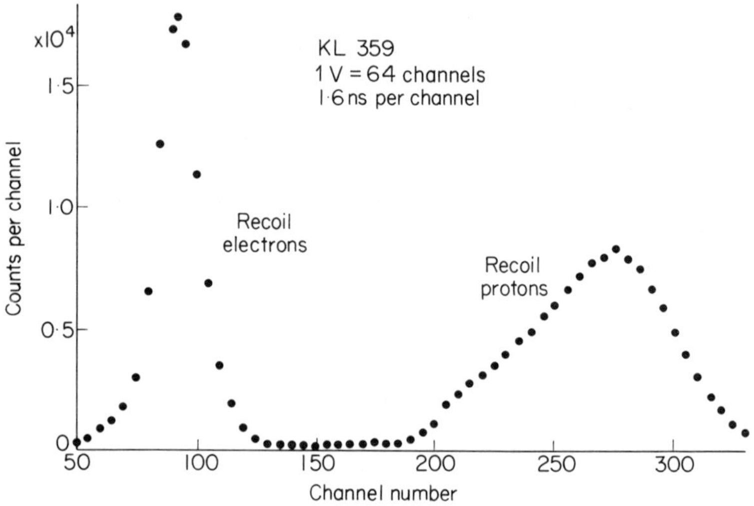

Figure 6.25 Zero crossing time distribution for KL 359 scintillator with a Am–Be source, according to McBeth, Lutkin and Winyard.[33] The pulses were selected in a height range equivalent to an electron energy range of 20 keV to 7 MeV

McBeth, Lutkin and Winyard[33] describe a system using CR shaping which gave almost 100% separation of neutron and gamma events over the energy range 20 keV to 7 MeV (equivalent electron energy). Typical results are illustrated in Figure 6.25 for KL 359, a xylene based liquid scintillator. Hollandsworth and Bucher[34] used delay line shaping and obtained a factor of rejection for gamma rays from neutrons of 200:1 for an energy range of 0·1 MeV to 1 MeV (equivalent proton energy).

The conventional zero crossing method described above approximately corresponds to sensing the time position of the instant of arrival of 50% of the total detector charge. Kuchnir and Lynch,[24] assuming Poisson statistics for the detection process in the scintillation counter claim that the optimum factor of merit should be obtained when the instant of measurement corresponds to the arrival of a fraction of 80–90% of the total collected charge.

This principle was realized practically by taking double delay line shaped pulse and adding to it the attenuated integrated detector signal so that an asymmetric bipolar pulse was obtained with an instant of zero crossing corresponding to 80–90 % of collected charge. The results obtained showed quite good agreement with theoretically predicted values. Depending on the setting of the discriminator it was possible to accept neutrons down to an energy of 350 keV with an efficiency of 70 %, with an interfering efficiency from gammas of 0·7 %, or alternatively a neutron efficiency of 90 % with a gamma efficiency of 1·37 %. The same principle was taken up by Johnson[35] who demonstrated that it was possible to distinguish Cerenkov events in the glass window of the photomultiplier from neutron and gamma events in stilbene.

For use with proportional counters where noise is an important factor Mathieson and Harris[36] have derived the optimum CR pulse shaping time constants in terms of a quantity they defined as the characteristic time constant of the counter and assuming a charge sensitive FET amplifier input stage. Sayres and Coppola[37] have described an application for discriminating between the products of the reaction ^{3}He $(n, p)T$ from ^{3}He recoils in a ^{3}He proportional counter. For surface barrier semiconductors Emmerich *et al.*[38, 58] have described a system suitable for discriminating between proton, deuteron and alpha induced events. Finally several workers[39, 56, 59, 60] have shown that it is possible to discriminate between γ, α, and proton induced events in CsI(Tl).

6.5.3 *Double integration method*

This method was originated by Brooks.[40] In essence it consists of integrating the charge contained in the early part of the detector pulse and comparing it to the total charge of the pulse. As implemented by Brooks a signal taken from the photomultiplier collector was passed into a RC filter of long time constant. A second signal from the last dynode was passed into a RC filter of short time constant. The output pulses from both filters were stretched by diodes and, after suitable weighting, the difference between the signals was taken. When the weightings were correctly adjusted it was possible to distinguish neutrons and gammas on the basis of whether the difference signal contained a positive going component. Figure 6.26 shows some of the results obtained with stilbene where, on the energy scale, $h_0 = 10$ corresponds to a gamma energy of $\sim$ 1·7 MeV.

Related methods have been published by many authors. Of the more recent examples, Sabbah and Suhami[41] claim that the method is capable of better performance than the zero crossing method for scintillators as it approximates more nearly to the theoretically optimum arrangement of Gatti and De Martini[25] discussed previously. In the system of Sabbah and Suhami the detector pulse was passed through a CR high pass filter of short time constant (~ 25 ns was calculated to be optimum) and the first lobe of the resulting bipolar pulse was removed with a diode. The remaining undershoot, which equalled the area of the first lobe, was integrated by a RC low pass filter and the difference was taken between this and the integrated whole detector pulse. The PSD was then based on whether this yielded a positive or a negative signal. The system appeared to give a very good performance, giving a rejection factor for γ-rays of 1000:1 for neutron energies down to 50 keV.

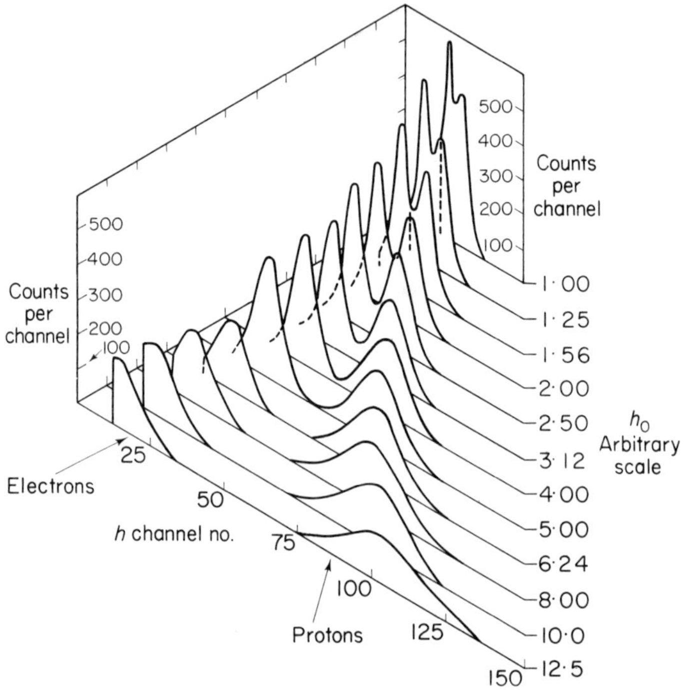

Figure 6.26 Experimental results of Brooks[40] using a stilbene crystal and a Po–Be source. The scale marked h_0 is the pulse amplitude where $h_0 = 10$ corresponds to a gamma energy of 1·7 MeV

Jones[42] used two CR high pass filters of time constants 60 ns and 60 μs to perform the two integrations. The amplitudes of the two resulting signals, after stretching, were differenced. As in most other systems, discrimination was based on whether the differenced signal was positive or negative and the system was capable of almost 100% separation of neutrons and gammas in the neutron energy range 0·5–15 MeV. Reid and Hummel,[28] St. Onge and Lockwood[44] and Czirr[61] have also described systems using simple RC filters to integrate the charge in the early part of the pulse and these systems are of interest as they quote their results in terms of the factor of merit, *M*, mentioned earlier. St. Onge and Lockwood give values of M of about 3 for the neutron energy range 1–30 MeV.

For use with Ge(Li) detectors Tamm, Michaelis and Coussieu[45] have used a variant of the double integration method. The detector pulse was clipped to a length of 200 ns by a delay line and, after a suitable delay, subtracted from another pulse obtained by clipping the detector pulse to a length of 4 μs by another delay line as shown in Figure 6.27. The criterion for discrimination was whether the resulting signal crossed the baseline.

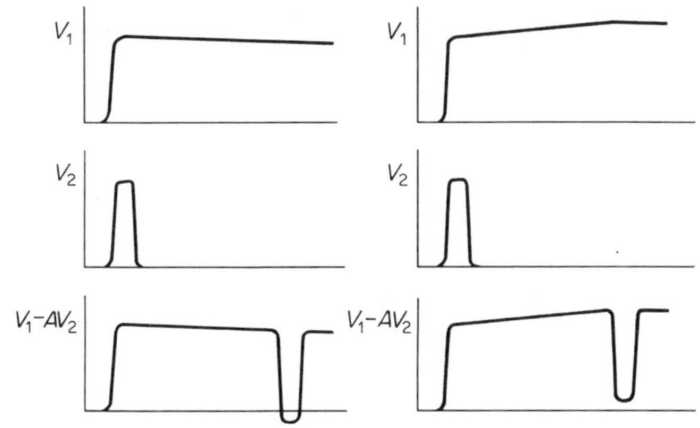

Figure 6.27 Principle of the pulse shape discrimination system of Tamm, Michaelis and Coussieu,[45] showing the differing pulse shapes for two different particle types

Using the technique for gamma spectroscopy it was possible to discriminate against events where the ionization was produced partly outside the depletion layer which consequently produced a pulse of slower rise time due to the lower collecting field in such regions. This enabled a reduction by a factor of 3 to be made in the background under the peaks of the pulse height spectrum, although the fwhm resolution was unaltered. The same technique was used by Swinth, Philipp and Hoitink[46] in combination with an anticoincidence shield. This technique resembled that used earlier by Ammerlaan, Rumphorst and Koerts[47] for discriminating between deuterons and alpha particles in a Si(Li) detector.

6.5.4 *Other techniques*

A very simple arrangement, described by Owen,[26] made use of the space charge saturation effect in photomultiplier tubes so that the last dynode causes secondary emission only during the latter part of the pulse when the amplitude is small. By suitable adjustment of the collector voltage it was possible to observe a positive overshoot on the last dynode with neutron events but not with gamma events. Many other earlier techniques were also reviewed in this publication by Owen.

Kinbara and Kumahara[48] used a technique based on a measurement of the rise time of the detector pulse. The time interval in Figure 6.28 was measured, the two instants being defined by integrated circuit differential amplifiers sensing the crossing points. The technique was used on a wide variety of detectors. For organic scintillators it was possible to discriminate neutron events down to 500 keV with a gamma rejection factor of 500:1.

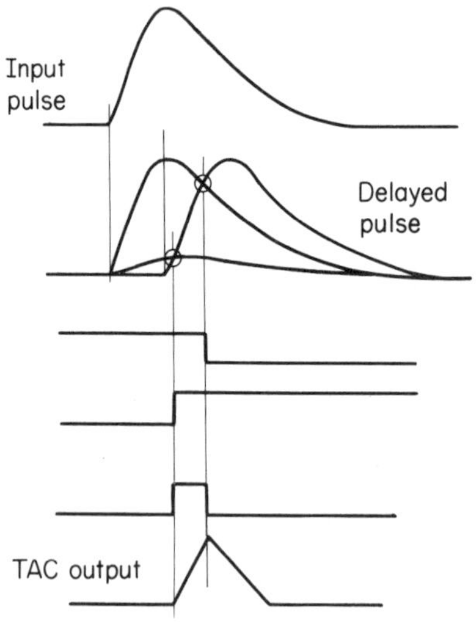

Figure 6.28 Principle of the rise time measurement method of Kinbara and Kumahara[48]

When used with a hydrogen filled proportional counter a gamma rejection factor of 50:1 was obtained at energies down to 60 keV. The technique was also used on a Ge(Li) detector to reduce the background from events outside the depletion layer and to improve the resolution. The improvement in resolution was possible since a significant proportion of the events did not occur entirely within the depletion layer and were subject to greater trapping and recombination. It was possible to reduce the fwhm of a 6346 keV peak from 86 keV (without PSD) to 69 keV. A further application described by the authors was the improvement of the resolution with a silicon surface barrier detector used for the detection of the products of the reaction ^{6}Li(n, α)T for thermal neutrons. The fwhm of the peak at 2·73 MeV was reduced from 160 keV to 71 keV with PSD, owing to the improvement of the effective collimation.

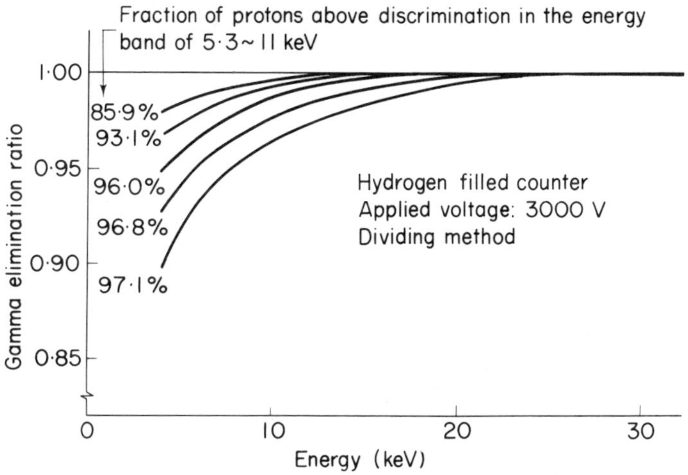

Figure 6.29 Gamma elimination ratios obtained with a hydrogen filled proportional counter by Obu, Ichimori and Shirakata.[49] Different curves relate to different settings of the threshold of the particle discriminator

Obu, Ichimori and Shirakata[49] have used a similar method for n/γ discrimination in a hydrogen filled proportional counter. In addition they have used another method based on the measurement of the ratio of the rate of rise of the pulse to the pulse amplitude. The rate of rise was derived as the height of the pulse after RC clipping (100 ns time constant) and the ratio of this to the pulse amplitude was formed by means of log and antilog function generator circuits. Both the methods gave similar performances and Figure 6.29 illustrates some of their results using the ratio method.

References

1. *Standard Nuclear Instrument Modules* Report No. TID20893, Rev. 3 (1969), U.S. Atomic Energy Commission.
2. *Electronic Counting,* Mullard Ltd., London, 1967.
3. A. I. Nizan and E. Elad, *Nucl. Instr. and Meth.,* **51**, 270 (1967).
4. C. B. Ward and C. M. York, *Nucl. Instr. and Meth.,* **23**, 213 (1963).
5. A. Barna, J. H. Marshall and M. Sands, *Nucl. Instr. and Meth.,* **12**, 43 (1961).
6. S. Rozen and E. Altman, *Nucl. Instr. and Meth.,* **17**, 337 (1962).
7. J. L. Le Comte and R. Allemand, *Nuclear Electronics,* Vol. 2, p. 399, Proceedings of international conference at Belgrade, IAEA, Vienna, 1962.
8. S. S. Klein, L. Hustman and J. Blok, *Nucl. Instr. and Meth.,* **60**, 88 (1968).
9. H. Verweij, *Nucl. Instr. and Meth.,* **10**, 308 (1961).
10. M. Coli, *Nucl. Instr. and Meth.,* **39**, 297 (1966).
11. L. M. Welter, *Rev. Sci. Instr.,* **36**, 487 (1965).
12. J. Lauch and H. U. Nachbar, *Nucl. Instr. and Meth.,* **73**, 292 (1969).

13. S. J. Rudnick, J. J. English and R. H. Howard, *Rev. Sci. Instr.*, **40**, 9 (1969).
14. S. Gorodetzky, A. Muser, J. Zen and R. Armbruster, *Nucl. Instr. and Meth.*, **13**, 282 (1961).
15. R. Van Zurk, *Nucl. Instr. and Meth.*, **16**, 157 (1962).
16. L. J. Herbst, *Nucl. Instr. and Meth.*, **70**, 189 (1969).
17. P. K. F. Grieder, *Nucl. Instr. and Meth.*, **56**, 229 (1967).
18. R. Van Zurk, *Nucl. Instr. and Meth.*, **46**, 125 (1967).
19. R. Van Zurk, *Nucl. Instr. and Meth.*, **53**, 45 (1967).
20. P. Bernard, J. Chambon, J. Mey and R. Van Zurk, *Nucl. Instr. and Meth.*, **60**, 213 (1968).
21. R. Sugarman, W. A. Higinbotham and A. H. Yonda, *Nuclear Electronics* Vol. 3, p. 3, Proceedings of international conference at Belgrade, IAEA, Vienna, 1962.
22. E. De Blust, V. Mandl and I. De Lotto, *Électronique Nucléaire*, p. 785, Proceedings of international symposium, Paris, S.F.E.R. 1963.
23. A. De Volpi, J. J. English, A. J. Metz, K. G. Porges and S. J. Rudnick, *Rev. Sci. Instr.* **37**, 1592 (1966).
24. F. T. Kuchnir and F. J. Lynch, *IEEE Trans. Nucl. Sci.*, NS **15**, No. 3, 107 (1968).
25. E. Gatti and F. De Martini, *Nuclear Electronics*, Vol. II, p. 265, Proceedings of international conference at Belgrade, IAEA, Vienna, 1962.
26. R. B. Owen, *IEEE Trans. Nucl. Sci.*, NS **9**, No. 3, 285 (1962).
27. G. W. McBeth, R. A. Winyard and J. E. Lutkin, *Pulse Shape Discrimination with Organic Scintillators*. Koch–Light Laboratories Ltd., Colnbrook, England, 1970.
28. W. B. Reid and R. H. Hummel. *Can. Nucl. Tech.* **5**, 36, Jan/Feb 1966.
29. J. Kalyna and I. J. Taylor, *Nucl. Instr. and Meth.*, **88**, 277 (1970).
30. T. K. Alexander and F. S. Goulding, *Nucl. Instr. and Meth.*, **13**, 244 (1961).
31. M. L. Roush, M. A. Wilson and W. F. Hornyak, *Nucl. Instr. and Meth.*, **31**, 112 (1964).
32. L. J. Heistek and L. Van Der Zwan, *Nucl. Instr. and Meth.*, **80**, 213 (1970).
33. G. W. McBeth, J. E. Lutkin and R. A. Winyard, *Nucl. Instr. and Meth.*, **93**, 99 (1971).
34. C. E. Hollandsworth and W. P. Bucher, *Rev. Sci. Instr.*, **39**, 165 (1968).
35. F. A. Johnson, *Nucl. Instr. and Meth.*, **78**, 1 (1970).
36. E. Mathieson and T. J. Harris, *Nucl. Instr. and Meth.*, **88**, 181 (1970).
37. A. Sayres and M. Coppola, *Rev. Sci. Instr.*, **35**, 431 (1964).
38. W. D. Emmerich, K. Frank, A. Hofman, A. Dittner, J. W. Klein and R. Stock, *Nucl. Instr. and Meth.*, **83**, 131 (1971).
39. G. Mathé, *Nucl. Instr. and Meth.*, **39**, 356 (1966).
40. F. D. Brooks, *Nucl. Instr. and Meth.*, **4**, 151 (1959).
41. B. Sabbah and A. Suhami, *Nucl. Instr. and Meth.*, **58**, 103 (1968).
42. D. W. Jones, *IEEE Trans. Nucl. Sci.*, NS **15**, No. 3, 491 (1968).
43. W. Meiling and F. Stary, *Nanosecond Pulse Techniques*, Gordon and Breach, 1968.
44. R. N. St. Onge and J. A. Lockwood, *Nucl. Instr. and Meth.*, **69**, 25 (1969).
45. U. Tamm, W. Michaelis and P. Coussieu, *Nucl. Instr. and Meth.*, **48**, 301 (1967).
46. K. L. Swinth, L. D. Philipp and N. C. Hoitink, *IEEE Trans. Nucl. Sci.*, NS **15**, No. 3, 486 (1968).
47. C. A. J. Ammerlaan, R. F. Rumphorst and L. A. Ch. Koerts, *Nucl. Instr. and Meth.*, **22**, 189 (1963).
48. S. Kinbara and T. Kumahara, *Nucl. Instr. and Meth.*, **70**, 173 (1969).
49. M. Obu, T. Ichimori and K. Shirakata, *Nucl. Instr. and Meth.*, **89**, 131 (1970).

50. V. E. Church, *Nucl. Instr. and Meth.*, **102**, 305 (1972).
51. H. A. Cole, *Nucl. Instr. and Meth.*, **84**, 93 (1970).
52. J. Harms, *Nucl. Instr. and Meth.*, **83**, 221 (1970).
53. A. Barna and E. L. Cisneros, *Nucl. Instr. and Meth.*, **75**, 261 (1969).
54. M. Bertolaccini and S. Cova, *IEEE Trans. Nucl. Sci.*, NS **19**, No. 3, 423 (1972).
55. E. Gatti, C. Cottini, S. Donati, V. Svelto and F. Vaghi, *Energia Nucleare*, **17**, 34 (1970).
56. R. A. Winyard, J. E. Lutkin and G. W. McBeth, *Nucl. Instr. and Meth.*, **95**, 141 (1971).
57. R. A. Winyard and G. W. McBeth, *Nucl. Instr. and Meth.*, **98**, 525 (1972).
58. K. Frank, W. D. Emmerich, A. Hofman and F. Vogler, *Nucl. Instr. and Meth.*, **96**, 247 (1971).
59. M. Bormann, V. Schröder, W. Scobel and L. Wilde, *Nucl. Instr. and Meth.*, **98**, 613 (1972).
60. M. F. D'Cunka and G. Joseph, *Nucl. Instr. and Meth.*, **95**, 515 (1971).
61. J. B. Czirr, *Nucl. Instr. and Meth.*, **88**, 321 (1970).

7

Timing Circuits

7.1 Characteristics of timing circuits

7.1.1 Introduction

There are several instances when it is necessary to have time information about the detection of one or more nuclear particles. Examples are the detection of coincidences, or anticoincidences, in two detectors, or in the measurements of times of flight or of short-lived nuclear states. In all cases it is first necessary to use a time pick-off to generate a logic pulse which ideally is precisely related in time to the nuclear event. Meiling and Stary[1] have comprehensively reviewed techniques used in high speed timing applications.

In a practical system, the time of the logic pulse relative to the nuclear event will always show statistical variations, or 'time jitter', from event to event; the magnitude of this will depend on the properties of both the detector and the time pick-off. One important component of time jitter originates from the action of the pick-off in giving a dependency of the relative timing of the output pulse with the amplitude of the detector signals, the so-called 'time walk' effect which was discussed in Section 6.2. Pulse height discriminators have a large component of time walk and hence give a large jitter. For most applications, therefore, a pulse height discriminator makes an unsatisfactory substitute for a time pick-off.

7.1.2 Coincidence detection and time jitter

The effects of time jitter may best be elucidated by consideration of the simple coincidence arrangement of Figure 7.1. A true coincidence event is defined as an event when both particles originating from a single nuclear event are each detected by their respective detector. As discussed in Section 6.2 the coincidence circuit designed to register such true coincidences will have a finite resolving time τ, so that it recognizes a coincident event when presented with the pair of pulses spaced less than a time τ apart.

A certain number of accidental coincidences will be passed by the coincidence circuit. If N_A and N_B are the mean pulse rates from the two detectors, the accidental rate is given by $2N_AN_B\tau$ (see Appendix B.6); the observed rate at the output will be the sum of this and the true rate. The essential problem is that the accidental rate may have a comparable magnitude to the true rate, since a true coincidence is the outcome of the comparatively unlikely situation of both the two nuclear particles from a single event entering, and being detected by, their respective detectors. Although the observed rate may be corrected for the content of accidental coincidences to give the true rate, this limits the statistical accuracy obtainable in the calculated true rate (see Appendix B.6). It is therefore essential in most cases to make the accidental rate small by making τ as small as possible.

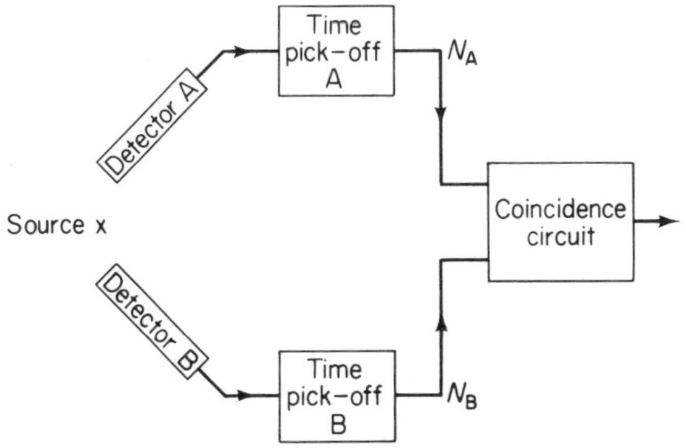

Figure 7.1 Simple coincidence arrangement

If, however, τ is made very small, inefficient collection of true coincidences occurs. Although the original nuclear particles may have entered their detectors simultaneously, owing to time jitter the logic pulses arriving at the coincidence circuit would not be simultaneous and may fail to be recognized as coincident if τ is short. The time jitter of the system is therefore the ultimate factor of performance of the system, since it determines how small a value τ may be efficiently used and, hence, how small an accidental rate may be obtained.

Confining our attention to true coincidences only, the time jitter may be characterized, as in Figure 7.2, by the time jitter distribution of the pulses from the time pick-off of one detector with respect to those from the other detector time pick-off. It may be noted in passing that the time fwhm of such a distribution ξ may be simply shown to be given by

$$\xi^2 = \xi_A^2 + \xi_B^2$$

where ξ_A and ξ_B are the fwhms of each of the two detector pick-off distributions (with respect to the original nuclear event), provided these distributions are of Gaussian shape. It will be observed that to obtain efficient recognition of most coincidences it is necessary to use a coincidence resolving time τ comparable to the fwhm ξ.

In practice a commonly used index of the performance of a system is a plot of counting rate versus relative delay. This is obtained by inserting various lengths of delay cable, prior to the coincidence circuit, in the signal path of one of the detectors shown in Figure 7.1. The resulting delay curve is

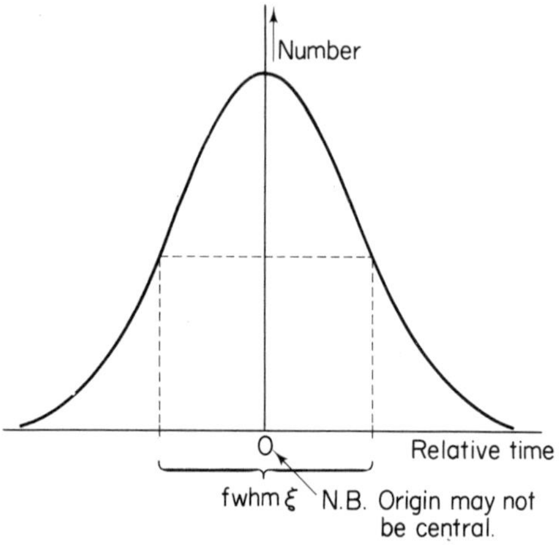

Figure 7.2 Time distribution of the output pulse at one time pick-off output, arising from a true coincidence, relative to the instant of the corresponding output pulse from the other time pick-off

dependent partly on the time jitter distribution of the pulses from the detector/time pick-offs and partly on the coincidence resolving time. Figure 7.3 shows delay curves obtained with various values of τ for the time distribution of Figure 7.2; it will be observed that the shape of the delay curve approximates to the time jitter distribution when τ is small. As noted earlier, τ should be comparable to the time jitter fwhm ξ for reasonably efficient registration of the true coincidences; in practical terms this may be sufficiently realized if τ is made rather less than that value which gives a flat topped delay curve. Most published descriptions of detector/time pick-off arrangements give only the delay curve (or the slope of the sides of the delay curve) as opposed to the time jitter distribution. This is not a disadvantage for the present purposes since, for a properly selected value of τ, giving no flat top to the delay curve, the two are similar and so the delay curve fwhm is usually an adequate factor of merit for the performance of a pair of detector/time pick-offs.

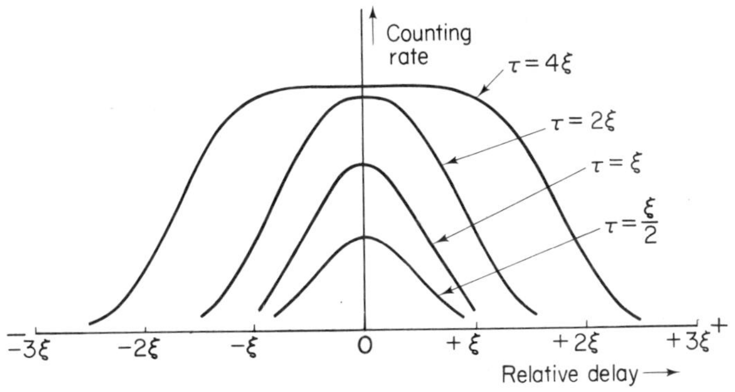

Figure 7.3 Typical delay curves for various values of the coincidence resolving time τ. ξ is the fwhm of the time distribution (see Figure 7.2)

It may here be noted that if we now include the contribution from accidental coincidences, the delay curves of Figure 7.3 become as in Figure 7.4. The accidental rate in an experiment may therefore be measured by placing sufficient delay cable in the signal path of one of the detectors so that all the true coincidence events are not recognized by the coincidence circuit. Since the delay cable attenuates the signal to some extent, and possibly changes the pulse shape, before using this method for accidental rate determination it is advisable to take a delay curve with a source emitting no true coincidences (or, alternatively with two separate sources each emitting particles into one

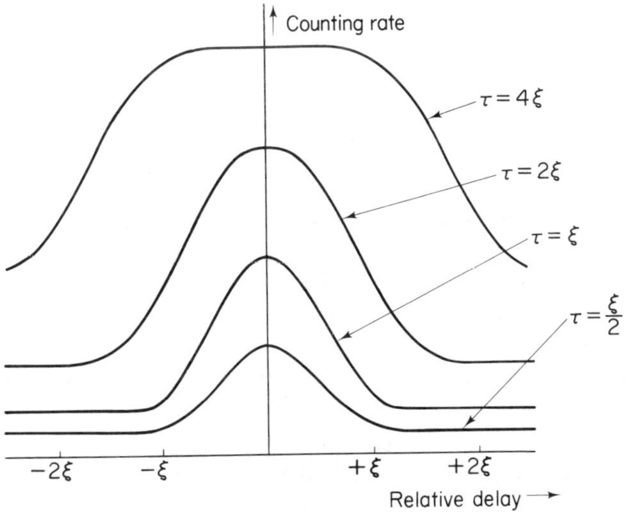

Figure 7.4 Delay curves as in Figure 7.3 but with the addition of accidental coincidences

detector only). Ideally the delay curve would then be flat, but if there is significant deviation from flatness a correction can be applied to the measured accidental rate in the experiment.

Where it is necessary to select the detector pulses on the basis of pulse height, in order to preserve the small time jitter attainable with time pick-offs the pulse height selection must be performed in parallel side channels or 'slow' channels, and some form of gating must be used which only delivers an output pulse if the event satisfies all the criteria of coincidence and pulse height. The arrangement of Figure 6.5, the slow–fast arrangement, achieves this and it may be noted that the timing of each input pulse to the coincidence circuit should derive from that of the corresponding time pick-off, i.e. the output of the time pick-off should 'strobe' the output from the discriminator.

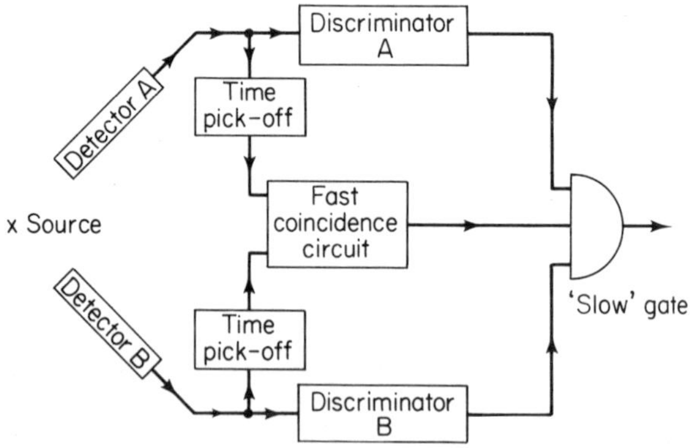

Figure 7.5 Fast–slow coincidence system

This will be the case in an AND gate if the strobing input pulse falls within the time duration of the other input. An alternative arrangement that is often used is the fast–slow arrangement of Figure 7.5. In both the fast–slow and the slow–fast arrangements there is a possibility of accidental coincidences at any of the gates, although usually only those occurring at the main fast coincidence circuit are important. Both the fast–slow and the slow–fast arrangements are widely used and De Volpi *et al.*[2] have compared the merits of the two arrangements.

Finally the arrangement discussed above may be alternatively synthesized using a linear gate. Unlike all the gating arrangements discussed so far which are intended to operate with logic pulses, a linear gate allows the transmission of a linear pulse (hopefully) undistorted during the presence of a logic gating pulse. Figure 7.6 shows a possible arrangement. Since the pulse height discrimination of one channel may be performed after the linear gate, the arrangement has the advantage of allowing a multichannel analyser to be

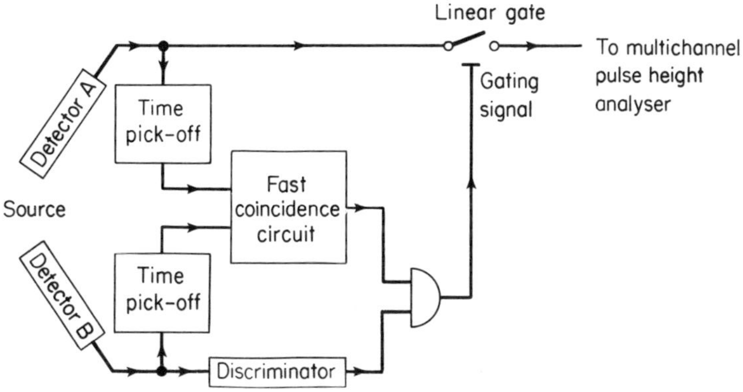

Figure 7.6 Fast coincidence selection with a pulse height discriminator in channel B and the use of a linear gate to allow the use of a multichannel analyser

used on the output so that the whole of the coincidence pulse height spectrum of one channel may be obtained simultaneously.

7.1.3 Other aspects of time jitter

The time jitter contribution of the detector will be dependent on several factors. In a Cerenkov counter the main source is the statistical variations in the electron transit times in the photomultiplier. In scintillation counters an additional component will come from the light emission mechanism; this will be small for fast scintillators, such as most organic scintillators, and larger for scintillators such as NaI(Tl). For semiconductor detectors the speed of the collection of the ionization will be the major factor and the time jitter will therefore be small for surface barrier types but larger for detectors with deep depletion layers such as Ge(Li) types. Unlike scintillation counters the shape of the detector pulse in a semiconductor detector is dependent on the position of the track of ionization in the detector. Other things being equal, the time jitter in any system will be less for high energy events, for two reasons: the detector signal shows less statistical variation, as it is made up from more charge carriers, and the relative contribution of noise is less.

Most pick-offs show a dependence of the timing of the output pulse on the amplitude of the input pulse and hence on the energy of the detected particle. This time walk may constitute a large component of the overall time jitter, depending on the ranges of pulse amplitude of interest. In a practical coincidence set up it is often desirable to count coincidences between, say, two particles of specified energies corresponding to two specific pulse height windows; this would be done by, for instance, the fast–slow arrangement of Figure 7.5 or the slow–fast system of Figure 6.5. In this case, only time walk corresponding to the variations of pulse heights within their respective windows need to be considered, since all other events are blocked by the slow gate(s).

In the case of measurements of lifetimes of short lived states whose birth and death is marked by the emission of particles, it is necessary to obtain a delay curve and compare it with that obtained with a prompt emitter. The half life of the delayed emitter can then be obtained either by comparing the difference in shape of the sides of the delay curves or alternatively by measuring the shift in the centroids of the two delay curves (see for example Meiling and Stary[1] or Schwarzschild).[3] For the measurement of the shortest life times it is important that the fwhm of the prompt delay curve should be small. With the centroid shift method, which is the preferred method for the shortest lifetimes, a problem arises in the choice of the prompt emitter. Since the prompt emitter would generally not emit particles of the same energy as the delayed emitter, and since the time pick-offs will have some degree of time walk, it follows that the observed centroid shift will contain an error due to time walk. This is a major source of error in the measurement of short lifetimes and it is therefore important that the time walk should be reduced to a minimum.

The time pick-offs to be described vary considerably in their time walk properties. For scintillation counters the zero crossing pick-offs show a very small time walk but derive the timing of the output pulse from a statistically uncertain part of the detector signal. On the other hand the leading edge pick-offs show a large time walk but derive the output signal from the leading edge which, for signals having a narrow range of amplitude, shows less variation in its time position. Since the time walk is dependent on the range of pulse amplitudes of interest, it is therefore difficult to compare the merits of these two types of time pick-off. The third main type of pick-off used with scintillation counters, the constant fraction of pulse height pick-off, in many ways combines the virtues of both the leading edge and zero crossing pick-offs; it gives a low time walk and yet derives the timing information from the statistically more certain part of the detector signal.

For semiconductor detectors the position is not as clear since the effects of variation in detector pulse rise time and of noise complicate the situation. The high energy resolution means that, in some cases, the pulses of interest can be confined to those of nearly constant amplitude; for such a case the time walk is of little account.

It may be noted that, for any detector and time pick-off, it is possible to compensate for the effects of time walk, and of rise time variations, if a time to amplitude converter is used. This is discussed in Subsection 7.7.3.

A further problem in the use of timing circuits concerns the effects of pulse pile up. If the pulse from a coincident event is closely preceded by another pulse the time pick-off will only 'see' the first pulse, since it will still be paralysed when the second pulse arrives. Even when the two pulses separately operate the time pick-off, the second pulse may fall on the tail of the first, or the pick-off may not be fully recovered; in all these cases the relative timing of the output pulse may be perturbed. These factors have the effect of distorting the delay curve and this may be important in delayed coincidence and time of flight measurements.[1,3,4,5] In this case it is often necessary to use some form of pile up rejector (Section 7.9).

7.2 Time pick-offs

7.2.1 Limiter type

The action of the limiter type pick-off is illustrated in Figure 7.7; the function is usually performed by causing a valve or transistor to cut off. The length of the output pulse from the limiter must usually be shortened and standardized before passing into the coincidence circuit; this is usually done by a

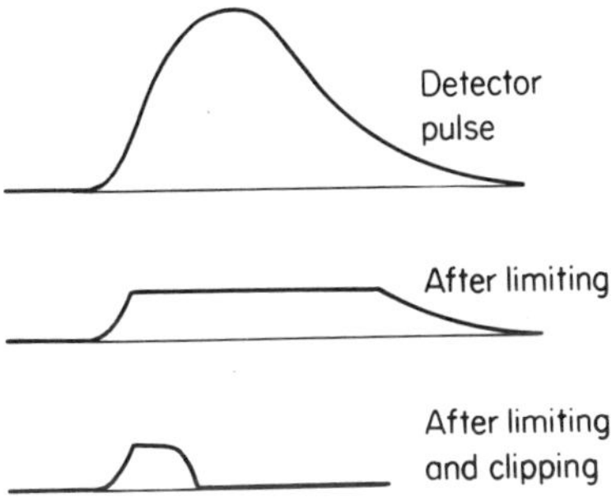

Figure 7.7 Action of a limiter type pick-off

delay cable shorted at one end (a 'clipping line'). Figure 7.8 shows the classic arrangement of Bell, Graham and Petch[6] which has been widely used and which combines the functions of limiting, clipping and coincidence selection in a very simple circuit. The high slope pentodes, which are quiescently conducting, are cut off by the first part of the signal from their respective detector. The anode load R of each is made to match the characteristic impedance of the cables A and B. If detector A delivers a pulse to its pentode, the limited pulse travels down cable A to the junction J where it is partly reflected back, partly propagated down cable B and partly down the clipping line. The parts reflected back down cable A and propagated down cable B are of no further interest as they are absorbed when they reach the anode loads, since these are perfect terminations. The signal which is propagated down the clipping line is reflected at the shorted end with a reversal of polarity and arrives back at junction J. Since the clipping line has a characteristic

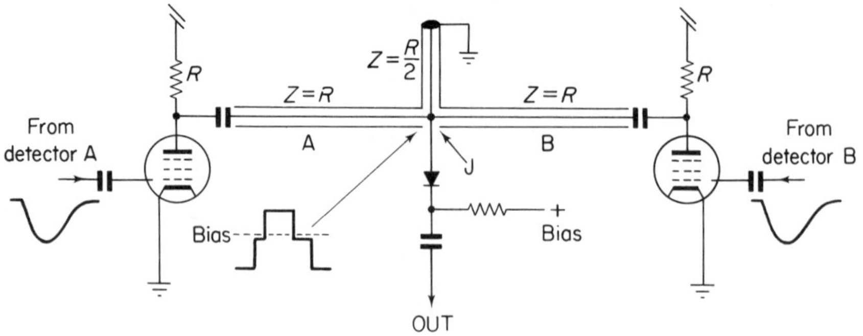

Figure 7.8 Combined limiter-type pick-off and coincidence circuit of Bell, Graham and Petch[6]

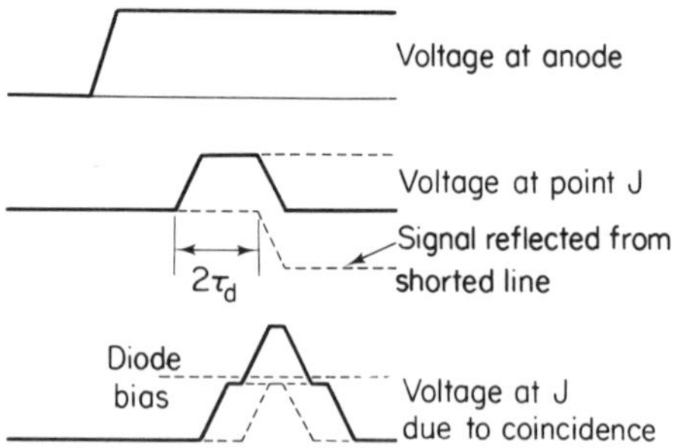

Figure 7.9 Voltage waveforms in the circuit of Bell, Graham and Petch[6] shown in Figure 7.8

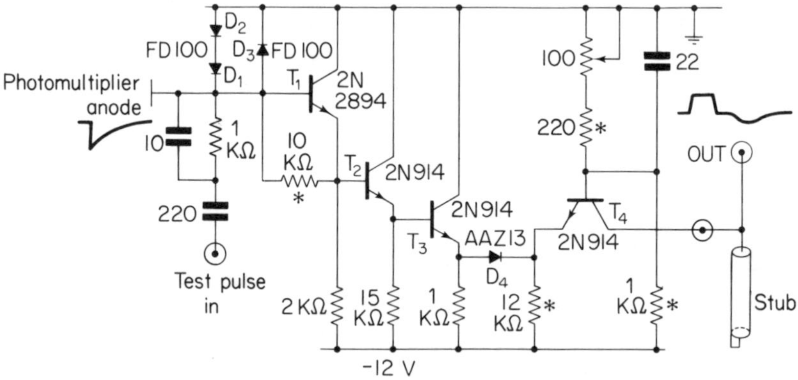

Figure 7.10 Time pick-off of Bertolaccini, Cova and Bussolati[7]

impedance of $R/2$, the signal encounters a perfect match with the cables A and B and so divides down A and B with no reflection. The voltage at J, due to this signal, therefore cancels out the remaining flat part of the incoming signal, as shown in Figure 7.9. The resultant clipped pulse at J therefore has a width $2\tau_d$, where τ_d is the propagation delay (one way) of the clipping cable.

Coincidence selection is performed by the reverse biased diode. Two coincident signals cause the diode to be lifted into conduction and it will be seen from Figure 7.9 that the coincidence resolving time is given by $2\tau_d$ approximately. Since the diode has a small capacitive feedthrough for subthreshold pulses, it is desirable to follow the diode with a secondary discriminator which also serves to shape the output pulse into a logic pulse. It is not necessary for the circuitry following the diode to have a fast response, since considerable pile up of the subthreshold feed-through pulses can be tolerated before these become comparable to the magnitude of the true coincidence pulses.

Ideally the limiter should deliver an output pulse with a steeply rising front. This requires minimal integration of the detector signal on the input capacitance and the output capacitance of the limiter, and the use of a high gain valve. More recent variations of the limiter have employed transistors in various modes. The common base mode gives minimum integration of the signal, but has a low gain, whereas the common emitter and emitter-follower modes give better gain at the expense of a higher input capacitance.

A typical arrangement using transistor limiting was described by Bertolaccini, Cova and Bussolati,[7] and is shown in Figure 7.10, where limiting takes place on both T_1 and T_4. With two NaI(Tl) scintillation counters the delay curve had a fwhm of $\sim$6 ns when used in a fast–slow system with the slow channel discriminator windows set on the 511 keV photopeaks of ^{22}Na.

One disadvantage of the limiter type circuits is that they do not discriminate absolutely against noise; it is therefore necessary to arrange the following circuitry to be insensitive to noise pulses. Another disadvantage is that the output pulse shape from a limiter is dependent on the input pulse shape and magnitude so that the optimum adjustment of the coincidence circuit is dependent on the input pulse characteristics. If the two detectors are dissimilar, the resulting difference in detector pulse rise times can be troublesome and may necessitate a large clipping time and a critical adjustment of the coincidence circuit. The delay curve in this instance is often found to be assymetric.

7.2.2 Leading edge trigger

In this case, when the front edge of the detector signal reaches a certain level it causes a trigger circuit to operate. The output signal rise time is then determined primarily by the characteristics of the trigger circuit, and, provided the triggering level is set high enough, the circuit is insensitive to noise. The length of the output signal should be short enough to operate the coincidence circuit, and this is usually arranged by using a monostable trigger circuit and clipping the output with a shorted delay line. To eliminate a slight dependence of output pulse shape on input pulse shape, some designs use a second trigger circuit after the first. Alternatively the same result can be achieved by approximately standardizing the detector pulse by limiting and clipping before passing it on to the trigger circuit.

The timing of a front edge trigger inevitably shows some time walk. The situation depicted in Figure 6.3 is an idealization of the time walk for a trigger which responds instantaneously. A practical trigger will show a delay before triggering, which is partly dependent on the rate of rise and the magnitude of the input signal, so that to some extent it behaves as a charge controlled device. To minimize the contribution of this variable trigger delay to the overall time walk, interest has centred almost exclusively on the two types of device, avalanche transistors and tunnel diodes, which have short trigger delays.

An example of an avalance transistor time pick off is given in the design of Grabari.[8] Using NaI(Tl) scintillation counters, each output was first limited with a common base stage and then passed to the avalanche stage. In the fast–slow system using a ^{22}Na source, and with the discriminator windows spanning the 511 keV photopeaks, he obtained a delay curve fwhm of 1·1 ns.

Tunnel diode time pick-offs have been widely used. In principle many of the tunnel diode discriminator circuits already described may be used as time pick-offs. For instance Clausen and Nainan,[9] using a tunnel diode monostable, obtained a fwhm of 1 ns when used with plastic scintillation counters on a ^{60}Co source, and the upper fourth of the pulse height spectrum was accepted. As with many other designs the detector pulse was pre-limited before passing it to the tunnel diode circuit.

Several authors have considered the switching behaviour of the tunnel diode as a charge controlled device. In particular Grunberg and Tepper[10] have used a snap-off diode to generate the pulses for a tunnel diode coincidence circuit. The snap-off diode was normally conducting, and an input pulse caused it to turn off. The resultant pulse of stored charge is then released very rapidly and its magnitude is dependent primarily on the forward current magnitude before turn off. The resultant pulse of charge was passed on to a tunnel diode coincidence circuit. Using inputs supplied with a pulser it was found that a 10:1 change in pulse amplitude in one detector caused the centre of the delay curve to shift by only 160 ps. With two plastic scintillation counters and a ^{60}Co source, arranged in a fast–slow configuration accepting ranges of pulse heights of 10:1, a delay curve fwhm of 0·8 ns was obtained.

Tamminen and Jauho[11] have used a snap-off diode as a preliminary to a tunnel diode trigger so that the tunnel diode is presented with standardized pulses and so triggers with consistent delay. With plastic scintillator counters on ^{60}Co they obtain a fwhm of 210 ps with the side channel discriminator windows set at 10% at 900 keV.

7.2.3 *Zero crossing type*

If the detector signal is converted into a bipolar signal by a linear circuit, such as a CR differentiator or by double delay lines, then the timing of the point of zero crossing will be invariant to changes in the input pulse amplitude. If an ideal trigger circuit is arranged to trigger at the zero crossing, the time pick-off would be walk free. A practical circuit is not perfectly walk free, since there will be a delay in triggering, which will be dependent on the steepness of the waveform near the zero crossing point and hence on the amplitude. Although this type of time pick-off has found wide application, it has a number of disadvantages. If the detector pulse shape varies, as in the case of a semiconductor detector, the time position of the zero crossing will vary (Section 7.4). In the case of a scintillation counter, for pulses of a small range of amplitude, the statistics of the detector pulse shape are such that zero crossing timing gives a larger time jitter than leading edge timing (see Section 7.3). The advantages of zero crossing timing are that it can be simply set up for a wide range of pulse amplitudes and that, on account of its low time walk, pulses of widely varying amplitude can be processed.

Most zero crossing time pick-offs use as a trigger element a tunnel diode which is switched from the low to the high voltage state at the point of zero crossing, since switching in this direction is more rapid than in the reverse direction. The tunnel diode is usually arranged in a monostable configuration as, for example, in the circuit of Whittaker.[12] With this circuit a change in pulse amplitude of 125:1 in one of the detectors caused a time walk of 200 ps. With a pair of organic scintillation counters detecting the products of the reaction $T(D, n)^4He$ (Q value + 17·6 MeV) the delay curve had a fwhm of 1·70 ns.

A different arrangement was used by Metz[13] whereby the tunnel diode was provided with a non-linear load line and was quiescently in the high voltage state. The first part of the signal switched the tunnel diode to the low voltage state and at the zero crossing the tunnel diode switched back again. The time information was therefore contained in the trailing edge of the output pulse which, after clipping, operated a further tunnel diode monostable. Using a light pulser, an input pulse amplitude variation of 70:1 in one detector caused a time walk of <200 ps; organic scintillation detectors gave a fwhm of 0·822 ns with a ^{22}Na source and 0·548 ns with a ^{60}Co source where, in both cases, practically the whole (>6 keV) of the pulse height spectra were accepted.

7.2.4 *Constant fraction of pulse height pick-off*

Although in the literature this type of time pick-off is occasionally termed zero crossing, the term 'zero crossing' will be used here to apply to the methods of the previous section which utilize a doubly differentiated signal. Constant fraction of pulse height timing has, to some extent, the advantages of both leading edge and zero crossing timing in that it is nominally walk free but utilizes the statistically more significant information of the leading edge of the pulse. As used by Gedcke and McDonald[14, 15] the principle of the method is illustrated in Figure 7.11. The detector pulse is delayed and a fraction of the

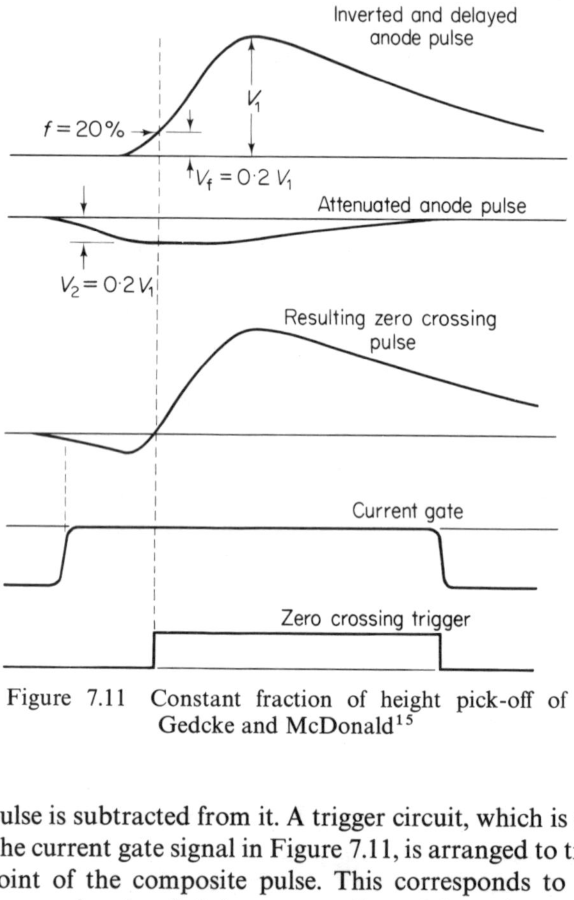

Figure 7.11 Constant fraction of height pick-off of Gedcke and McDonald[15]

undelayed pulse is subtracted from it. A trigger circuit, which is enabled (i.e. primed) by the current gate signal in Figure 7.11, is arranged to trigger at the crossover point of the composite pulse. This corresponds to sensing the arrival of the same fraction *f* of charge regardless of the pulse amplitude, and is therefore nominally walk free. The factor *f* can be chosen to give minimum time jitter. Earlier results by McDonald and Gedcke[14] on organic scintillators had shown that the optimum value of *f* was in the region of 10–15 % and some of these results are given in the following section. The circuit later used by Gedcke and McDonald[15] used a tunnel diode as trigger element and showed a time walk of ±120 ps for a pulse height variation of 100:1 from a light pulser.

The same principle was used by Maier and Sperr,[17] but using an integrated circuit differential amplifier to form the difference of the two pulses illustrated in Figure 7.11. Using a plastic scintillation counter on ^{60}Co and with the side channel discriminator windows set to accept 1 MeV $\pm 20\%$, they obtained a delay curve fwhm of 200 ps.

Karlsson[18,68] has used a rather different approach. The basic trigger element was a tunnel diode whose biasing was adjusted differently for each input pulse, so as to cause triggering at a constant fraction of pulse height. The biasing was derived from the integrated detector pulse, and the arrival of the main pulse was suitably delayed to allow time for the biasing to be established. As mentioned earlier, any trigger element behaves to some extent as a charge controlled device so that the instant of triggering depends on the rate of rise of the trigger voltage. This means that the tunnel diode would not trigger at a true constant fraction of pulse height and would show a slight time walk effect. In the arrangement of Karlsson this residual time walk was partly compensated out by using a non-linear relationship between the pulse amplitude and the biasing voltage. In this way it was possible to accommodate a tenfold change in pulse amplitude with a time walk of less than 200 ps Abbattista *et al.*[71] have also described an arrangement using the same principle.

7.3 Timing for scintillation counters

A certain amount of experimental and theoretical work has been done on the timing characteristics of various scintillator–photomultiplier combinations. The most relevant work for the present purposes has been the measurements of the time jitter fwhm of the instant when the charge on the collector reaches a fraction f of the total charge. This is roughly the instant at which the time pick-off circuits already discussed respond, since the detector current signal is integrated on the stray capacitance. Thus a 50 ohm load and a capacitance of 5 pF, due to the tube anode and the time pick-off input capacitance, act as an integrator of time constant 0·25 ns, which is comparable to the time taken for the signal to reach triggering level.

The form of the curves of time jitter fwhm against f all show a minimum for a small value of f for any type of scintillator, and this is also the case when the tube is coupled to a light pulser. For organic scintillators, the values of f giving minimum time jitter appear to be $\sim 10\%$, and a representative plot from McDonald and Gedcke[16] for NE 213 liquid scintillator and an XP 1040 photomultiplier tube are shown in Figure 7.12. These curves were obtained by irradiating the scintillator with a brief pulse of neutrons and give the fwhm of the delay curve of the output of the time pick-off with respect to the neutron pulse. Each curve corresponds to a narrow range of pulse amplitudes, so that the effects of time walk should be negligible.

For NaI(Tl) and a 56 AVP tube, Braunsfurth and Körner[19] found that the optimum values of f were much smaller, $<1\%$, and for a NaI crystal cooled to 80° K they were $\sim 8\%$. Using leading edge pick-offs in a coincidence arrangement with narrow windows centred on 1250 keV with a ^{60}Co source, NaI(Tl) crystals gave a delay curve fwhm of 0·80 ns. The use of a NaI crystal cooled to 80° K improved this value by 35 %.

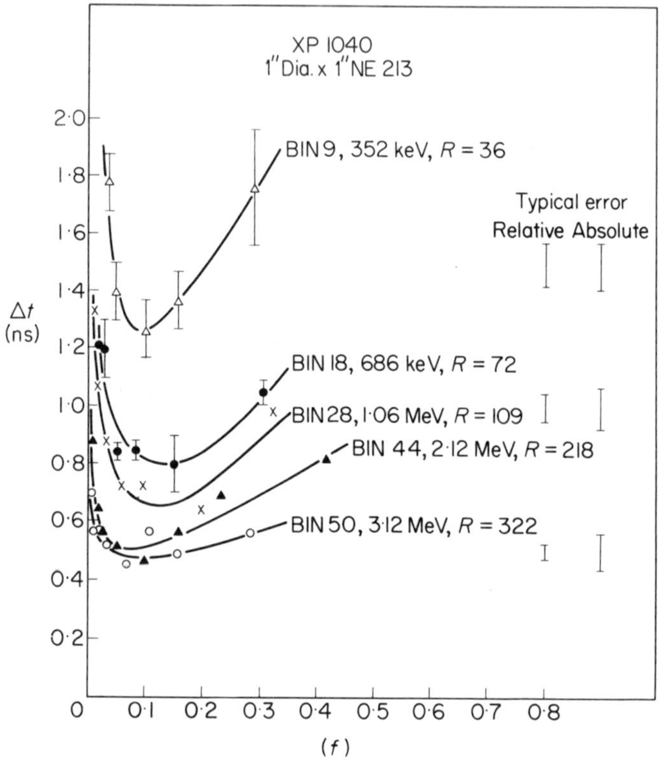

Figure 7.12 The fwhm time resolution Δt for a XP 1040 photomultiplier and NE 213 scintillator plotted against f, the fractional triggering level, according to McDonald and Gedcke[16]

A certain amount of theoretical work has been done on the timing performance of scintillation counters, and this has been reviewed by Donati, Gatti and Svelto[20] and by Bengston and Moszyński.[21] Until recently there was considerable disagreement between the predictions of the theory and experiment. In particular, the curves of fwhm against f, of the type shown in Figure 7.12, were predicted to have different minima and different energy dependences from those observed. Many of the difficulties have been resolved by the work of Bengston and Moszyński.[21] They used two XP 1021 photomultiplier tubes with NE 111 or Naton 136 scintillators in a coincidence arrangement with leading edge time pick-offs. By confining the pulse heights to a narrow range by side electronics, and by varying the threshold of one of the pick-offs, they could construct the fwhm against f curves. They showed that the dependence of the fwhm on the energy E was correctly predicted by the theory, provided the window width of the side electronics was adjusted to be proportional to $\sqrt{E}$, so as to allow acceptance of the same fraction of the distribution of photoelectrons from monoenergetic nuclear particles. They also showed that the height of the scintillator, and the area of the photocathode which was illuminated, had a profound effect on the magnitude of the fwhm and on the shape of the fwhm against f curves. This led them to amend the earlier theoretical models by the inclusion of a simple statistical model of the light emission characteristics of the scintillator. It was then possible to fit the theory to the experimental data in a self-consistent way.

Of the two scintillators examined, NE 111 had a much faster response and the fwhm obtained was limited essentially by the phototube. For Naton 136, the light spread in the scintillator was essentially the limiting factor. The authors also investigated a constant fraction of pulse height time pick-off and found that it was possible to reduce the fwhm by 30% of that obtained with a leading edge pick-off when a window of 20% was used. It therefore appears that, even with quite narrow window widths, time walk in the leading edge pick-off is still a significant factor and the use of a constant fraction pick-off can effect a considerable improvement. A fwhm of 132 ps for ^{60}Co with a window width of 15% at 930 keV was obtained with NE 111. Bialkowski and Moszyński[69] have extended this work to include the use of NE 111 with various types of photomultiplier tube.

In the work of Present and Scarl[22] two photomultipliers were arranged to view Cerenkov light flashes, so that each effectively received no more than one photon from the common source. They investigated the use of photomultiplier tubes with GaP first dynodes which gave a multiplication factor of the order of 50 as against 4–5 for more conventional dynodes. When used with a constant fraction pick-off set to 20%, the delay curve fwhm was reduced from 1·5 ns to 0·9 ns by the use of the tubes with GaP dynodes. It therefore appears that in this case the sampling of the greater number of paths between the first and the second dynodes had a significant effect. Krall and Persyk[70] have reviewed recent work on fast photomultipliers using GaP(Cs) dynodes. They suggest that the better timing performance of such tubes may result partially from the 'thermalization' of the emitted secondary electrons, so that when they arrive on the next dynode they have a nearly uniform velocity distribution.

The critical dependence of the fwhm on f explains the variability of the delay curve fwhms obtained in the earlier work. In particular it shows that zero crossing timing using a doubly differentiated signal cannot give as small a time jitter as leading edge timing, if the effects of time walk are not considered. For the case of double delay line differentiation, this may be seen from Figure 7.13 which shows the impulse response and the detector current pulse. The signal after differentiation may be obtained by evaluating the convolution integral (Subsection 2.2.2) between the impulse response and detector current. The point of zero crossing, i.e. when the convolution integral is zero, is depicted in Figure 7.13(c), and occurs at an instant which, apart

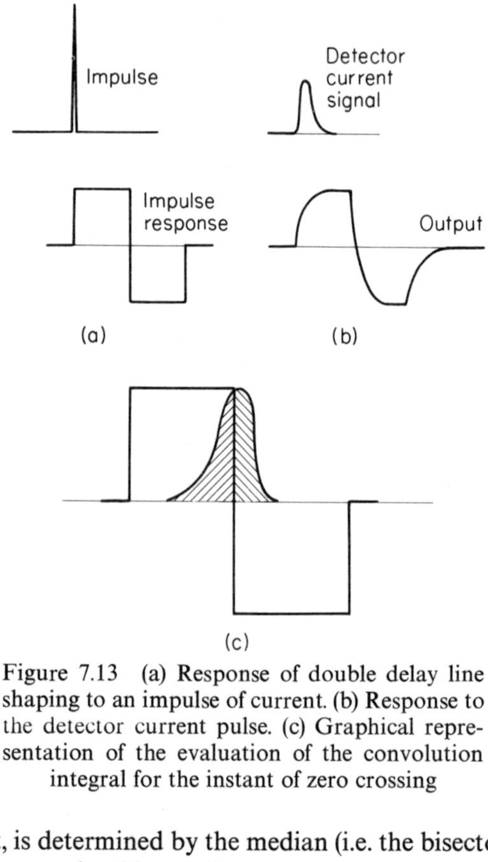

Figure 7.13 (a) Response of double delay line shaping to an impulse of current. (b) Response to the detector current pulse. (c) Graphical representation of the evaluation of the convolution integral for the instant of zero crossing

from a constant, is determined by the median (i.e. the bisector of the area) of the detector current pulse. The median of the current pulse corresponds to the instant at which $f = \frac{1}{2}$ and so the timing derives from the equivalent of $f = \frac{1}{2}$. Since the optimum values of f are always $\gtrsim 15\%$, it is clear that zero crossing timing of such delay line shaped signals cannot attain the ultimate accuracy of other methods when time walk is not considered. A similar conclusion seems to hold for CR differentiated signals. If the impulse response in the region of zero crossing approximates to a straight line, the evaluation of the convolution integral, to give the instant of zero crossing for the signal (Figure 7.14), indicates that the timing then derives from the centroid of the detector current pulse, and this also might be expected to give inferior timing to the methods using small values of f.

At the moment, the constant fraction of pulse height method should yield better results since the value of f can be chosen to be optimum and it is walk free when used with an ideal trigger (i.e. one that responds instantaneously to the true zero crossing level). However, any method relying on detection of a fraction of the total charge does not use all the time information in the detector pulse and it should be possible, in theory at any rate, to design a better system.

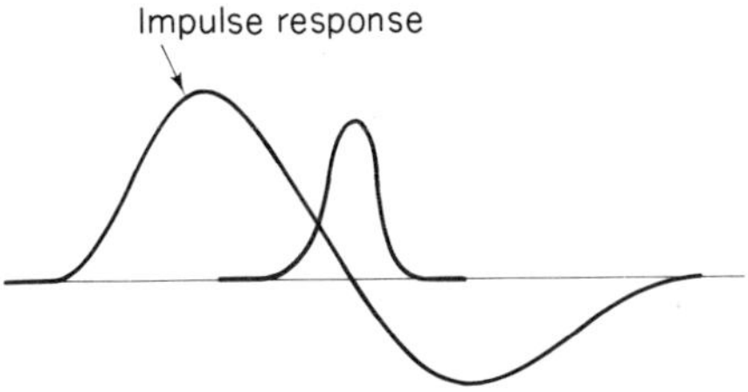

Figure 7.14 Impulse response for CR differentiation. The detector current signal is shown reversed in time and superimposed on the impulse response to illustrate the evaluation of the zero crossing instant

Finally it may be noted that while the stopping time of a nuclear particle in a scintillator is brief (10^{-11} to 10^{-12} seconds), in the case of a large scintillator the variations in transit time of the particle and of the light emitted may be the cause of some time jitter (e.g. for light over 10 cm in vacuo, transit time $= \frac{1}{3}$ ns).

7.4 Timing for semiconductor detectors

The derivation of a timing signal from a semiconductor detector differs in at least two respects from the case of the scintillation counter. Firstly, the shape of the detector pulse is dependent on the initial distribution of ionization in the active volume and, secondly, the contribution of noise is often important. For the surface barrier types, the collection time is of the order of 1 ns and such detectors give a timing performance similar to fast scintillation counters. With lithium drifted types the collection time may be of the order of 100 ns and it is with this type that most effort has been expended in deriving precise timing information.

Figure 7.15 shows the voltage pulse induced on the collector for the initial ionization at alternative point locations. The pulses are constructed on the assumption of equal electron and hole mobilities and for uniform electrical field. Since for gamma rays over 200 keV most of the events derive from Compton interactions, and may therefore be multiple interactions, and the range of both Compton and photoelectrons is not negligible compared to the active region, the pulse shapes encountered in practice will not be as simple. It does follow, however, that if a timing signal is derived when the leading edge of the pulse at an instant corresponding to a fraction f of the

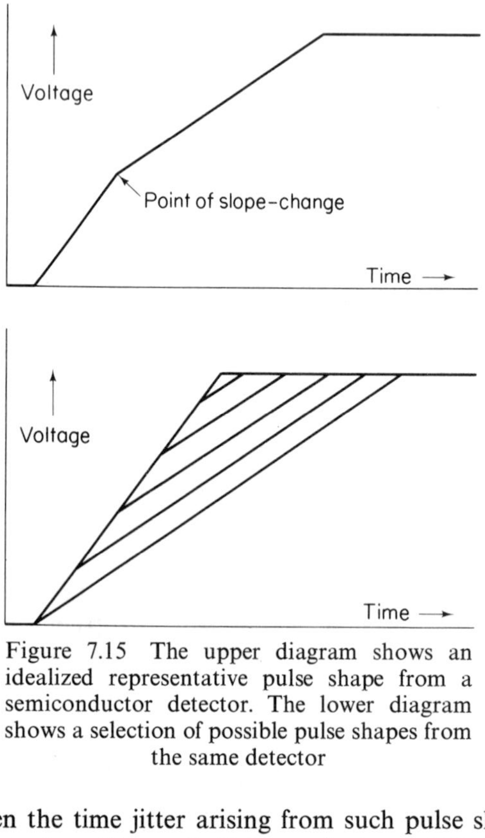

Figure 7.15 The upper diagram shows an idealized representative pulse shape from a semiconductor detector. The lower diagram shows a selection of possible pulse shapes from the same detector

total charge, then the time jitter arising from such pulse shape variations would be smallest when f is small. This conclusion, which has ignored the statistical nature of the ionization process, is borne out by the work of Strauss, Larsen and Sifter[23] who, with a 1 cm planar Ge(Li) detector, examined the pulse shape variations from the detection of the 662 keV γ of ^{137}Cs. Their results indicated that a timing signal from $f = \frac{1}{2}$ would give a time jitter fwhm of 30 ns, whereas for $f = 5\%$ the fwhm would be 5 ns. It follows that zero crossing timing of the type which is equivalent to $f = \frac{1}{2}$ will yield very poor results.

A further problem concerns the positioning of the time pick-off circuit. Although ideally the time pick-off would derive its input directly from the detector itself, it appears that this to some extent degrades the energy resolution when used in a spectrometry system. The reason for this is that, in order to obtain maximum signal to noise ratios in a spectrometry system, it was seen in Subsection 5.3.1 that the resistance at the collector of the detector should be $\gtrsim 10^9$ ohms. Even where it is possible to maintain such a high resistance, the time pick-off circuit itself will introduce some excess noise at the input. Williams and Biggerstaff[24] have used a transformer arrangement at the input of the charge sensitive preamplifier to give the fast timing output. Other workers[25, 26] have used two FET preamplifiers in parallel, one the main charge sensitive preamplifier, and the other to deliver the fast timing pulse. For a Si(Li) detector, Vogler *et al.*[76] have used a normal charge sensitive preamplifier connected to one detector lead and have used a current sensitive amplifier connected to the other detector lead to furnish the timing signal. However most workers have used the output of the charge sensitive preamplifier itself to operate the time pick-off. The rise time of the signal is somewhat degraded due to the finite preamplifier rise time (10–100 ns currently), but it is not yet clear what effect this has on timing performance. A compromise solution was used by Sherman and Roddick.[27] They used a charge sensitive preamplifier, from which a fast timing signal could be taken at a point early in the preamplifier. Another compromise solution was used by Millard, Blalock and Hill.[72] The normal charge sensitive configuration was modified so that the 'integrating' action of the feedback capacitor was cancelled by the 'differentiating' action of a later circuit element.

All the work to be described was done using a Ge(Li) detector in coincidence with a fast scintillation counter and it may be assumed that the time jitter of the scintillation counter is small, so that the delay curve fwhms are due essentially to the time jitter of the Ge(Li) detector.

The work of Graham, MacKenzie and Ewan[28] illustrates some of the variables. With a limiter type pick-off limiting at $f = 10\%$, located after a charge sensitive amplifier of rise time 30 ns, and with the slow channel set to accept the photopeak of the 511 keV gamma ray of ^{22}Na, a planar Ge(Li) detector (5 mm at 1000 V) gave a fwhm of 7 ns. A coaxial Ge(Li) detector (radial depth 8 mm at 1500 V) gave a fwhm of 8 ns when the radiation was collimated into the coaxial region and 35 ns when it was collimated in the closed end of the detector. For the coaxial region, but with limiting set at 50%, the fwhm was 30 ns. These values are generally in accord with the observations of many other workers. It is a common fact that the best timing characteristics are obtainable with detectors of planar geometry. Of the non-planar geometries, the true right circular cylindrical geometry appears to be better than the closed end cylindrical geometry or the trapezoidal coaxial geometry. These differences are mostly accounted for by the differering degrees of variation of the charge collection times due to the different degrees of inhomogeneity of the electric field.

In distinction to the scintillation counter, the fact that the signal from a Ge(Li) detector has a leading edge of variable slope, as illustrated by the variable position of the point of slope-change in the idealized case of Figure 7.15, has stimulated development of time pick-offs which are invariant to this effect. Fouan and Passerieux[29] used a technique shown in Figure 7.16, which sensed the time instants t_1 and t_2 at which the detector signal crossed the threshold levels E and $2E$ respectively. The value of $t_2 - t_1$ was then, in

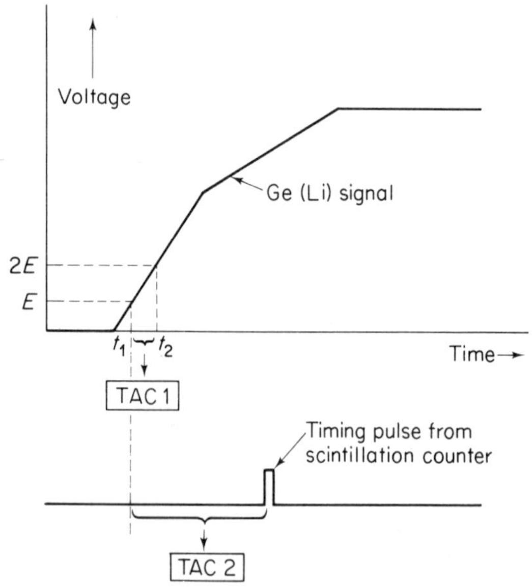

Figure 7.16 The pulse shape from a semiconductor detector showing the principle of the extrapolated leading edge timing method of Fouan and Passerieux[29]

essence, subtracted from the time t_1 so as to extrapolate to the instant of the beginning of the pulse. This system should therefore be invariant both to changes in pulse amplitude and to changes in the initial slope of the signal. The system was implemented by using two time to amplitude converters (TACs). TAC 1 delivered an output pulse of amplitude proportional to the time interval $t_2 - t_1$. The main TAC, TAC 2, responded to the time interval between t_1 and the timing signal from the other detector (a NaI(Tl) scintillation counter). The equivalent of the extrapolation procedure was performed by adding the output of TAC 1 to that of TAC 2.

It will be apparent that this technique of extrapolated leading edge timing is inaccurate if the detector signal undergoes a slope change before the level $2E$ is reached. For this reason $2E$ should ideally be small and a value the equivalent of 180 keV was used. With a 25 cm^3 coaxial Ge(Li) detector used to detect the γ rays of ^{60}Co (1·17 MeV and 1·33 MeV), all events having an equivalent energy of over 200 keV were accepted. The fwhm obtained was 7 ns as against 12 ns with TAC 1 disconnected. In the absence of TAC 1 the delay curve showed an extended tail on the 'late' side. Inclusion of TAC 1 almost completely eliminated this and yielded a nearly symmetrical delay curve.

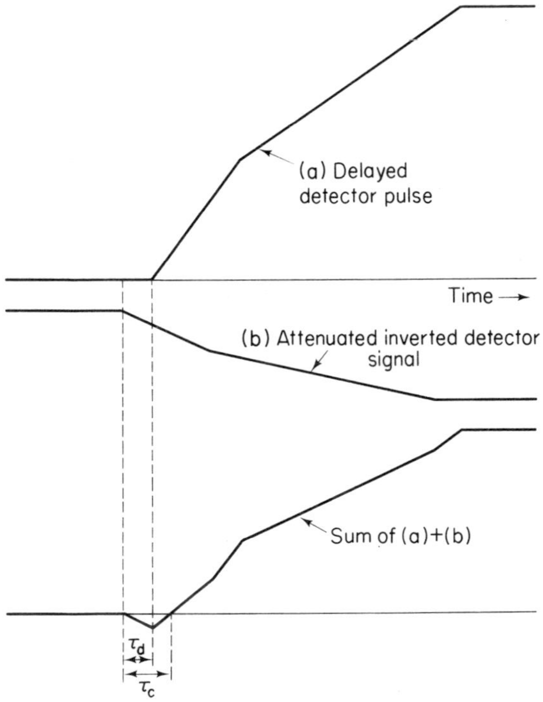

Figure 7.17 The principle of the amplitude and rise time compensated (ARC) method of Chase[30]

Chase[30] has described a time pick-off which incorporates both amplitude and rise time compensation (ARC). The technique resembles that of the constant fraction of pulse height pick-off of Gedcke and McDonald.[15] As shown in Figure 7.17, the detector pulse is delayed by an amount τ_d and is summed with the attenuated, inverted but undelayed pulse. The timing signal is derived from the zero crossing of the composite pulse. Such a timing should be invariant to pulse amplitude; it should also be invariant to changes of initial slope of the detector waveform provided the point of slope-change occurs at least a time τ_c after the pulse beginning. In order to make τ_c as small as possible the delay τ_d was chosen to be only large enough (3–10 ns) that the signal had risen above the noise by this time.

The inverted pulse was attenuated by a factor of 3 compared to the delayed pulse. It will be observed that, because of the small value of τ_d, the timing signal does not derive from the arrival of a certain fraction of detector pulse charge, and in this respect it differs from the system of Gedcke and McDonald. In essence the system is equivalent to sampling two instants on the detector waveform, spaced a time τ_d apart, and extrapolating to give the instant of the beginning of the pulse. In this light it may be seen as a variant of the system of Fouan and Passerieux described above.

With a 8 cm^3 planar detector, and with the pulse height window set to accept the photopeaks of either the 1·37 MeV or 2·75 MeV γ ray of ^{24}Na, the system gave a fwhm of 1·2 ns or 0·8 ns respectively. With a 30 cm^3 coaxial detector the 2·75 MeV photopeak gave a fwhm of 1·1 ns.

In a later modification of Chase's ARC system, Cho and Chase[73] extended the lower energy limit at which the system could be used. The original system used a true zero crossing detector which, in order to avoid spurious triggering by noise, was only primed (i.e. enabled) during the period τ_c in Figure 7.17. For small values of τ_c (and hence of τ_d) this priming became unreliable for low energy events. The modification consisted of using a second ARC pick-off, operating with a larger τ_d ($\sim$10 ns), to generate the priming signal for the main ARC pick-off. The main ARC pick-off could then be used with the small value of τ_d ($\sim$3 ns) desirable for the reasons outlined earlier. This dual primed system could then operate reliably with events of energies down to 20 keV, as against 200 keV for the original single primed system.

An alternative method of extending the low energy limit of Chase's original ARC method was described by Akimov *et al.*[74] In this case the zero crossing detector was a NAND gate and spurious outputs from this arising from noise were suppressed by a low pass filter. It appears that this method gives comparable results to that of the dual primed method of Cho and Chase.

Since the accuracy of any time pick-off appears to be highest when the early part of the detector waveform is used, the contribution of noise is important. Douglass and Williams[31] have therefore given some consideration to the use of different filters placed before the time pick-off with the object of reducing the effect of noise. They have also[32] experimentally compared some of these filters together with the ARC system of Chase. Their results indicated that the ARC system gave fwhms a factor of two smaller than those of the other systems tested.

At this point it is useful to attempt to consider the relative merits of some of the time pick-off systems used with Ge(Li) detectors. The basis for this is provided by the extensive theoretical and experimental studies by Cho and Chase.[75,77] We will assume that pulses of constant amplitude are selected and will consider the effect of changes of slope of the detector signal. For the idealized planar detector of Figure 7.15, the point of slope-change may occur anywhere during the leading edge of the signal with approximately equal probability. A simple leading edge (LE) pick-off adjusted to have a low threshold would then deliver an accurate timing signal for the majority of events. The waveforms of a small proportion of events would, however, have slope changes before the threshold, and the timing signals would then be late. This will cause a broadening of the delay curve on the late side, giving the so-called late tailing effect. The extrapolated leading edge (ELE) method, provided the point of slope change does not occur below $2E$ (defined in Figure 7.16), would respond correctly. However the minority of events with slope changes before the $2E$ level would cause the system to respond early, so causing some

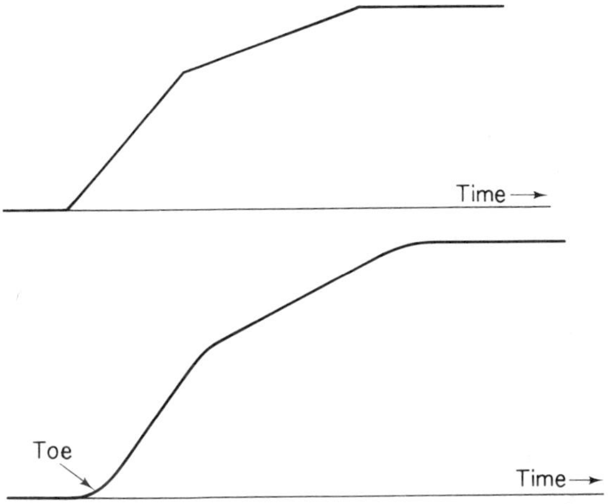

Figure 7.18 The idealized pulse shape (upper diagram) of a semiconductor detector and the modified shape (lower diagram) as observed at the preamplifier output

early tailing on the delay curve. Finally the amplitude and rise time compensation (ARC) system would respond correctly provided no slope-change occurred within the time interval τ_c (defined in Figure 7.17) of the pulse beginning. If the point of slope-change occurs before τ_c, the system will respond early, so causing early tailing on the delay curve.

The pulse shapes encountered in practice will differ from those of Figure 7.15 in several respects. As already mentioned, the fact that the ionization does not originate from a point will mean that the pulse shapes will have more rounded contours than those of Figure 7.15. The finite rise time of the preamplifier will further degrade the pulse shape; in particular the pulses will have a 'toe' region as shown in the example of Figure 7.18. This particular aspect was considered in detail by Cho and Chase, as it has a profound effect on the timing accuracy of any method deriving information from the early

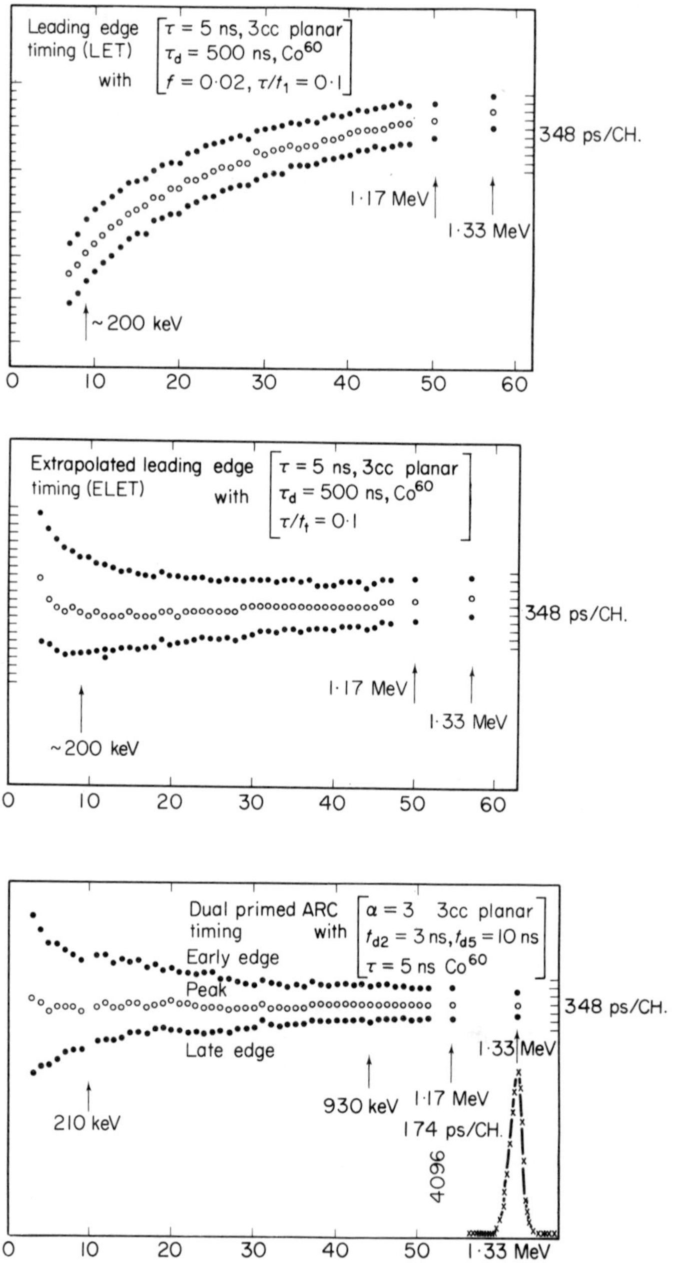

Figure 7.19 Timing performance of a 3 cm^3 planar Ge(Li) detector used with various timing systems, according to Cho and Chase[75]

part of the signal. For a detector of non-planar geometry, the electric field in the active volume is not constant and the pulse shapes would show further departures from the simple case. Finally, most detectors of all types give a minority of slowly rising pulses that appear to originate from edge effects.

Cho and Chase[75] compared the timing performance of three types of pick-off, a simple LE pick-off with a threshold just above the noise, a ELE system and, thirdly, the dual primed ARC system. The other detector was a plastic scintillation counter coupled to a constant fraction of pulse height pick-off which in a separate experiment was shown to contribute a time jitter of less than 400 ps. Figure 7.19 shows a selection of their results for a 3 cm^3 planar detector. Each trio of points represents the peak and fwhm positions of the delay curve for events of a single energy. The time walk over any energy range may therefore be inferred from the locus of the peak points over that energy interval. The results of Figure 7.19 may be summarized as follows. For a single energy, the LE pick-off generally gives the smallest fwhm, but, as expected, over a wide energy range shows considerable time walk compared with the other two methods. The ELE system gives a good time resolution over a wide energy range, but for a single energy is not as good as either the LE or ARC system. The ARC method gives a good time resolution over a wide energy range, but is poorer than the LE and ELE methods at low energies. These observations were generally in accord with the theory.

When used with a 35 cm^3 true coaxial detector a rather different picture emerged. The LE system gave both a large time walk and a poor single energy performance. This was thought to arise from the variability of the initial slope of the detector waveform. Both the ELE and ARC systems gave rather similar performances over a wide energy range, the ARC system having a slightly better single energy performance.

In a later paper Cho and Chase[77] have further considered the case of the LE pick-off with a planar detector. For events of a single energy most of the time jitter originated from variations of the slope of the detector waveform. An additional circuit was therefore used whose purpose was to sense whether the detector waveform underwent a change of slope during approximately the first half of the voltage change. If this was the case the event was rejected. Although this caused rejection of approximately 50 % of all events, it removed one of the main sources of variability in the timing. A further advantage was that it was then possible to set the threshold of the LE pick-off at quite a high level (20 % of the total pulse amplitude) so as to be clear of the toe of the detector waveform. For events in the photopeak of the 511 keV of ^{22}Na the system gave a fwhm of 0·77 ns. A theoretical analysis showed that this fwhm could be satisfactorily accounted for on the basis of the effect of noise from the FET of the preamplifier. It therefore appears that the performance is near to the theoretically optimum for this type of system.

Moszyński and Bengston[78, 79, 80] have also used a form of pulse shape discrimination to improve the timing performance. When a simple LE pick-off was used it was found that the delay curve had a broad tail on the late side of the curve. They demonstrated that this was due to a minority of pulses with a slow rise time that originated from the boundary regions of the active volume of the detector. The slow rise was the consequence of temporary charge carrier trapping and low electric field in these regions. A pulse shape discriminator, operating on the zero crossing method (Subsection 6.5.2), was used to block the analysis of such events. For γ rays in the 1·33 MeV photopeak, the resulting delay curve showed a slightly improved fwhm but gave a delay curve shape with steeply descending sides. The same pulse shape discrimination technique was also applied to the ARC system. This had previously yielded delay curves with a late tailing effect. Provided the circuit was adjusted to reject the slower 50% of all pulses the delay curve exhibited a shape near to the ideal Gaussian shape. As in the LE system the fwhm was slightly improved (1·1 ns for the 511 keV photopeak with a 0·25 cm^3 planar detector).

7.5 Fast coincidence circuits

7.5.1 Additive type

The time pick-off associated with each of the detectors will be assumed to deliver an output pulse of fixed height having a brief and fixed duration. It is the function of the coincidence circuit to deliver an output pulse only when it receives inputs from all the time pick-offs within a time $\pm\tau$, the resolving time, of one another. The proper choice of the value of τ to take account of the time jitter has already been discussed.

Probably the simplest type of coincidence circuit is of the additive type. The pulses from the two time pick-offs are added together and, if they overlap (Figure 7.20), the resulting large pulse exceeds the threshold of a following circuit, causing an output pulse to be delivered. The resolving time is therefore approximately equal to the width of the input pulses, although it is partly dependent on the input pulse shape and the setting of the threshold level. The simplest threshold circuit uses a reverse biased diode, as in the circuit of Bell, Graham and Petch[6] already discussed (Subsection 7.2.1). The diode has a finite junction capacity which causes subthreshold pulses to be transmitted slightly. To eliminate these, and to standardize the shape and height of the coincidence output pulses, it is usual to place a trigger circuit after the diode. This secondary trigger often need not have a fast response since the coincidence rate is usually low, the feedthrough singles are too small for pile up to be important, and the coincidence output from the trigger is usually only required to operate a slow gate.

In more recent variants of the additive type circuit the diode has been replaced by a tunnel diode, which in some cases may remove the need for a secondary discriminator. Thus, for example, the circuit of Iaci and Lo Savio[33] added current pulses from the two time pick-offs and passed them into a tunnel diode arranged in a LR monostable configuration. A coincidence caused the combined current to exceed the peak current of the tunnel diode and so triggered it to the higher voltage state.

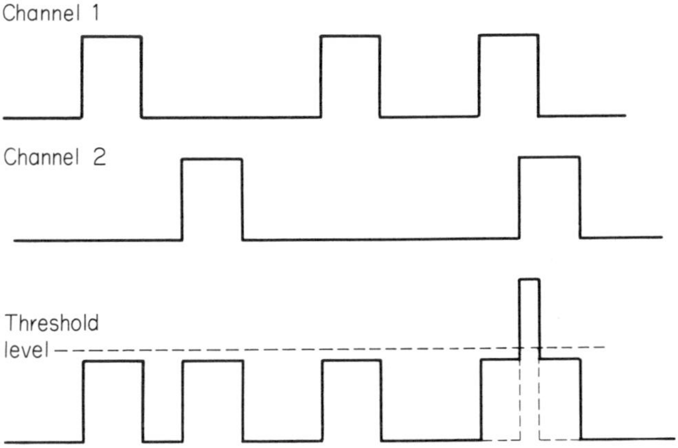

Figure 7.20 Principle of the additive type of coincidence circuit

An anticoincidence circuit (i.e. an AND NOT gate) may also be fabricated using the additive principle. This would be achieved in the example of Figure 7.20 by arranging the veto input pulses to have negative polarity. Due to the effects of time jitter, the veto pulses would have to be wider than the other positive pulses in order to efficiently block coincident events. The anticoincidence resolving time may be then seen to be approximately equal to the difference in pulse widths.

The additive circuits rely on discrimination between single pulses and pulses of double height due to coincidences, and so the setting of the threshold level is fairly critical. If it is required to detect coincidences when three or more time pick-offs all deliver simultaneous pulses, then the threhold setting would but even more critical and it is not practicable to use an additive type circuit.

7.5.2 Diode AND gate

Probably the most commonly used coincidence circuit uses the diode AND gate or Rossi circuit, after his original design using triodes.[34] All the n diodes in Figure 7.21(a) are normally each conducting a current I. If $n-1$ input pulses appear simultaneously, all but one diode are cut off, but the output voltage changes very little since one diode is still conducting. Assuming $R \gg r$ the voltage change at the output is approximately $(n-1)Ir$. If n

inputs are present however, all the diodes are cut off and the output voltage rises to E. Thus it is only necessary to have minimal discrimination following the output to exclude voltage changes of up to $(n - 1) Ir$. This discrimination may be performed by a diode, a tunnel diode or a suitably biased transistor of the following amplifier. The resolving time is given approximately by the width of the input pulse, although the setting of the discriminator will influence it slightly. The diode AND gate may be modified for anticoincidence, if the diode at the anticoincidence (veto) input is arranged to be quiescently non-conducting, as in Figure 7.21(b), and the veto input is of opposite polarity to the coincidence inputs, so that it causes the diode to become conducting. Any number of coincidence and anticoincidence inputs may be accommodated in the same way. As before, it is necessary to have the anti-coincidence pulse longer than the coincidence pulses so that, allowing for time jitter, the rejection of coincidences is efficient.

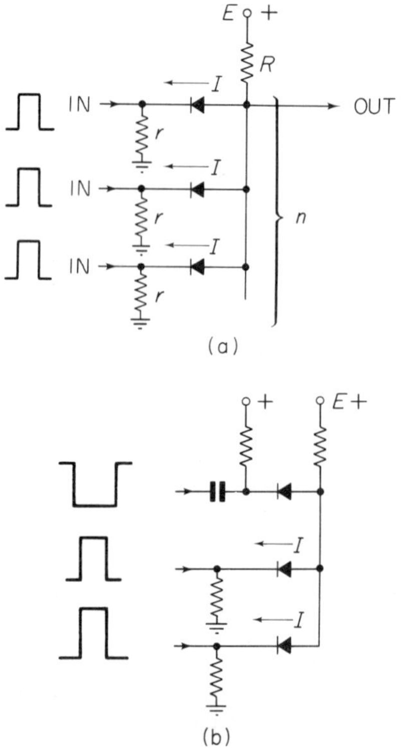

Figure 7.21 Principle of diode AND gates used (a) as a coincidence circuit and (b) as a mixed coincidence and anticoincidence circuit

An example of a variant of the diode AND gate may be seen in the arrangement of Tomlinson and Brown[35] shown in Figure 7.22(a). Diode D3 is normally conducting and remains so when only one input is present. When both inputs are present together D3 becomes non-conducting and an output pulse is obtained. Other circuits have used an output diode in series, as in the circuit of Simms[36] (Figure 7.22b). The complete circuit could accommodate up to four coincidence or anticoincidence inputs and it incorporated electronically generated variable delays and pulse clipping for all the inputs. By means of the switches in the circuit each channel could be switched 'in' or 'out'. Since each input diode was driven from a transistor monostable, any channel could be converted to anticoincidence by changing its input connection to the collector of the complementary transistor of the monostable pair.

Verweij[37] has described a circuit which is entirely DC coupled and can accommodate up to five coincidence or anticoincidence channels. In this case the secondary discrimination was performed by a tunnel diode and it is

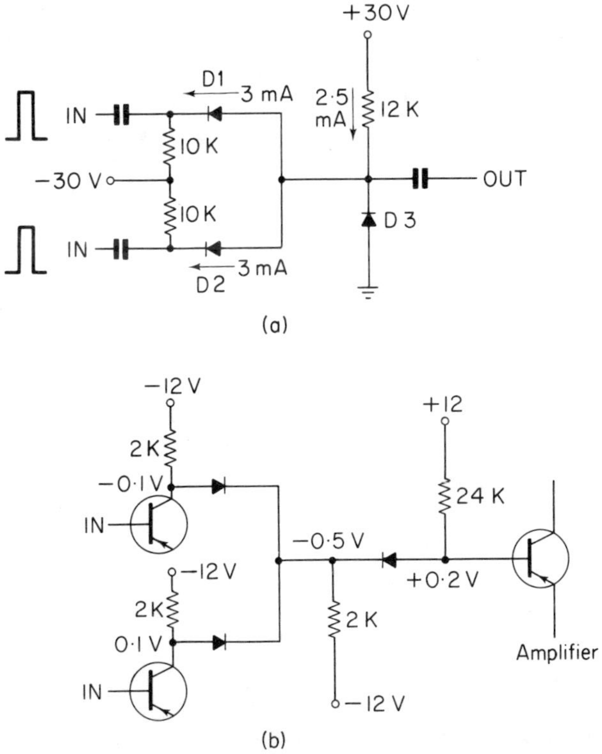

Figure 7.22 Principle of the coincidence circuits used by (a) Tomlinson and Brown[35] and (b) Simms[36]

claimed that this circuit and the associated discriminators can function at pulse rates up to 100–200 MHz.

With four detectors and time pick-offs there are six possible combinations of pairs of coincidences. The diode AND circuit lends itself easily to the detection of all these modes of coincidence, and Caini and Tove[38] have described such a circuit.

7.5.3 Differential coincidence circuits

All the circuits mentioned above require brief input pulses if a small resolving time is to be achieved. The clipping of the pulse to such a short length may present problems if the detector pulse has a long rise time. The differential coincidence system of Bay[39] can give a resolving time much shorter than the width of the input pulses. The principle of the Bay method is illustrated by Figure 7.23. The input pulses are fed to two coincidence circuits A and B, but with a mutual delay of $2\tau_x$. Figure 7.23(a) therefore shows the two delay curves that would be obtained from these two coincidence circuits. Figure 7.23(b) shows the variation of pulse height obtained at the output of the differential amplifier for various delays between the detector signals and Figure 7.23(c) shows the variation of pulse height after the phase rectifier. If the discriminator following the phase rectifier is set to accept small pulses as shown in Figure 7.23(c), then no output will be delivered for relative delays of $\pm < \tau$. If this output is now used to operate an anticoincidence gate from the auxiliary coincidence circuit, it follows that a final output will only be obtained when the detector pulses are within a time interval of $\pm\tau$ of one another. By setting the discriminator to a low level this effective resolving time may be made much smaller than the resolving time achievable by the individual coincidence circuits. It may be noted that the auxiliary coincidence circuit need not have a particularly short resolving time. In the original circuit of Bay, the two coincidence circuits were incorporated in a balanced diode bridge arrangement.

Another differential coincidence circuit was described by Meunier[40] and the principle is shown in Figure 7.24. The two inputs are arranged to be of opposite polarity. Single pulses from input 1 are blocked by the diodes and cause no response. Single pulses from input 2 charge all three capacitors equally and the difference amplifiers give no output (apart from a small transient) to the coincidence circuit. Simultaneous inputs from both detectors cause the voltage at point B to be constant and zero, while point A will be first positive and then negative so charging up the capacitor negatively. Point C will go first negative and then positive and will also charge the capacitor negatively. Both difference amplifiers will therefore deliver outputs and a coincidence output will be obtained. If the detector inputs are not quite simultaneous the three capacitors are charged asymmetrically and one of the differential amplifiers will not deliver an output. The coincidence

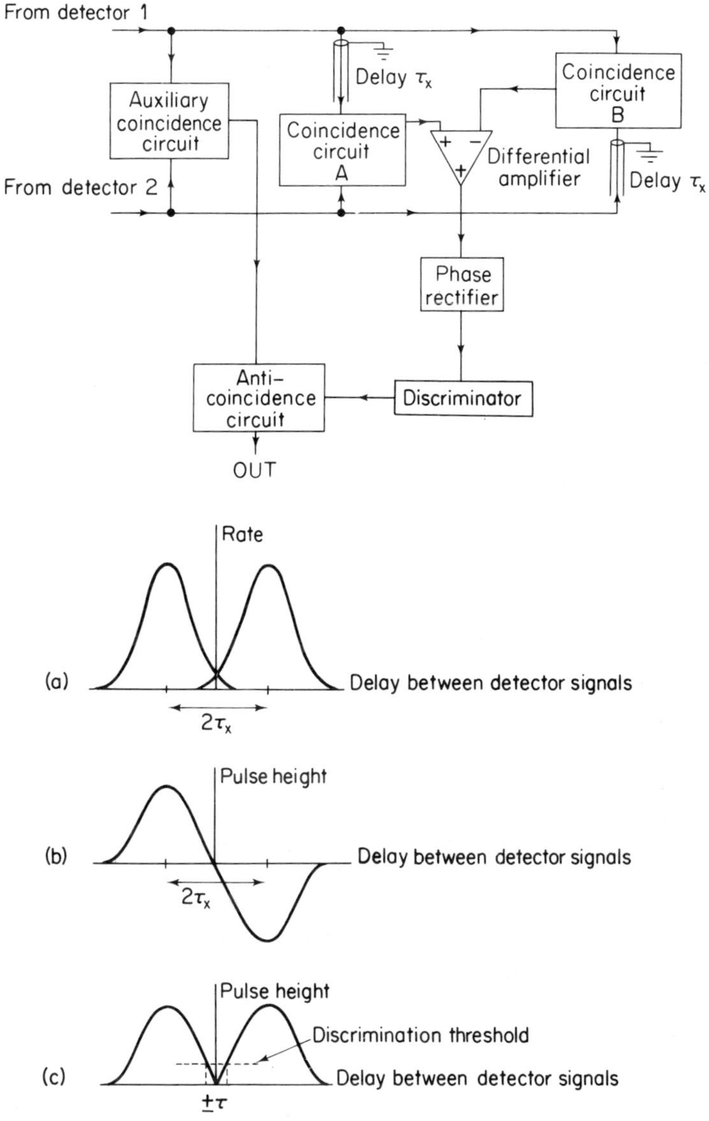

Figure 7.23 Differential coincidence circuit of Bay.[39] (a) Delay curves for coincidence circuits A and B. (b) Plot of pulse height versus delay at the output of the differential amplifier. (c) Plot of pulse height versus delay at the output of the phase rectifier

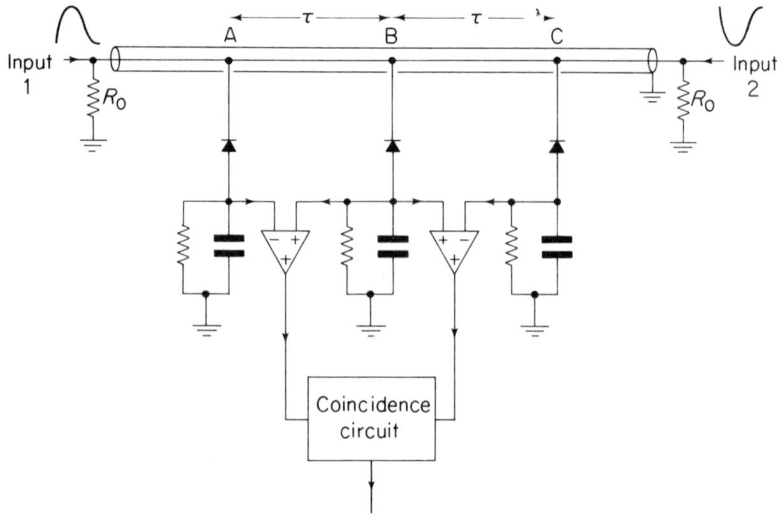

Figure 7.24 Differential coincidence circuit of Meunier[40]

resolving time is therefore approximately equal to τ, the propagation time between points A and B and between B and C.

7.5.4 Other types

Many other types of coincidence circuit have at times been described in the literature. Several years ago techniques using gated beam tubes[41] such as the type 6BN6 were fairly common. This tube has two control grids with similar control characteristics and it is normally used with both grids quiescently cut off, so that anode current only flows when positive pulses are applied simultaneously to both grids. Other techniques have used cathode ray tubes with a suitably masked screen which is viewed by a photomultiplier. Bergström *et al.*[42] have described several of these techniques based on the use of a high speed oscilloscope.

Some progress has been made in the development of techniques where the coincidence gating is done in the photomultiplier tube itself, usually by means of an additional grid in the tube.[43, 44, 45] The advantage of this procedure is that the tube and the various circuits are only required to process pulses which originate from coincident events. This is sometimes desirable, since the maximum pulse rate acceptable in a conventional fast–slow coincidence set-up is often limited, not by the accidental rate at the fast coincidence circuit, but rather by the maximum pulse handling rate of the slow channels. This is particularly so when the coincidences are relatively numerous and the experimental time is limited as in the case of a pulsed accelerator.

7.6 Multichannel time analysers

7.6.1 Introduction

Where it is only necessary to detect coincidences between a pair or more of detector pulses, the simple coincidence circuits already described will suffice. In other types of experiment however the particles may not be emitted simultaneously and it may be necessary to acquire information about the distribution of their relative times of detection. Examples of this are in time of flight experiments or in measurements of the lifetimes of radioactive nuclear states or of unstable particles. While information about the relative time of detection of a pair of particles may be obtained with a simple coincidence circuit by taking a delay curve, this is a time consuming procedure and has other disadvantages. A better arrangement is often to use a multichannel time analyser which senses the relative time of arrival of the pair of pulses from the two detector time pick-offs and accordingly directs the output pulse into one of several time channels, each of which has an associated scaler. Thus each time channel may be seen as equivalent to a coincidence circuit of resolving time equal to the width of the time channel. The relationship of a time analyser to the simple coincidence circuit may be seen to be analogous to that of a multichannel pulse height analyser to the single channel pulse height analyser. The time analyser can be arranged so that the time channels cover the overall range of the time intervals of interest and so the whole of the time interval histogram may be obtained simultaneously by examining the contents of the scalers. This has the additional advantage that changes in the source intensity will affect all channels equally; the same will be true for some sources of apparatus drift. It is, however, important that the widths of the time channels should be stable. In many designs all the time channels are arranged to have the same width and to be contiguous, but this is not always necessary or desirable. In practice the type of experiment will dictate the range of time intervals of interest and the number of time channels required to span this range. For slow neutron time of flight measurements the range may be of the order of microseconds, whereas for the measurement of short-lived states the lower limit will be set by the time jitter of the detector-time pick-off and may be of the order of 100 ps. The corresponding channel widths for these two extremes will therefore be of the order of 100 ns and a few picoseconds respectively.

Three main techniques are used in time analysis. The first consists of feeding the pair of input pulses to several coincidence circuits, one for each time channel, with a successively increased relative delay prior to each circuit. The second method consists of using one of the input pulses to start a high frequency clock, which is stopped by the other input pulse, and counting the number of clock cycles which occurred in the interval. The third method, which is probably the most common, is to use a time to amplitude converter (TAC) which delivers an output pulse of height proportional to the time interval between input pulses. The output pulse heights can then be analysed by means of a multichannel pulse height analyser and the resulting pulse height spectrum gives the time interval histogram.

7.6.2 *Multiple coincidence circuit type*

This type of circuit may be seen as an extension of the simple coincidence circuit. The circuit of McLusky and Moody,[46] for example, is shown in Figure 7.25 and each coincidence circuit is, in essence, of the additive type and uses a

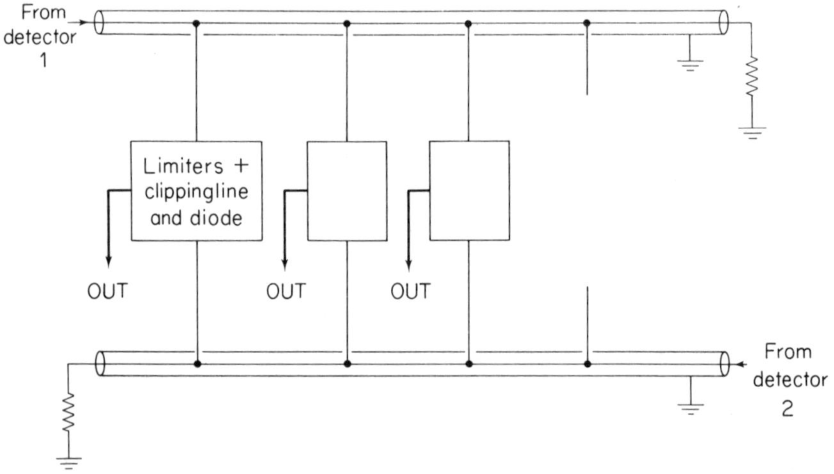

Figure 7.25 Principle of the time analyser of McLusky and Moody[46]

biased diode. In this case the width of each of the 10 time channels is set primarily by properties of the coincidence circuits (clipping line length and diode bias) and the spacing of the channels is set by the delay between coincidence circuits.

A more recent variant of this type of circuit has been described by Venable.[47] The two input pulses are fed to two series of tunnel diode bistable stages, with successively increasing relative delay (Figure 7.26). Each bistable stage, in switching to the high voltage state, generates a step function which switches the next bistable. The tunnel diode stage bridging each pair of tunnel diode bistables only switches if the upper bistable is switched first, i.e. if the signal in the upper line precedes the signal in the lower line. It follows that the only anticoincidence circuit to deliver an output will be the one corresponding to that section in which the signals in the upper and lower lines are coincident, since only that anticoincidence circuit receives an input from the tunnel diode switch on the left but not from the switch on the right. The tunnel diode bistables are all eventually reset, in readiness for the next event, by means of a pulse derived from the last pair of stages. The arrangement has the advantage over simpler circuits that the time channels are contiguous and Venable obtained channel widths down to 5 ps with a stability of 5 ps per °C. When supplied with pairs of pulses from a pulse generator and the relative time interval was varied, it was found that the boundaries of the time channels exhibited a jitter of less than 2 ps.

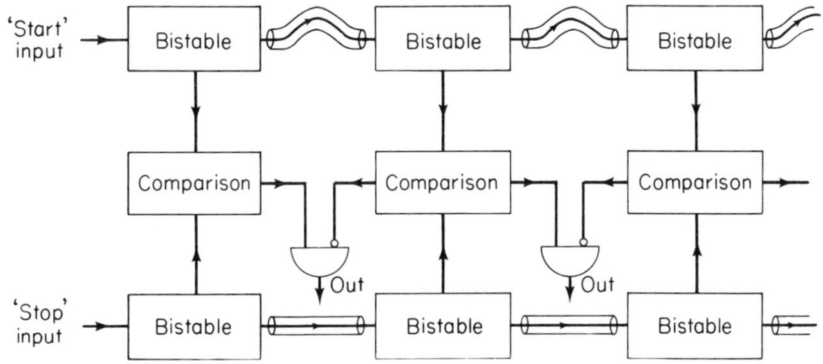

Figure 7.26 Principle of the time analyser of Venable[47]

7.6.3 *Time to digital converters*

It is clearly impracticable to use the multiple coincidence circuit type of analyser for a large number of time channels and in this case it is necessary to use either a time to digital converter or a time to amplitude converter (TAC).

In a time to digital converter one of the input pulses, the 'start' pulse, causes a high frequency clock oscillator or pulse generator to be started or gated on. The second pulse, the 'stop' pulse, is arranged to stop or gate off the clock. The number of clock cycles is counted in a scaler and the result used to address the memory, and to deposit the count in the appropriate address. It may happen, of course, that the pulse applied to the 'stop' input precedes that arriving at the 'start' input in which case the circuit will not respond. Hence it is usual to provide a certain amount of delay in the path of the 'stop' pulse so that the range of time intervals of interest are shifted into the region where the analyser can respond in the majority of events. A further contingency arises when a single pulse arrives at the 'start' input, but the second detector does not give an output and so no 'stop' pulse is provided. In this case the clock would not be switched off for a considerable time and the analyser would be unable to process any other events in this time. Hence it is usual to provide some type of supervisory circuit which either switches the clock off after the maximum time interval of interest or, alternatively, examines the 'start' and 'stop' inputs for coincidence within the time interval of interest and prevents the clock from starting if no 'stop' pulse is seen within this interval. The requirement of such a supervisory circuit is common to all converters using the start–stop principle. It will be apparent that much of the circuitry of a time to digital converter is common to that of a multichannel pulse height analyser (MCA) and, for that reason, some commercial MCAs have facilities for alternate operation as a time to digital time analyser.

The channel width obtainable in such a system will be equal to the period of one oscillator cycle. Thus to obtain a channel width of 1 ns it would be necessary to use a clock frequency of 1 GHz and the problems of operating the necessary logic circuitry at this frequency are, at the moment, formidable. Certain other problems concern the definition of the channel boundaries since, if the clock is gated on rather than switched on, there is a possibility that the gate will open at any part of the oscillator cycle. In any case the clock will usually be switched off in mid-cycle and the circuit registering the number of cycles must respond consistently to such part cycles.

Many designs of time to digital converters have been published. Ivanov,[48] for instance, described a 256 channel analyser with a clock frequency of 200 MHz. The 'start' pulse, of negative polarity, triggered a tunnel diode bistable into its high voltage state and so allowed a tunnel diode multivibrator, of frequency 200 MHz, to operate. The 'stop' pulse was of positive polarity and this returned the bistable to its initial state and so stopped the multivibrator. The pulses from the multivibrator went to an 8 digit binary counter using tunnel diodes. The contents of this counter are transferred, in parallel binary code, by an interrogation pulse 1·5 μs after the 'start' pulse, and these operated the address system of a 256 channel ferrite core memory. The total dead time for the register and address system was 2 μs. The time for the transfer to the memory was 30 μs but it was possible to use the register to accept another conversion while the memory was being read in so that the effective dead time was only 2 μs.

To obtain smaller channel widths it is possible to interpolate between oscillator cycles. Meyer,[49] for instance, used a clock frequency of 125 MHz and passed the clock signal down an 8 ns delay line (equal to one clock cycle) along which 8 coincidence circuits were spaced. The other input to all the coincidence stages was the 'stop' pulse and the circuit was designed so that an output was obtained which indicated to which of the 8 parts of the cycle the 'stop' pulse most nearly corresponded. In this way an effective channel width of 1 ns was obtained.

7.6.4 Vernier chronotrons

A different principle is used in the vernier chronotron of Lefevre and Russell[50] and is illustrated in Figure 7.27. Two clock oscillators are employed, of slightly different frequency, and the 'start' pulse starts the clock of lower

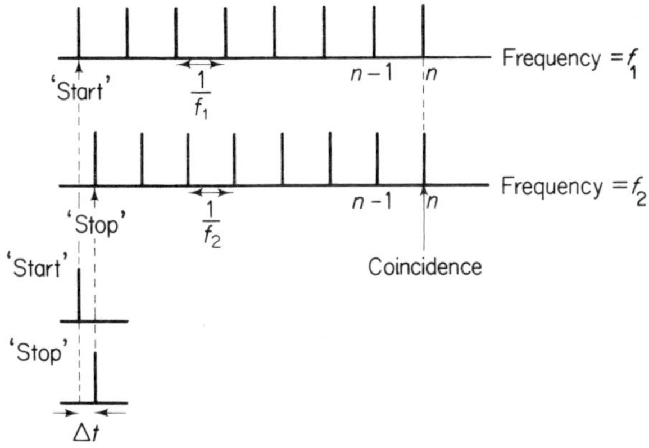

Figure 7.27 Principle of the vernier chronotron

frequency. Some time later the 'stop' pulse starts the higher frequency oscillator. A coincidence circuit between the two clock outputs detects coincidences between the two clocks and switches them off when this occurs. By counting the number of cycles n (which is the same for both clocks) prior to coincidence it is possible to infer the time interval Δt between the 'start' and 'stop' pulses, since $\Delta t = (n - 1)(1/f_1 - 1/f_2)$ where f_1 and f_2 are the two clock frequencies. In this arrangement the period of the clocks must be larger than the maximum time interval to be measured. By making the clock frequencies very near to one another it is possible to realize small channel widths without having to use a high clock frequency, although this is obviously achieved at the expense of a long measuring time and hence a long dead time.

Figure 7.28 shows the vernier chronotron of Lefevre and Russell,[50] which consists of two pulse generators having periods of 300 ns and 299 ns. The 'start' pulse caused a pulse to be passed into the left delay loop which, on arriving back at the gate circuit, was regenerated and continued round the

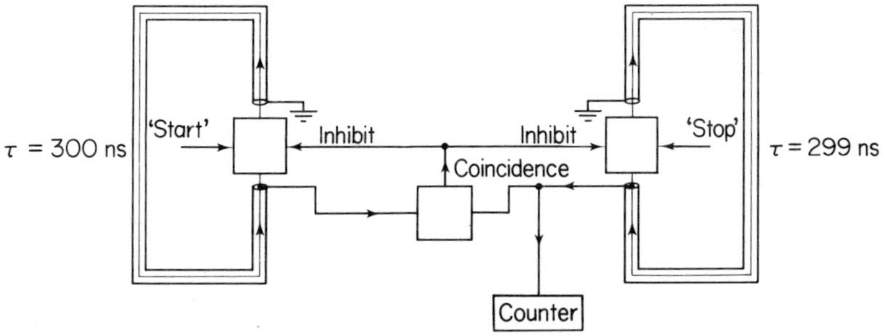

Figure 7.28 Principle of the vernier chronotron of Lefevre and Russell[50]

loop. A similar process was initiated in the right loop by the 'stop' pulse. The coincidence circuit caused the regeneration circuits to be blocked when coincidence occurred. It will be seen that the channel width of 1 ns was obtained, and with a total number of channels of 300 gave corresponding dead times of up to 90 μs.

The vernier chronotron imposes strict conditions on the frequency stability of the two oscillators. If the periodic time varies slightly from period to period this produces a large variation in the apparent channel number. An improvement on the original design in this respect was made by Kindlemann and Sunderland[51] who stabilized the oscillator periods by a quartz crystal oscillator. Since only the relative stability of the two circulating clock pulses is of importance, Aveynier and Van Zurk[52] used an ingenious arrangement whereby the two circulating signals shared common circuit elements (i.e. delay lines) for the most part.

It is possible to extend the vernier technique to measure time intervals greater than one clock cycle. For instance Thenard and Victor[53] included additional circuitry that counted the number of cycles of both clocks; the clocks were only switched off when coincidence occurred and when both clocks had executed the same number of cycles. Clock periods of 50 ns and 49 ns were used, giving channel widths of 1 ns, but up to 3000 channels were available (up to 3 μs time interval) so giving dead times up to 150 μs. Barton and King[81] described a fast converter using the same principle, and which was assembled almost entirely from ECL integrated circuits.

Cresswell and Wilde[54] used a rather different method. They counted the number of cycles n_1 of the first clock which occurred before the second clock was switched on, as well as counting the number of cycles n_2 that subsequently occurred before coincidence of the clocks. In this case the time interval is given by $(n_1 - 1)/f_1 + (n_2 - 1)(1/f_1 - 1/f_2)$. Thus the arrangement may be seen to be equivalent to a time to digital converter, which is combined with a vernier technique for interpolation between clock cycles. The measurement time in this case is much less than that using the principle of Thenard and Victor. Using clock periods of 5 ns and 4·5 ns, giving a channel width of 0·5 ns, up to 100 channels were available (50 ns altogether) but with a maximum measuring time of up to 100 ns.

7.6.5 *Time to amplitude converters* (*TACs*)

A large number of time to amplitude converters have been described. The popularity of this method arises partly from the usual ready availability of a multichannel pulse height analyser. An important advantage of the TAC is that the number of channels, and the range of time intervals analysed, may be varied over a wide range simply by changing the scale of conversion, whereas most of the other types of time analyser are designed to operate with time channels of fixed width. A disadvantage of the TAC is that the speed is limited by the measuring time of the multichannel analyser, which may be up to 60 μs. TACs employ one of two main principles of operation, the pulse overlap type or the start–stop type.

For a TAC coupled to a multichannel analyser (MCA), the width of the time channels can be varied down to zero by choosing a large enough conversion gain (in volts per channel). A more important criterion is the stability of the system and the excess time jitter introduced by the TAC, since these will limit the minimum useful channel width in a practical system. As with other types of time analysers, these can be measured by taking the input pulses from a pulse generator so that each pair of pulses are separated by an exact time interval. Another consideration is the non-linearity of the TAC. In most designs this is less than 1 % but may be considerably less than this. Although the start–stop converters may seem to give better linearity in principle than the pulse overlap type, this is not always borne out in practice.[4]

In the use of MCAs for pulse height analysis, the use of servo stabilization arrangements is becoming more common. These usually operate on the basis of periodic reference pulses of known amplitude being supplied at the detector input; by sensing the channel number after each has been processed by the MCA, a correction is made to the gain of the system when necessary. This type of arrangement is fully discussed in Section 8.5. For the case of a TAC coupled to a MCA, it is desirable that any such stabilization system should include the TAC as well. Lapshin *et al.*[55] have described such a system. Two pairs of reference pulses were supplied to the two scintillation counters by means of light emitting diodes, the first pair having a small relative time interval and followed by the second pair having a larger relative time interval. The resulting channel classifications in the two cases were then used to control the zero setting and the gain of the system in a similar manner to the examples discussed in Section 8.5. Using the system it was possible to stabilize the centroid of the delay curve within ± 3 ps over an 8 hour period, compared to a variation of 200–300 ps over the same period without stabilization.

Karlsson[82] has described a similar arrangement. In this case, to avoid the instabilities associated with light emitting diodes, the reference pulses were fed directly into the time pick-offs.

7.7 Time to amplitude converters (TACs)

7.7.1 Pulse overlap type

This type of TAC is really an extension of the simple coincidence circuit. Figure 7.29 shows the output of such a circuit for varying degrees of overlap of the input pulses, and it will be seen that the area under the output pulse is proportional to the time interval between input pulses. If, therefore, the output pulse is integrated, the arrangement forms a TAC. To obtain a strictly linear relationship between output pulse amplitude and input pulse time interval, it is necessary to use input pulses of rectangular shape, although a limited region of linearity will result from any input pulse shape with a flat top. The arrangement is, of course, unable to distinguish events in which input pulse 1 precedes pulse 2 from those in which pulse 2 precedes pulse 1, and so it is usual to delay one of the pulse inputs so that the time intervals of the events

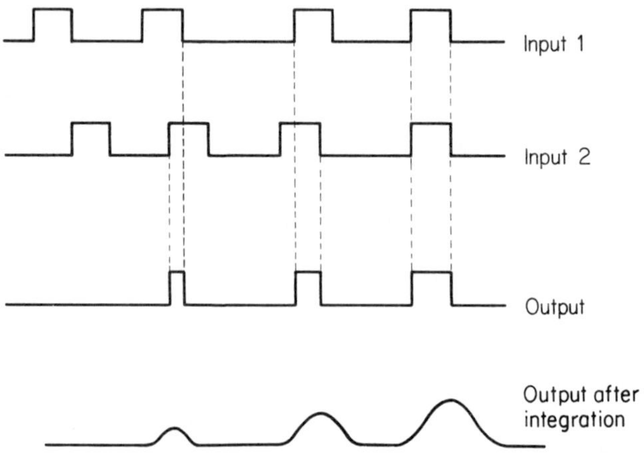

Figure 7.29 Principle of the overlap TAC

of interest are shifted to a region that permits a mostly unambiguous interpretation of the output of the TAC. In some cases a supervisory circuit may be used to exclude events where, say, input 1 precedes input 2, and it may be noted that this will reduce the accidental coincidence rate by a factor of two.

Circuits have been described based on coincidence circuits of both the additive type and the Rossi type. Thus the circuit of Rodda, Griffin and Steward[56] is similar to the arrangement of Bell, Graham and Petch (Figure 7.8), except that the diode threshold detector is replaced by a junction transistor which is quiescently cut off. As with the circuit of Bell *et al.*, the time pick-offs were of the limiter variety. The stability of this circuit was 55 ps per °C.

Bell and Khan[57] have described a TAC based on the Rossi type circuit. The coincidence circuit, a diode AND gate, was fed with pulses of 30 ns length giving a time analysis range of 20 ns.

7.7.2 Start–stop type

In this type of converter the 'start' pulse causes some storage device to accummulate charge or current and this is terminated by the 'stop' pulse.

Most commonly the storage device is a capacitor which is charged or discharged with a constant current, so the change in voltage is proportional to the start–stop interval.

The arrangement of Dardini *et al.*[58] used the circuit having the principle shown in Figure 7.30. The 'start' pulse, a step function of voltage, was applied to the base of T_2, cuts off T_2 and causes a constant current from T_1,

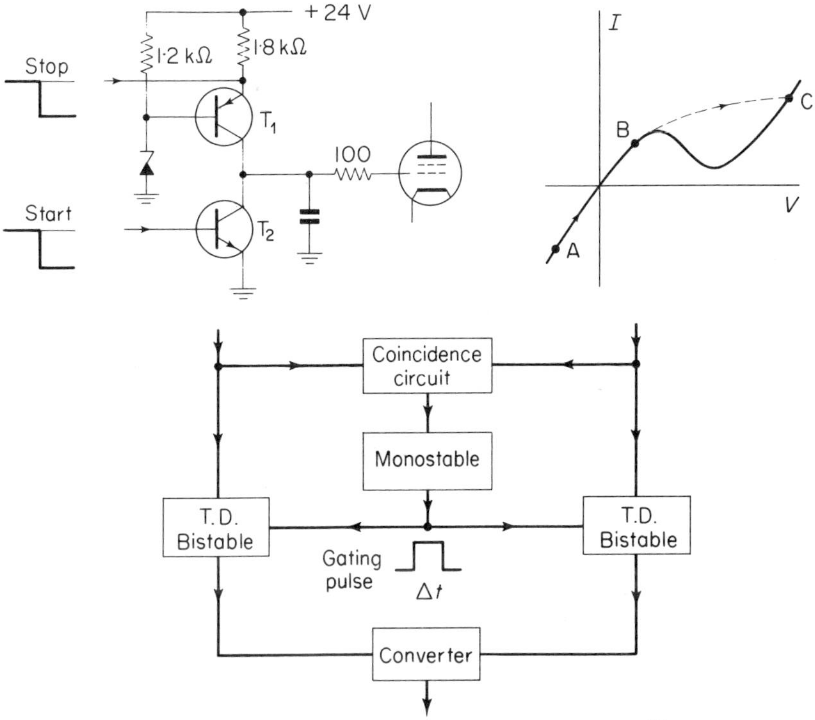

Figure 7.30 Start–stop converter of Dardini *et al.*[58]

which is in the grounded base mode, to flow into the capacitor. The stop pulse, also a step function of voltage, is applied to the emitter of T_1 and causes this current to be cut off. The voltage developed across the capacitance is amplified by a valve in order to preserve a high input impedance. The means by which the start and stop step function pulses are generated is also illustrated in Figure 7.30. If a pair of input pulses occur within the time interval range of interest, the coincidence circuit output causes the monostable to operate, and this delivers a gating pulse of duration Δt to the bistables. The two bistables are tunnel diodes biased quiescently at point A on the characteristic (Figure 7.30). The arrival of the gating pulse causes the operating point to move to B, so each bistable is now enabled to permit triggering to point C by its input pulse, and so generates the 'start' or 'stop' step function for use by the converter. The gating pulse duration Δt must be longer than the maximum time interval it is required to process and at the end of the gating pulse the bistables are reset. Thus the arrangement supervises the incoming events and the circuit senses the first pair of pulses within the region Δt. The dead time of the arrangement is therefore nearly Δt; this could be as small as 50 ns. The TAC had a stability of 10 ps per °C, and the time jitter introduced by the TAC when fed with input pulses from a pulse generator was ~10 ps fwhm.

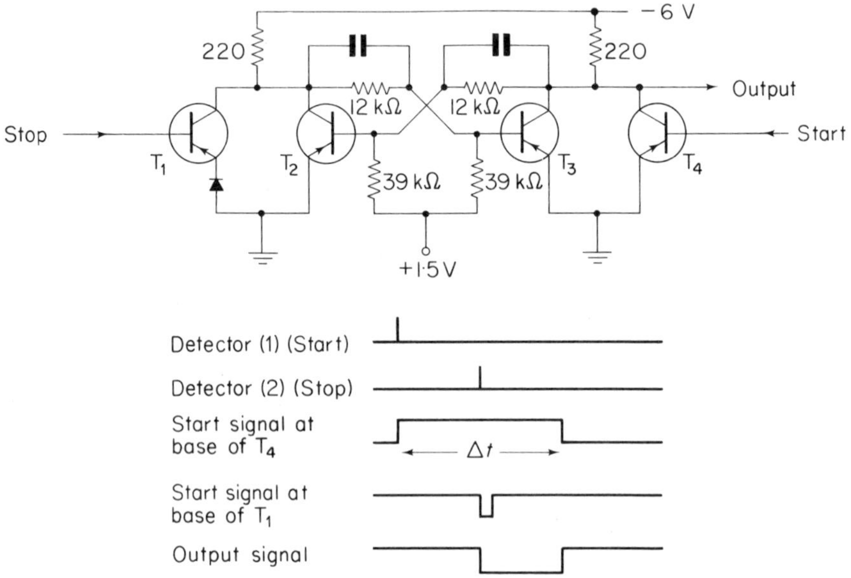

Figure 7.31 Principle of the converter of Balaux and Boulay[59]

In the design by Balaux and Boulay,[59] a transistor bistable (Figure 7.31) was used. The 'start' pulse is shaped to be a rectangular pulse of duration Δt, equal to the maximum time interval of interest (Figure 7.31), and is applied to the base of T_4. Since T_4 is quiescently in the saturated condition, this causes T_4 to be cut off. This enables the bistable to trigger to the second state when the 'stop' pulse arrives on the base of T_1, and it remains in this second state until the rectangular 'start' pulse is removed. Figure 7.31 shows the sequence of voltage changes, and it will be seen that the output signal has a length equal to the time interval between the stop pulse and the end of the start pulse. This is then integrated by further circuitry to give a pulse of height proportional to the interval. It should be noted that the bistable can only trigger to the second state when the start pulse is present at T_4 base, so a single 'stop' pulse has no effect. A single 'start' pulse on the other hand enables the bistable for a period Δt but then the circuit returns to quiescent condition in readiness for another event. Thus no other supervisory circuits are necessary and the measurement time of the arrangement is approximately equal to Δt. The dead time was essentially limited by the following circuits, including the discharging of the capacitor, and amounted to 4·5 μs. The inherent time jitter of the converter was 2·8 ps fwhm, and time interval ranges between 50 and 400 ns could be accommodated.

7.7.3 Compensation of amplitude time walk in TACs

All time pick-offs show some degree of time walk, as already discussed in Subsection 7.1.3. With leading edge pick-offs, a large amplitude detector pulse will cause the time pick-off to respond earlier. If this corresponds to the first of the two pulses (i.e. the 'start' pulse) at the TAC input, this increases the time interval between the pair of input pulses; if it corresponds to the 'stop' pulse it will decrease the time interval as seen by the TAC. The consequent error in the amplitude of the TAC output will therefore be in different directions, according to whether the large detector pulse occurred in the detector corresponding to the 'start' pulse or the 'stop' pulse. With overlap type TACs, although the 'start' and 'stop' pulse terminology was not used, nevertheless one input is usually delayed so as to follow the other input for the majority of events of interest, so that the two inputs may be seen as equivalent to 'start' and 'stop' inputs and the same conclusion is valid.

Since TACs are widely used in measurements of lifetimes of short lived states, the existence of time walk is troublesome in two respects. Firstly the fwhm of the delay curve is increased and, secondly, lifetime measurements derived from observation of the centroid shift contain an error due to displacement of the delay curves, due to differing time walks for the delayed emitter and the prompt emitter (Subsection 7.1.3.).

Most compensation circuits operate on a linear compensation principle. If A_1 and A_2 are the amplitudes of the two detector pulses, then a compensating voltage of $k_1A_1 - k_2A_2$, where k_1 and k_2 are constants, is added to the TAC output. This form of compensation therefore assumes that, in the case of leading edge timing, the pulse shape at the input of the time pick-off has a linear rise. For example, Schwarzschild[3] used this principle on a system employing organic scintillation counters and tunnel diode leading edge trigger time pick-offs. With the slow channels set to accept most of the pulse height spectrum of ^{60}Co, the addition of the compensation circuit caused the fwhm to be reduced from 0·7 ns to 0·4 ns approximately.

For a combination of a 8 mm planar Ge(Li) detector and plastic sctintillation counter, Gorni *et al.*[60] used a rather different principle. They applied compensation to the TAC output by adding a voltage $k_1R_1 - k_2R_2$, where k_1 and k_2 are constants and R_1 and R_2 are the times the leading edges of detector pulses took to rise from 10% to 90% of the mean pulse height. Using limiter type pick-offs, and with the slow channels set to accept the 511 keV photopeaks of ^{22}Na, they obtained fwhms of 4·0 ns and 6·0 ns with and without compensation.

Theiberger[61] used a different principle. The pulses from the time pick-offs were fed to a second TAC, but with their time sequence reversed by means of a delay inserted before one input of the second TAC. Thus the detector which provided the 'start' pulse for the main TAC also provided the 'stop' pulse for the second TAC, and vice versa. In this way the dependence of the two TAC outputs on time walk in the pick-offs was in opposite directions and a fraction of the output of the second TAC was therefore added to the output of the main TAC to provide the compensation. The second TAC was comparatively 'slow', being designed to operate on input pulses whose rise times were longer than those in the main TAC but whose shapes were similar on a longer time scale. With this system, using plastic scintillation counters and tunnel diode leading edge trigger time pick-offs, and with the slow channels set to accept between 400 and 930 keV of the pulse height spectrum of ^{60}Co, the fwhm was 0·4 ns and 0·9 ns with and without compensation.

For most systems, the time walk is linearly related to the pulse amplitudes only over a limited range. Hence some workers have sought to apply non-linear techniques in order to compensate more accurately for the time walk over a larger range of amplitude. Barton and King[83] describe an arrangement where the amplitudes of the two detector pulses are passed into two non-linear networks made up from resistors and a diode. The sum of the outputs of these was added to the TAC output to effect the compensation. Burchall *et al.*[84] have described a compensation technique which is applicable to systems using a digital computer. In this instance the computer was being used to process and store the experimental information, and this included the storage of timing information. The starting points for the compensation technique were the curves of time walk versus pulse amplitude of each detector, which had been determined experimentally. These curves were approximated by simple analytical expressions; the computer was then programmed so that, using these analytical expressions, the timing information of any event could be appropriately corrected. When used with Si detectors, a shift in particle energy from 0·5 to 10 MeV caused a time walk of 1 ns, as against 50 ns without the compensation technique.

7.8 Linear gates

A linear gate is required when it is necessary to gate a pulse prior to pulse height analysis. This is often required when a multichannel pulse height analyser is used, since it is impractical to do coincidence or anticoincidence selection after such an analysis. Thus a linear gate should transmit the linear input pulse with minimum distortion when a logic type gating pulse is present at the gating input. It is necessary to ensure that the gating pulse has a duration and time position so that it overlaps either the whole of the linear pulse or at least the portion of interest, such as the peak region.

Linear gates may be seen as equivalent to a series or parallel switch as in Figure 7.32(a), or a combination of both as in Figure 7.32(b), where the switches are operated for the duration of the gating pulse. Since in practice the switches are imperfect, a number of sources of possible malfunction arise. When the gating pulse only is present but there is no linear input pulse, there is usually a shift in the voltage at the output, the so-called pedestal, and possibly some other transients. When both linear and gating pulse are present, the output pulse will be seen superimposed on this pedestal and, in addition, there may be some non-linearity between input and output pulse heights. When the gate is nominally closed because of the absence of the gating pulse, there will be a small amount of transmission, or feedthrough, of the linear pulse.

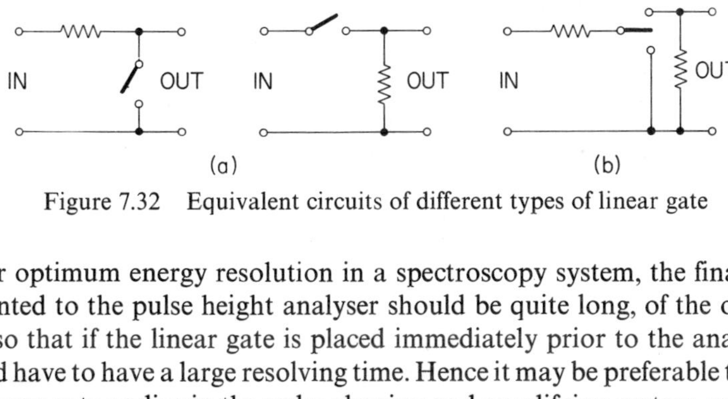

Figure 7.32 Equivalent circuits of different types of linear gate

For optimum energy resolution in a spectroscopy system, the final pulse presented to the pulse height analyser should be quite long, of the order of 1 μs, so that if the linear gate is placed immediately prior to the analyser it would have to have a large resolving time. Hence it may be preferable to place the linear gate earlier in the pulse shaping and amplifying system, so that it operates on a shorter linear pulse and consequently can have a shorter resolving time. In this case the pedestal may prove troublesome since the subsequent lengthening of the output pulse will also serve to partially integrate the pedestal, giving a large shift in final pulse height.

In many cases, although the shift in pulse height due to the pedestal may be important, a much more critical factor is the stability of the height of the pedestal. In some linear gate circuits the pedestal is suppressed by the use of a second linear gate circuit fed with the gating pulse only, so as to generate the pedestal only, and whose output is subtracted from the output of the main linear gate. In other circuits the use of a balanced configuration causes the pedestal to be cancelled out. In both types of circuits the suppression of the pedestal also serves to improve the stability of the height of the small residual pedestal.

Linear gate circuits have been published using diodes, junction transistors or FETs as the active elements. As pointed out by Mills,[62] to switch a germanium diode so its conductance changes by a factor of 10^3 requires a voltage change, and hence a pedestal, of about 0·2 volts. Mills used a junction transistor, which gave a conductance change between collector and emitter of 10^6 for a change in V_{ce}, and hence a pedestal, of 0·05 volts. The main gate of his circuit, Figure 7.33, is made up of Q_{11}, where the gating pulse is applied to the base, and is therefore of the parallel switch type. Identical gates, made up of Q_{13} and Q_{12}, are respectively supplied with the gating pulse only and

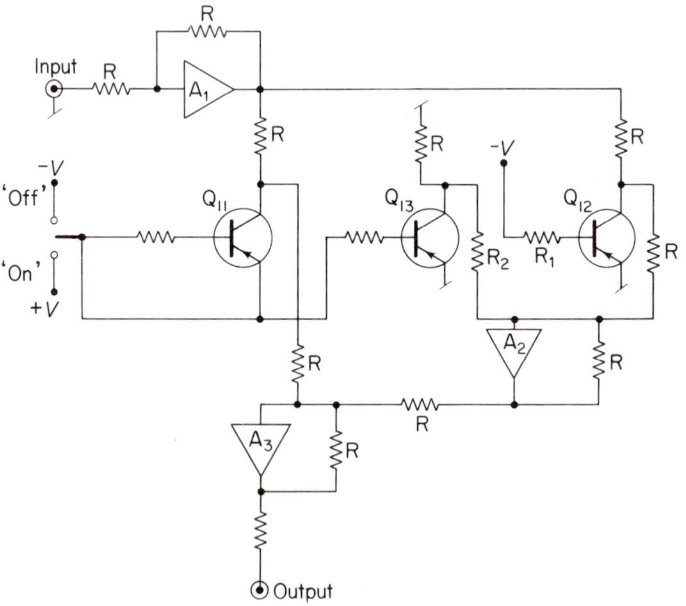

Figure 7.33 Principle of the linear gate of Mills[62]

the input pulse only and so generate the pedestal only and the feedthrough pulse only. The outputs of these two gates are then subtracted from the output of the main gate, by means of the operational amplifier configurations, to give cancellation of the pedestal and of the feedthrough in the final output. The overall circuit had a pedestal of $3 \times 10^{-4} + 5 \times 10^{-5}/°C$ of the full scale output. The gate accepted input pulses in the range ± 10 V, 1–5 μs long with a switching time of 70 ns, and a transfer stability of $10^{-4}/°C$ with the gate open. With the gate closed the feedthrough was 2×10^{-4}.

Smith[63] has described a DC coupled linear gate using diodes in a balanced series–parallel switch arrangement (Figure 7.34). In addition, special stabilization circuits were incorporated to give a stable pedestal. The linear input signal is converted into two identical current signals I_{sig}, which are applied to points A and B. In the quiescent condition D_3 and D_4 are conducting and D_1 and D_2 are blocked. The signal currents are not large enough to alter this state of affairs and flow only through D_3 and D_4, so that the resultant change in voltage at the output is negligible. The gating pulse is applied to the bases of the transistors with opposite polarity, and causes diodes D_3 and D_4 to block and D_1 and D_2 to conduct, so that the input signal appears at the output. The circuit would operate with gating pulses as short as 10 ns and for output pulses up to 22 mA into 50 ohms (i.e. ~1 volt) gave a pedestal of <0·5 mV.

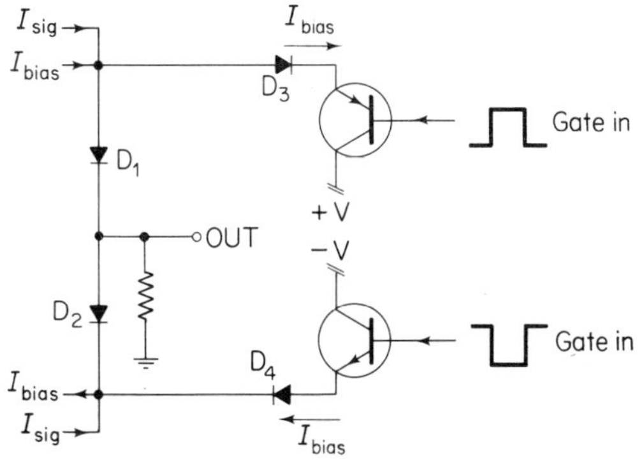

Figure 7.34 Principle of the linear gate of Smith[63]

Other workers have employed the balanced series–parallel type switch circuits using a diode bridge arrangement. A recent example of this was published by Schuster[64] (Figure 7.35). With no gating pulse present, diodes D_1 to D_4 are blocked and diodes D_5 and D_6 are conducting. A linear input pulse does not alter this state of affairs, and so negligible output voltage is produced. The gating pulse is applied as complementary currents to points A and B of the bridge and causes diodes D_1 to D_4 to conduct and diodes D_5 and D_6 to be blocked, so that a linear input pulse appears practically unattenuated at the output.

The opening of the gate took about 20 ns; the pedestal could be adjusted to zero, but had a stability of <2 mV for a temperature change of 20 °C to 35 °C (full scale output 10 V), and the feedthrough of the closed gate was 10^{-3}.

Manfredi, Maranesi and Rimini[65] have described a variant of the simple diode gate (Figure 7.36). The voltage to current converter circuit is arranged to have a transconductance g_m equal to the reciprocal of R when the gating pulse is absent, and so the resultant changes at the output are very small, especially so since the diode is just biased in the forward conducting state. The gating pulse causes the output transistor of the voltage to current converter to be cut off and so disables the converter. The positive linear input signal is therefore transmitted to the output via R and cuts off the diode so that it appears at the output with negligible attenuation. To suppress the pedestal

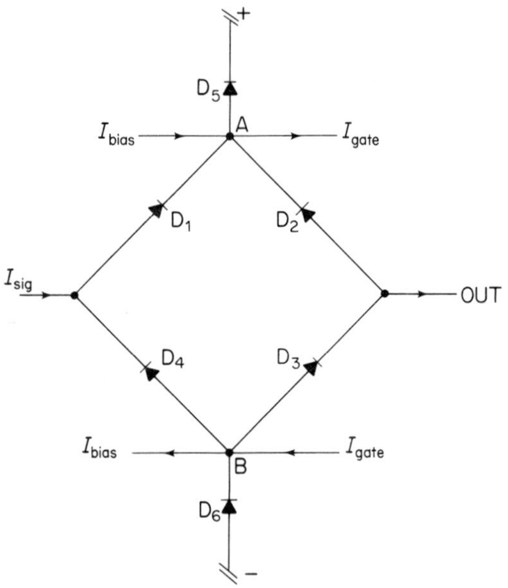

Figure 7.35 Principle of the linear gate of Schuster[64]

an identical circuit was provided with the gating pulse only, and thus generated the pedestal only, which was then subtracted from the main output. For positive input pulses in the range 0–10 V the feedthrough was <0·1 %. The stability of output pulse height was <0·5 mV/°C. The gate would function with the above performance with gating signals down to 50 ns in length.

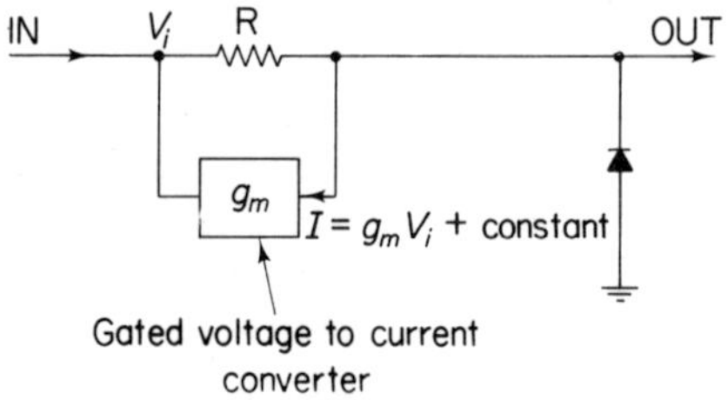

Figure 7.36 Principle of the linear gate of Manfredi, Maranesi and Rimini[65]

7.9 Pile up rejector circuits

The effect of pulse pile up distorting the pulse height spectrum (Subsection 3.1.2 and Section 4.5) and distorting the form of the delay curve (Subsection

7.1.3) have already been discussed. Pile up rejectors act by sensing whenever two pulses from a detector arrive with less than a certain minimum time interval, here termed the pile up rejection time, and accordingly produce a gating signal which, in most cases, operates a linear gate placed in the main signal path. For pulse height analysis, the pile up rejection time will be chosen to be sufficiently large so that the tail of the first pulse has decayed to a low enough level to preserve the required energy resolution. In timing applications it is the recovery time of the time pick-off, and perhaps the time analyser, which dictate the pile up rejection time to be chosen.

A pile up rejector circuit will inevitably fail to recognize pairs of pulses which are very closely spaced, the minimum detectable spacing being usually termed the resolving time. The ability of any circuit to recognize two closely spaced pulses inevitably depends on how brief the two pulses are. While it is comparatively simple to derive brief pulses from an organic scintillation counter, either by keeping the integration of the detector current signal to a minimum or by suitably clipping the partially integrated pulse, the detector current pulse from a NaI(Tl) scintillation counter, for instance, cannot be clipped to as short a length. Thus the type of detector influences the resolving time attainable. In addition, it is easier to recognize two closely spaced large pulses than a large pulse closely followed by a small pulse. Hence the resolving time is a variable quantity depending on the magnitudes of both pulses. The efficiency of pile up rejection in a given case will therefore depend on the range of pulse amplitudes present and, to some extent, on the form of the pulse height spectrum. The influence of all these factors on the performance of a pile up rejector means that it is difficult to compare the relative merits of the various circuits that have been described.

Several principles have been used in pile up rejectors. Probably the most common arrangement is to form a bipolar pulse and detect the time between the leading edge and the point of zero crossing. For single pulses this should be invariant with respect to pulse amplitude, provided the detector pulse shape is constant. Two closely following pulses will cause the time interval to be prolonged and this forms the basis of the method. The same principle is used in pulse shape discrimination (Subsection 6.5.2). A recent example of this technique was given by Monier and Tripard.[66] The pulse was double delay line shaped and the first lobe limited (Figure 7.37). The area in the resultant positive lobe was integrated by means of a pulse width to height converter circuit.

Moszyński, Jastrzebski and Bengston[5] have described a rather different system. As shown in Figure 7.38 the pulses from the detector, after differentiating and limiting, are fed to one of the inputs of a coincidence circuit. The delay is necessary to prevent self-coincidence of single pulses. When the circuit was optimized for use with an organic scintillation counter, a resolving time of <60 ns was obtained for a dynamic range of 10:1. When optimized for a

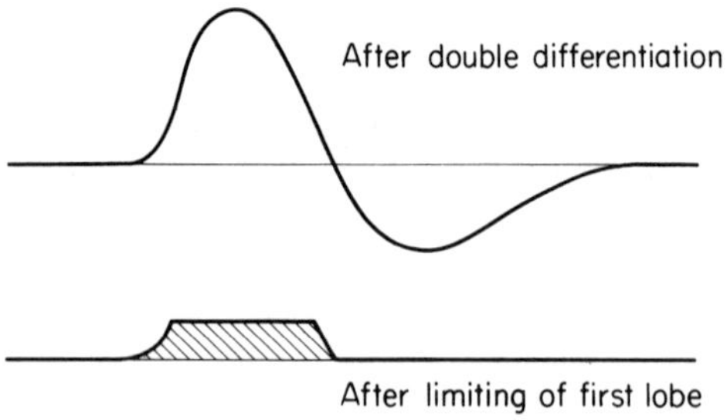

Figure 7.37 Pile up rejection by the detection of the zero crossing time. Principle used by Monier and Tripard[66]

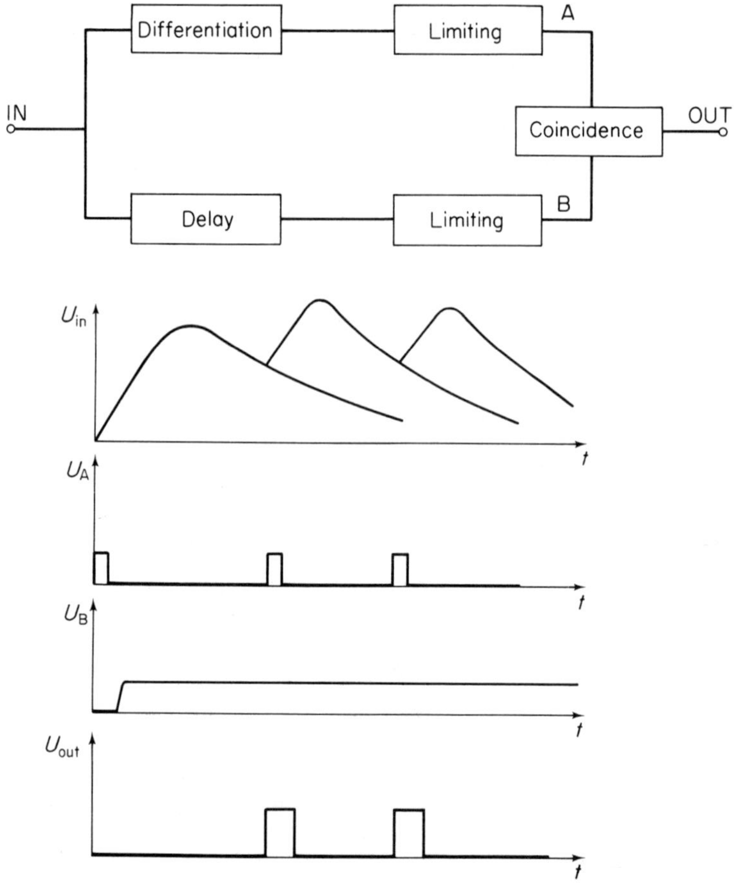

Figure 7.38 Principle of the pile up rejector of Moszyński, Jastrzebski and Bengston[5]

NaI(Tl) scintillator, a resolution time of < 10 ns was obtained for the 662 keV photopeak of the ^{137}Cs spectrum.

Yet another technique has been described by Sabbah, Klein and Arbel.[85] The pulses were first shaped into near-Gaussian form, which we will assume were of voltage $v(t)$. The voltage

$$v(t) + k\frac{d^2v(t)}{dt^2}$$

was then formed, where k is a constant. As shown in Figure 7.39, k was chosen so that, for a single pulse (i.e. no pile up), the central dip in voltage of the

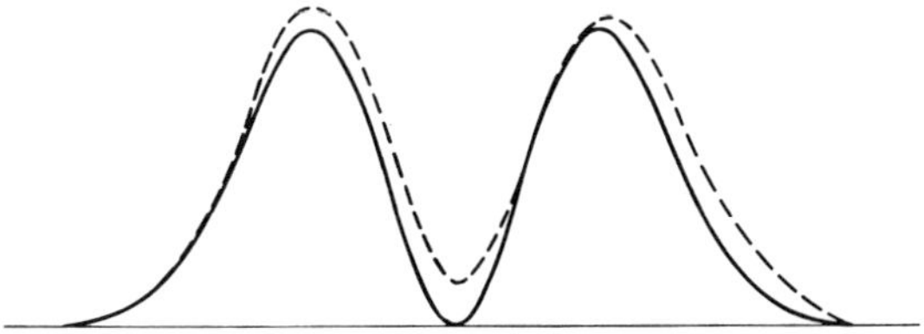

Figure 7.39 Pulse shape used by Sabbah, Klein and Arbel[85] for use in a pile up rejector. The broken line shows the waveform resulting from two signals piling up

resultant waveform just touched zero. If, now, two pulses pile up, the central dip does not reach zero (broken line in Figure 7.39) and this was the basis of the pile up rejection. The voltage level at the dip was sensed by a voltage comparator circuit, whose output was so gated that it could only deliver an output during the passage of a detector pulse. For pulse rates up to 9×10^4 per second the authors give representative Ge(Li) pulse height spectra, obtained with and without the use of the pile up rejector.

Blatt, Mahieux and Kohler[67] have considered in some detail the design requirements of a pile up rejector. The zero crossing method cannot recognize the pile up of a small pulse preceding a large pulse and, for detectors such as a NaI(Tl) scintillator, it is necessary to use a large resolving time, so that the statistical variations arising from the slowly decaying fluorescence do not cause faulty operation. Other types of system often suffer from a poor dynamic range and/or large dead times. The system they evolved has the advantage that it only examines events for pile up if they fall within the prescribed amplitude range. The system is shown in Figure 7.40. The detector pulse, of suitably short duration, must exceed the bias level of the high level discriminator and satisfy any coincidence or anticoincidence requirements for the pile up inspection to take place. Assuming it is a single pulse (no pile up), the pulse will not activate the low discriminator since its path is blocked by the fast linear gate which is operated via the signal in the upper line. If,

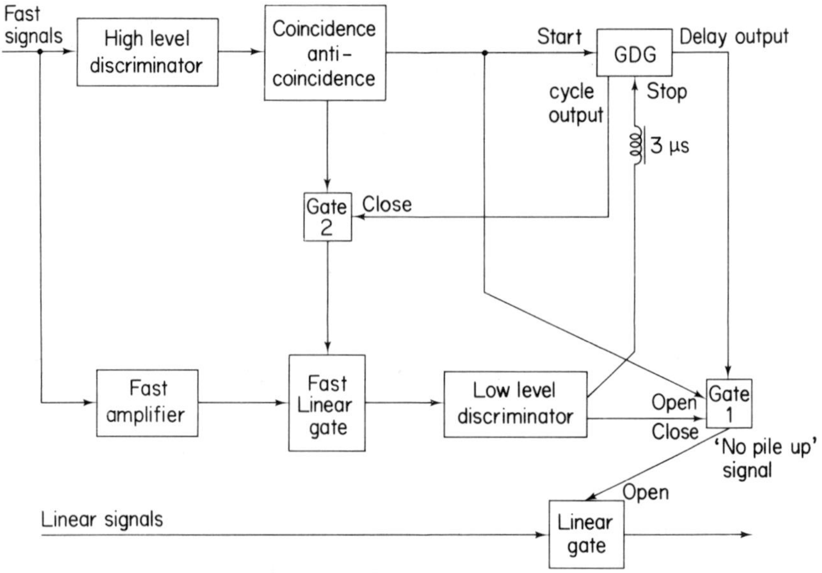

Figure 7.40 Principle of the pile up rejector of Blatt, Mahieux and Kohler[67]

however, a second pulse is present shortly before or after, it will pass the fast linear gate and will cause gate 1 to deliver a signal which closes the main linear gate. In this way it was possible to realize a resolving time of 20 ns over a pulse height spectrum extending from 0·5 to 30 MeV, using a NaI(Tl) scintillator. It was possible to accommodate a dynamic range from 0·3 to 40 MeV by the use of a resolving time of 50 ns.

References

1. W. Meiling and F. Stary, *Nanosecond Pulse Techniques*, Gordon and Breach, 1968.
2. A. De Volpi, J. J. English, A. J. Metz, K. G. Porges and S. J. Rudnick, *Rev. Sci. Instr.* **37**, 1592 (1966).
3. A. Schwarzschild, *Nucl. Instr. and Meth.*, **21**, 1 (1963).
4. A. Ogata, S. J. Tao and J. H. Green, *Nucl. Instr. and Meth.*, **60**, 141 (1968).
5. M. Moszyński, J. Jastrzebski and B. Bengston, *Nucl. Instr. and Meth.*, **47**, 61 (1967).
6. R. E. Bell, R. L. Graham and H. E. Petch, *Can. J. Phys.*, **30**, 35 (1952).
7. M. Bertolaccini, S. Cova and C. Bussolati, *Nucl. Instr. and Meth.*, **37**, 297 (1965).
8. V. Grabari, *Nucl. Instr. and Meth.*, **53**, 136 (1967).
9. K. D. Clausen and T. D. Nainan, *Nucl. Instr. and Meth.*, **49**, 22 (1967).
10. J. Grunberg and L. Tepper, *IEEE Trans. Nucl. Sci.*, NS **13**, No 1, 389 (1966).
11. A. Tamminen and P. Jauho, *Nucl. Instr. and Meth.*, **65**, 132 (1968).
12. J. K. Whittaker, *IEEE Trans. Nucl. Sci.*, NS **13**, No 1, 399 (1966).
13. A. J. Metz, *Rev. Sci. Instr.*, **38**, 1445 (1967).
14. D. A. Gedcke and W. J. McDonald, *Nucl. Instr. and Meth.*, **55**, 377 (1967).

15. D. A. Gedcke and W. J. McDonald, *Nucl. Instr. and Meth.*, **58**, 253 (1968).
16. W. J. McDonald and D. A. Gedcke, *Nucl. Instr. and Meth.*, **55**, 1 (1967).
17. M. R. Maier and P. Sperr, *Nucl. Instr. and Meth.*, **87**, 13 (1970).
18. L. Karlsson, *Nucl. Instr. and Meth.*, **93**, 563 (1971).
19. J. Braunsfurth and H. J. Körner, *Nucl. Instr. and Meth.*, **34**, 202 (1965).
20. S. Donati, E. Gatti and V. Svelto, *Adv. Electronics and Electron Physics*, **26**, 251 (1969).
21. B. Bengston and M. Moszyński, *Nucl. Instr. and Meth.*, **81**, 109 (1970).
22. G. Present and D. B. Scarl, *Rev. Sci. Instr.*, **41**, 771 (1970).
23. M. G. Strauss, R. N. Larsen and L. L. Sifter, *Nucl. Instr. and Meth.*, **46**, 45 (1967).
24. C. W. Williams and J. A. Biggerstaff, *Nucl. Instr. and Meth.*, **25**, 370 (1963).
25. J. J. Samueli and A. Sarazin, *IEEE Trans. Nucl. Sci.*, NS **13**, No 3, 306 (1966).
26. I. S. Sherman, R. G. Roddick and A. J. Metz, *IEEE Trans. Nucl. Sci.*, NS **15**, No 3, 500 (1969).
27. I. S. Sherman and R. G. Roddick, *IEEE Trans. Nucl. Sci.*, NS **17**, No 1, 252 (1970).
28. R. L. Graham, I. K. MacKenzie and G. T. Ewan, *IEEE Trans. Nucl. Sci.*, NS **13**, No 1, 72 (1966).
29. J. P. Fouan and J. P. Passerieux, *Nucl. Instr. and Meth.*, **62**, 327 (1968).
30. R. L. Chase, *Rev. Sci. Instr.*, **39**, 1318 (1968).
31. T. D. Douglass and C. W. Williams, Int. Symp. Nucl. Electronics, Versailles, 1968, Vol. 1, 66.
32. T. D. Douglass and C. W. Williams, *IEEE Trans. Nucl. Sci.*, NS **16**, No 1, 87 (1968).
33. G. Iaci and M. Lo Savio, *Nucl. Instr. and Meth.*, **65**, 103 (1968).
34. B. Rossi, *Nature*, **125**, 636 (1930).
35. R. B. Tomlinson and R. L. Brown, *IEEE Trans. Nucl., Sci.*, NS **11**, No 2, 28 (1964).
36. P. C. Simms, *Nucl. Instr. and Meth.*, **70**, 311 (1969).
37. H. Verweij, Int. Symp. Nucl. Electronics, Versailles (1968), Vol. 1, 60.
38. V. Caini and P. A. Tove, *Energia Nucl.*, **11**, 186 (1964).
39. Z. Bay, *IEEE Nucl. Sci.*, NS **3**, No 4, 12 (1956).
40. R. Meunier, *Electronique Nucléaire*, Proc. Int. Symp. on Nuclear Electronics, Paris, Vol. 1, 263 (1959) (I.A.E.A.).
41. J. Fischer and J. Marshall, *Rev. Sci. Instr.*, **23**, 417 (1952).
42. I. Bergström, E. C. O. Bonacalza, A. Jech, M. Perez and P. Thieberger, *Nucl. Instr. and Meth.*, **8**, 151 (1960).
43. S. V. Muchin, H. J. Pohl, S. V. Richwizky, I. N. Semenyuskin and I. Foltin, *Nucl. Instr. and Meth.*, **33**, 339 (1965).
44. U. J. Roose, *Nucl. Instr. and Meth.*, **36**, 333 (1965).
45. P. Görlich, H. J. Pohl, S. V. Mukhin and I. N. Semenyuskin, *IEEE Trans. Nucl. Sci.*, NS **13**, No 3, 112 (1966).
46. G. R. J. McLusky and N. F. Moody, *Electronic Engineering*, **24**, 330 (1952).
47. W. H. Venable, *Rev. Sci., Instr.*, **37**, 1443 (1966).
48. A. A. Ivanov, *Pribory i Tekhnika Eksperimenta*, No 1, p. 111, Jan/Feb 1966. English translation in, *Instrumental and Experimental Techniques*, p. 113 (1966).
49. H. Meyer, *Electronique Nucléaire*, Proc. Int. Symp. on Nuclear Electronics, Paris, p. 633, S.F.E.R. Paris (1963).
50. H. F. Lefevre and J. T. Russell, *Rev. Sci. Instr.*, **30**, 159 (1959).
51. P. J. Kindlmann and J. Sunderland, *Rev. Sci. Instr.*, **37**, 445 (1966).
52. J. Aveynier and R. Van Zurk, *Nucl. Instr. and Meth.*, **78**, 161 (1970).
53. J. Thenard and G. Victor, *Nucl. Instr. and Meth.*, **40**, 318 (1966).
54. J. Cresswell and P. Wilde, *Proc. Conference on Automatic Acquisition and Reduction of Nuclear Data*, Karlsruhe, EANDC, 1964, p. 300.

55. V. G. Lapshin, M. N. Omel'yanenko, V. I. Rykalin, A. N. Sinaev, V. P. Khromov and Z. Tsisek, *Pribory i Teknika Eksperimenta*, No 3, p. 44, May/June 1968. English translation in, *Instrumental and Experimental Techniques*, p. 551 (1968).
56. J. L. Rodda, J. E. Griffin and M. G. Steward, *Nucl. Instr. and Meth.*, **23**, 137 (1963).
57. J. Bell and N. G. A. Khan. *Nucl. Instr. and Meth.*, **59**, 314 (1968).
58. C. Dardini, G. Iaci, M. Lo Savio and R. Visentin, *Nucl. Instr. and Meth.*, **47**, 233 (1967).
59. N. Balaux and R. Boulay, *Nucl. Instr. and Meth.*, **78**, 109 (1970).
60. S. Gorni, G. Hochner, E. Nadav and H. Zmora, *Nucl. Instr. and Meth.*, **53**, 349 (1967).
61. P. Thieberger, *Nucl. Instr. and Meth.*, **44**, 349 (1966).
62. A. P. Mills, *Nucl. Instr. and Meth.*, **50**, 132 (1967).
63. B. Smith, *Nucl. Instr. and Meth.*, **55**, 138 (1967).
64. H. J. Schuster, *Nucl. Instr. and Meth.*, **58**, 179 (1968).
65. P. F. Manfredi, P. Maranesi and A. Rimini, *Rev. Sci. Instr.*, **38**, 1253 (1967).
66. L. F. Monier and G. E. Tripard, *Rev. Sci. Instr.*, **37**, 316 (1966).
67. S. L. Blatt, J. Mahieux and D. Kohler, *Nucl. Instr. and Meth.*, **60**, 221 (1968).
68. L. Karlsson, *Nucl. Instr. and Meth.*, **100**, 193 (1972).
69. J. Bialkowski and M. Moszyński, *Nucl. Instr. and Meth.*, **105**, 51 (1972).
70. H. R. Krall and D. E. Persyk, *IEEE Trans. Nucl. Sci.*, NS **19**, No 3, 45 (1972).
71. N. Abbattista, B. Marangelli and V. L. Plantamura, *Nucl. Instr. and Meth.*, **95**, 327 (1971).
72. J. K. Millard, T. V. Blalock and N. W. Hill, *IEEE Trans. Nucl. Sci.*, NS **19**, No 1, 388 (1972).
73. Z. H. Cho and R. L. Chase, *IEEE Trans, Nucl. Sci.*, NS **19**, No 1, 451 (1972).
74. Yu. K. Akimov, K. Andert, A. I. Kalinin and H.-G. Ortlepp, *IEEE Trans. Nucl. Sci.*, NS **19**, No 3, 404 (1972).
75. Z. H. Cho and R. L. Chase, *Nucl. Instr. and Meth.*, **98**, 335 (1972).
76. F. Vogler, A. Hofman, G. Philip and K. Thomas, *Nucl. Instr. and Meth.*, **101**, 167 (1972).
77. Z. H. Cho and R. L. Chase, *Nucl. Instr. and Meth.*, **102**, 299 (1972).
78. M. Moszyński and B. Bengston, *Nucl. Instr. and Meth.*, **80**, 233 (1970).
79. M. Moszyński and B. Bengston, *Nucl. Instr. and Meth.*, **100**, 285 (1972).
80. B. Bengston and M. Moszyński, *Nucl. Instr. and Meth.*, **100**, 293 (1972).
81. R. D. Barton and M. E. King, *Nucl. Instr. and Meth.*, **97**, 359 (1971).
82. L. Karlsson, *Nucl. Instr. and Meth.*, **105**, 349 (1972).
83. R. D. Barton and M. E. King, *Nucl. Instr. and Meth.*, **100**, 165 (1972).
84. J. Birchall, B. Frois, R. Roy and R. J. Slobodrian, *Nucl. Instr. and Meth.*, **96**, 45 (1971).
85. B. Sabbah, I. Klein and A. Arbel, *Nucl. Instr. and Meth.*, **95**, 163 (1971).

8

Multichannel Pulse Height Analysers

8.1 Components of multichannel analysers (MCAs)

8.1.1 Analogue to digital converter (ADC)

A multichannel analyser (MCA) is in principle equivalent to many differential pulse height analysers with their windows arranged contiguously. The channels of an MCA may usually be arranged to span a selected region of the pulse height spectrum; thus a 400 channel MCA may be arranged to span the pulse height range of 2·2 to 6·2 volts, say, giving a channel width (i.e. the window width of each channel) of 10 mV. The process of assigning the input pulse height to one of many channels is therefore that of analogue to digital conversion (ADC) and in addition an MCA is usually provided with means for recording the number of events classified in each channel. By examining the number of events classified in all the channels at the end of an experiment, that part of the pulse height histogram between 2·2 and 6·2 volts is obtained. To achieve the same result with a single channel analyser would require 400 times the experimental time, since the window would have to be set sequentially at the 400 different channel positions in the pulse height spectrum. With Ge(Li) detectors an energy resolution of down to 0·05 % fwhm is obtainable at high energies and, in order that the pulse height histogram obtained should approximate to the pulse height spectrum, the number of channels necessary to span the whole of the pulse height spectrum would be $\sim$ 10,000. In such a case it would usually be impracticable to obtain the spectrum by means of a single channel analyser, as the large experimental time would be unacceptable. Thus the advantage of an MCA over a single channel instrument is the saving in experimental time and the fact that the whole of the pulse height histogram is accumulated concomitantly. The latter point means that the pulse height histogram will be more faithfully rendered when the intensity of the source or the detection efficiency vary during the experiment. The advantages of the MCA are offset by the fact that the dead time is longer than that of a single channel instrument and so has a lower maximum counting rate.

Ideally all the channels of a MCA should have the same width. If this were not the case and, say, one of the channels had a width 5% larger than the rest, then this channel would collect roughly 5% more counts than it would if its width were correct, and so the channel contents would be a distortion of the true pulse height histogram. In general, the channel widths should be the same within ±1%, this quantity being known as the differential non-linearity. It is this requirement that places a severe constraint on the design of the ADC, particularly if the number of channels is large, and it is this consideration that has led, until recently, to the virtual monopoly of ADCs using the Wilkinson method.

A further quantity of interest is the maximum integral non-linearity, which expresses the fractional maximum deviation from linearity between the channel boundaries and the corresponding channel numbers.

8.1.2 Storage of digital data

Once the input pulse has been classified or 'coded' by the ADC into, say, the kth channel, some method of recording this classification is necessary. Most MCAs use a computer type of ferrite core memory (see Subsection 2.4.2) for this purpose, which is built into the equipment. Each channel has a corresponding 'address' in the memory where a number of cores are allocated to it and where the number of events which have been classified in that channel is stored, in either binary code or binary coded decimal.

The information at the output of the ADC, conveying that the classification of the input pulse is in the kth channel, may be coded in various ways. In a Wilkinson ADC the coding would be in the form of a train of k pulses. In other types of ADC the information would most commonly be in the form of parallel binary coding of the number k. Some form of recoding may be necessary to give the address code, by which the memory can be addressed at the correct location corresponding to the kth channel. In the Wilkinson ADC it is necessary to count up the k pulses in a counter; outputs from this counter can be taken to give the address code.

Once the address code has been obtained the sequence of operations is as follows. The memory contents of the cores constituting that address are read out into a memory register, one digit is added, and the register contents are then read back into the cores. This cycle of operations takes a time of 1–10 μs typically. The total dead time of the MCA will therefore be the conversion time (of the ADC) plus the read–write cycle time of the memory. For MCAs with a small number of channels the memory cycle time may constitute a large proportion of the overall dead time. In this case a reduction in dead time may be made by using a buffer register to temporarily store the coding, so the memory cycle can then take place while the ADC is freed to accept another event. In addition, the use of thin film, plated wire and semiconductor memories offers the possibility of reduced memory cycle times and, for several years, at least one design[1] has been in use using a thin film memory.

The advent of the large scale manufacture of small general purpose digital computers has provided an alternative source of memory capacity, since the cost per bit of memory often compares favourably with that in a specially built instrument. For MCAs with a large number of channels this may be an attractive method, particularly if the computer cost is offset by its use in other applications.

Still another possibility for the fast recording of the output of the ADC lies in the use of a magnetic tape system. At the end of the experiment the tape can be replayed into a computer at a lower speed, or even replayed several times, to enable the channel classifications to be sorted. At the other end of the speed scale it is sometimes necessary to record the ADC output where the events are infrequent, as in the case of some cosmic ray experiments. The use of a fast memory is therefore wasteful and it is often sufficient to record the ADC output on magnetic or paper tape and perform sorting of channel classifications later when the tape is replayed at a higher speed.

8.1.3 Ancillary circuits

Besides the ADC and the memory, a number of ancillary circuits are usually required. Since the dead time of an ADC and memory is considerable, and regions of the whole pulse height spectrum are often of no interest, a saving in dead time may be made by preventing analysis of pulses in such regions. This may be accomplished by inspecting the input pulses with a single channel analyser whose upper and lower thresholds include the region of interest. The output from the single channel analyser is used to operate a linear gate place prior to the input of the ADC. In this way only pulses falling within the range of interest are committed to the ADC for the time consuming analysis, and the percentage dead time (i.e. the total dead time per second) is reduced. The same linear gate is usually supplied with an additional input so that it is closed immediately an analysis has begun, thus preventing possible disturbance from following pulses. An external input to the linear gate is often provided so that coincidence or anticoincidence function may also be incorporated.

Most designs require the use of a pulse stretcher before the input pulse can be satisfactorily processed by the ADC. As its name implies, this gives an output voltage of the same magnitude as the input pulse height but for a length of time comparable with the conversion time of the ADC. Pulse stretcher circuits are usually based on the principle of Figure 8.1(a). The capacitor acts as a storage device for the peak pulse height and, in the case of a Wilkinson ADC, it is usually the same capacitor which is subsequently linearly discharged to form the basis of the ADC method. In practice, rather than the simple diode, most designs use the amplified diode arrangement of Figure 8.1(b), which effectively decreases the diode impedance by a factor of $(A + 1)$, and so the capacitor voltage during charging more nearly follows the input voltage. The reverse leakage current, which causes an error through the progressive lowering of capacitor voltage, is unaffected. Several workers have considered the design of pulse stretchers.[2,3,4,5] As the precision of

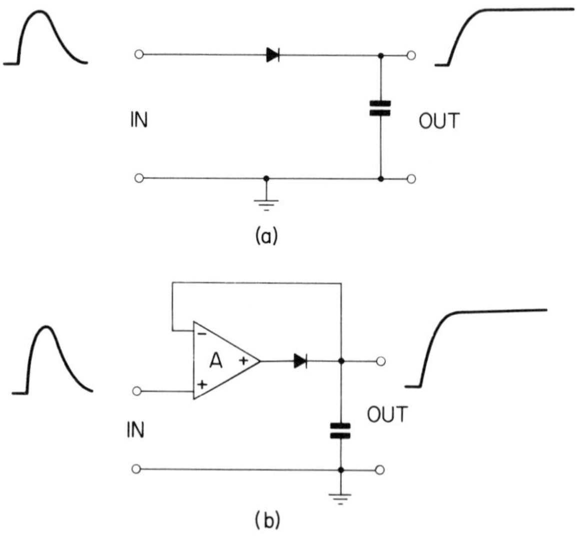

Figure 8.1 (a) Simple diode pulse stretcher. (b) Amplified diode pulse stretcher

ADCs has been improved over the years so the design of the pulse stretcher becomes more critical. In particular the effect of dielectric charge absorption causes the voltage on the capacitor to reduce with time and in many designs this factor has become the main limitation on the precision of MCAs. Recent designs of pulse stretchers have usually sought to compensate for dielectric charge absorption and Koeman,[5] for example, achieves this by injecting a small current into the capacitor during storage.

Another feature of importance is the means for setting the input pulse height corresponding to channel zero. This is useful to enable all the available channels of a MCA to be used to span a region in the middle of the pulse height spectrum. This facility is variously termed 'back biasing', 'zero channel threshold' or 'lower threshold'. Back biasing may be analogue or digital. In analogue back biasing a voltage is subtracted from the pulse height in the linear gate (i.e. applying a 'pedestal') or in the ADC itself. In digital biasing the full magnitude of the input pulse height is processed by the ADC but an appropriate number is digitally subtracted from the ADC coding before recording the event in the memory. With analogue back biasing the back biasing voltage must be highly stable as it directly affects the stability of the analysis as a whole. Digital back biasing does not present this problem but requires an ADC capable of coding input pulses larger than would be the case with analogue back biasing. This increases the conversion time and hence the dead time.

8.2 ADCs

8.2.1 Stacked discriminator type

This may be seen as an extension of the single channel analyser. The input pulse is fed simultaneously to *n* discriminators in parallel, each defining one channel boundary. By means of the anticoincidence circuits connected to the discriminator outputs (Figure 8.2) an output pulse in the relevant channel is

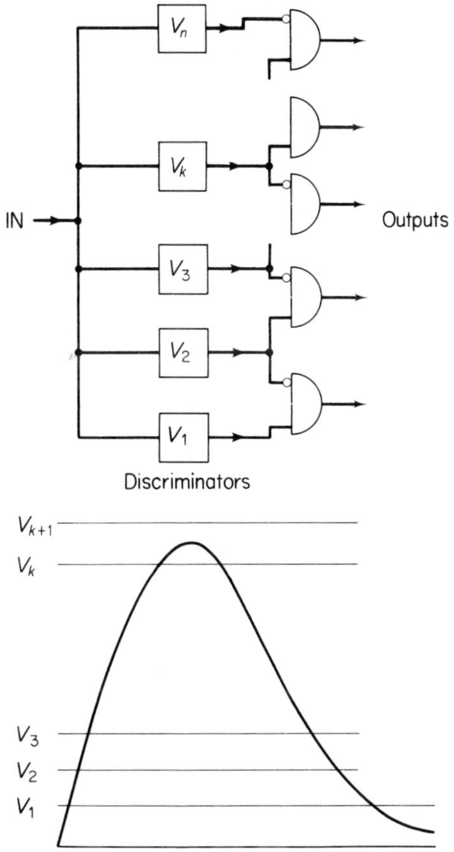

Figure 8.2 Stacked discriminator and its mode of action

obtained. Thus the outputs of the anticoincidence circuits may be seen as delivering a parallel digital code, and so the advantage of the stacked discriminator is one of speed.

The essential problems of the stacked discriminator arrangement are the cost and the difficulty of obtaining adequate differential non-linearity. The problem of differential non-linearity may be illustrated for the case of a 100 channel instrument. For a differential non-linearity of $<2\%$ (i.e. $V_{k+1} - V_k$ constant and stable to $<2\%$) the discriminator thresholds $V_1, V_2, \ldots, V_n$ must each be precise and stable to less than 1 part in 10^4. This difficulty essentially stems from the fact that the channel boundaries $V_1, V_2, \ldots, V_n$ are all derived independently.

A method of overcoming this problem was proposed by Gatti *et al.*[6,7,8] which consisted in superimposing on the input pulse an incremental pulse of magnitude rather less than the spacings $V_{k+1} - V_k$ between consecutive discriminator thresholds. Each discriminator output feeds into a coincidence circuit whose other input is the incremental pulse (Figure 8.3). In this way the only channel to give an output is that in which the discriminator did not

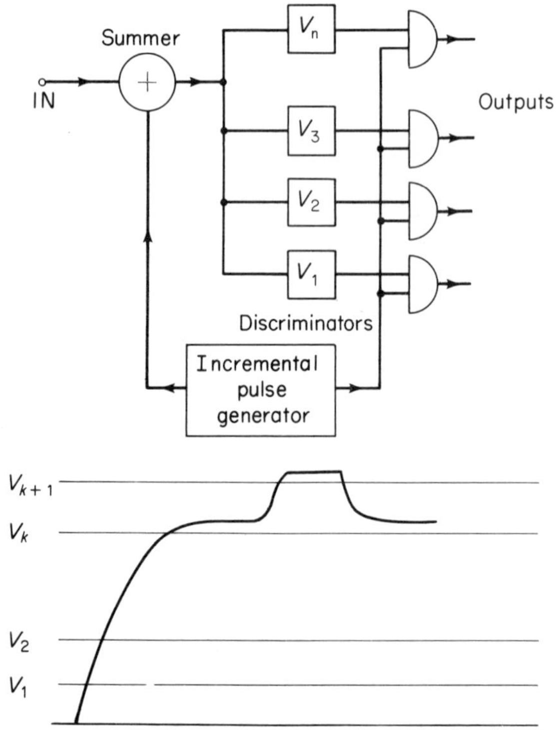

Figure 8.3 Incremental pulse technique of Gatti *et al.*[6,7,8]

fire prior to the incremental pulse but did fire when the incremental pulse arrived (the $(k + 1)$th in Figure 8.3). Thus, provided the height of the incremental pulse is less than the $V_{k+1} - V_k$, the channel width is equal to the height of the incremental pulse and is therefore constant for all channels; the lower boundaries of the channels are set by $V_1, V_2, \ldots, V_n$. It is true that such an arrangement does give channels which are not quite contiguous (there are small gaps between channels) but this is usually of little importance.

The few designs of stacked discriminators that have been published recently have been for a small number of channels and they were designed for high speed of operation working with fairly low precision. Thus the design of Lewyn[9] had ten channels made up of tunnel diode discriminators which gave a differential non-linearity of $\pm 3\%$ with a dead time of 10 ns. Gorodetzky *et al.*[10] made a feasibility study of a 50 channel analyser using tunnel diode discriminators. It was predicted that a differential non-linearity of $<10\%$ and a dead time of 50 ns could be achieved.

8.2.2 Wilkinson type

The principle of this type of ADC is that a storage capacitor is first charged up so that it has a voltage equal to the peak height of the input pulse. The capacitor is then linearly discharged by a constant current, so producing a ramp waveform (Figure 8.4(a)), and during this period a high frequency clock oscillator is switched or gated on. Thus the period of discharge, and the number of cycles of the clock, are proportional to the magnitude of the input pulse. An alternative arrangement is to charge up a second capacitor with the constant current until its voltage reaches that of the input pulse height. In either case the number of clock pulses during the ramp gives the channel number and after counting these in a register the classification can be recorded, usually in a ferrite core memory. To prevent ambiguities in channel assignment it is preferable to prevent the discharge starting in the middle of a clock cycle. Such a 'phase correlated' clock may be obtained by switching the oscillator on at the instant of initiation of the discharge, or alternatively by the use of a free running clock oscillator, such as a quartz controlled oscillator, and only allowing the discharge to begin at an instant when a clock pulse is present.

A variant of the Wilkinson method is the so-called staircase method. In this case the capacitor is successively discharged by discrete quanta of charge, one per clock pulse, so that the capacitor voltage has a staircase profile (Figure 8.4(b)). The number of quanta necessary to completely discharge the capacitor gives the channel number. Unlike the linear discharge method it is not necessary to have a stable clock frequency nor a phase correlated clock. It is, however, slower than the linear discharge method and has been rarely used of late.

From the principle of the Wilkinson method it will be seen that the channels are necessarily contiguous and, provided the clock frequency is stable and the discharge linear, the channels widths are all nearly identical. Thus the problem of differential non-linearity which was encountered in the stacked discriminator is avoided, since in a Wilkinson ADC the channel boundaries V_1 to V_n are all effectively generated by a repetitive procedure. The penalty

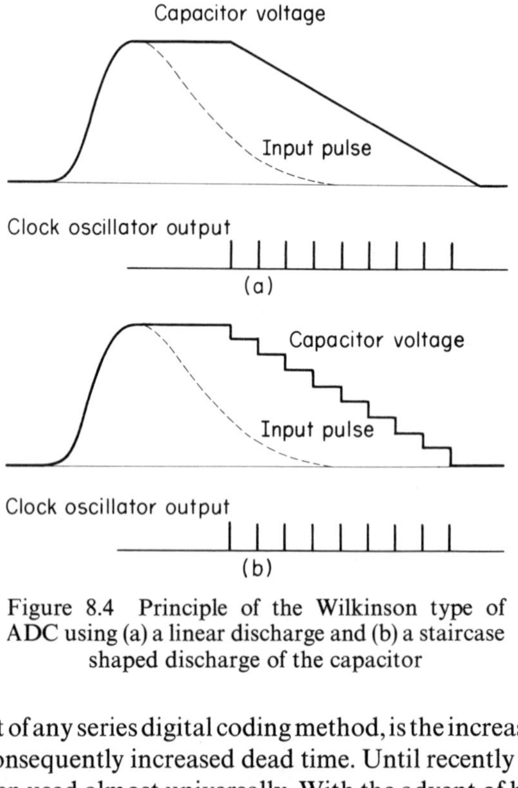

Figure 8.4 Principle of the Wilkinson type of ADC using (a) a linear discharge and (b) a staircase shaped discharge of the capacitor

of this, or in fact of any series digital coding method, is the increased conversion time and the consequently increased dead time. Until recently the Wilkinson method has been used almost universally. With the advent of high resolution Ge(Li) detectors the necessity of ADCs having a large number of channels has led to the design of other types of ADC which have a speed and differential linearity comparable if not better than the best Wilkinson type ADCs.

8.2.3 Successive approximation type

The method of successive approximation will be illustrated for the case of successive binary approximation as shown in Figure 8.5. In the first stage the input pulse height V is compared first with a 'weight' of voltage $E/2$ where E is the maximum pulse height of interest. If V is larger than $E/2$, a '1' is generated in the corresponding output and the remainder $E/2 - V$ is generated for the second stage. If V is smaller than $E/2$ then '0' is generated and the voltage V is passed unaltered to the second stage.

In the second stage the input voltage is compared to a weight $E/4$ and if larger, '1' is generated by this stage, otherwise a '0' is generated. In the first case $E/4$ is subtracted from the voltage; in the second case the voltage is passed unaltered and is presented to the third stage. Thus the process may be seen as similar to weighing with a balance and successively balancing out

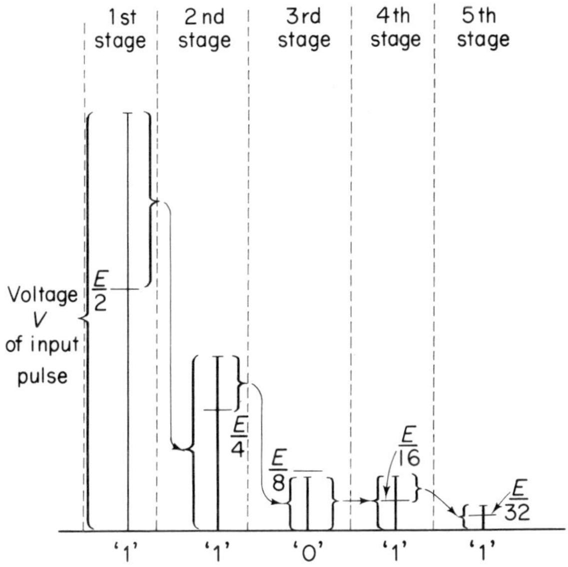

Figure 8.5 Schematic representation of a successive binary approximation ADC

the unknown weight by testing and, if necessary, adding standard weights of decreasing size. For the binary system of weights it is apparent that the outputs of the stages give the channel number in binary code. Thus the first stage gives the most significant bit, the second stage the next significant bit etc. For 1024 (2^{10}) channels ten successive stages and ten different weights $E/2$, $E/4, \ldots, E/1024$ are required. The main difficulty with the successive approximation method is to realize the weights accurately enough, especially the larger weights, so as to give an acceptable differential non-linearity. In general, if the differential non-linearity is to be less than $x\%$, then the largest weight should be accurate within an amount equivalent to $x\%$ of the smallest weight. In a 1024 channel ADC, for instance, if the $E/2$ weight is too small with an error of 1 part in 10^4, the other weights being perfectly accurate, then channel number 512 will have a channel width approximately 5% too small, the other channels having the correct width. This difficulty in realizing the larger weights with sufficient accuracy and stability has led to the use of methods of channel smoothing (Subsection 8.3.2) to improve the differential non-linearity.

Since the number of operations required for an n channel ADC in binary successive approximation is $\log_2 n$ compared to n in a Wilkinson ADC, the successive approximation ADC has found application where the number of channels is large, where it has a faster conversion time than is currently obtainable with the Wilkinson ADC.

8.2.4 Hybrid types

The conversion time of the Wilkinson ADC can, for a given clock frequency, be reduced by using two or more discharge currents of differing magnitude. Thus, in the two ramp Wilkinson the capacitor is discharged first by a large current and then when its voltage is nearly zero, a small discharge current is used. The fast and slow discharge periods may therefore be likened to a coarse and fine measurement. In the design of Patzelt,[11] to be described later, the fast discharge current is 64 times the slow discharge current. During the fast discharge each clock pulse is therefore made to correspond to increments of 64 channels, whereas during the slow discharge period each clock pulse corresponds to increments of one channel. It is important that the starting and stopping of the fast discharge current should be stably synchronized to the clock pulses so that the fall of the capacitor voltage during the fast discharge period is linearly related to the number of clock cycles.

The multi-ramp method may be regarded as a hybrid of a Wilkinson ADC and a successive approximation ADC using, in the above example, only two 'weights'. The problems of the stability of the weights in a true successive approximation design is alleviated in the hybrid version since the ratio of the largest weight to the smallest is less. In the above example the fast and slow discharge currents must be in the ratio of 64:1 with an accuracy of one part in 10^4 to give a differential non-linearity of less than 1 % in all channels.

Bonsignori *et al.*[12] have described a 4096 channel (12 bits) ADC which is a variant of the successive approximation type. Unlike the binary successive approximation method however, the first four significant bits of the conversion were determined by parallel coding by feeding the input pulse height V to a 16 channel stacked discriminator. Each discriminator which fires causes a voltage ΔE to be produced in a digital to analogue converter (DAC). If k discriminators fire, the voltage $k\Delta E$, less a small fixed amount, is substracted from V as shown schematically in Figure 8.6. The residue is then passed on to 8 stages of binary successive approximation to give the other 8 least significant bits. An interesting feature of this arrangement is that the stability of the thresholds of the stacked discriminators is not important. The weight ΔE must of course be accurately realized but as the ratio of this, the largest weight, to the smallest is only 256:1, its realization is not as critical as in a full binary approximation ADC of the same number of channels, where the ratio would have been 2048:1. A further interesting feature of the design was that the weights for the binary successive approximation part of the conversion were also generated by the same DAC. A sliding scale method of channel smoothing (described later) was employed to give a differential non-linearity of $<0{\cdot}5\%$. The conversion time was 10 μs.

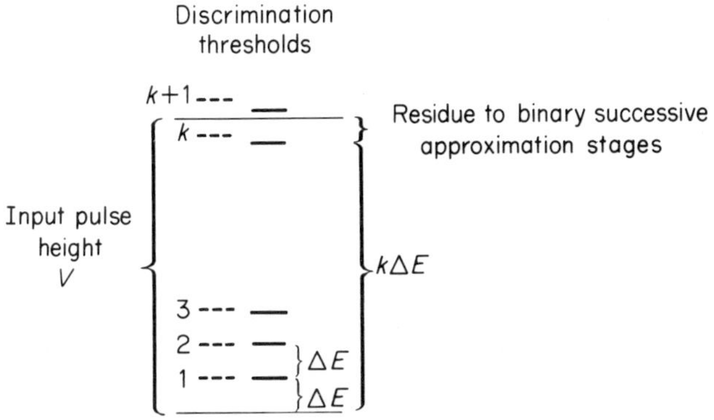

Figure 8.6 Schematic representation of the system used by Bonsignori *et al.*[12]

Lebail[13] has described a 64 channel ADC which uses four stages of successive approximation, each yielding four bits. The first stage consisted of three stacked discriminators with thresholds $E/4$, $E/2$ and $3E/4$. The largest weight that the input pulse exceeded was subsequently subtracted from the input pulse height and the residue passed on to the second stage. The second stage operates in a similar manner but has threshold levels of $E/16$, $E/8$ and $3E/16$. The residue from the second stage is passed to a third stage with threshold levels of $E/64$, $E/32$ and $3E/64$ to complete the coding into one of the 64 channels. The overall conversion time was 50 ns and the ADC had a differential non-linearity of $\sim \pm 1\%$.

8.3 Speed and precision of ADCs

8.3.1 Relative speed and precision of the different types

The precision of an ADC can be characterized by the differential and integral non-linearity, together with the time and temperature stability. In addition the ADC should itself introduce negligible loss of resolution in the pulse height histogram. The latter requirement means that, for instance, input pulses of any constant height should always be classified in a single channel. A useful criterion of this is the so-called channel profile, which is the distribution of the probability against pulse height for classification in the channel in question. The ideal channel profile would therefore be rectangular and of the same width for all channels and equal to the nominal channel width of the instrument. In practice the channel profiles will overlap or underlap slightly due to various causes. In the Wilkinson ADC, for example, if the clock phase is not correlated (i.e. stably synchronized) to the start of the capacitor discharge then, with constant input pulses, the number of clock cycles, and hence the channel number assignment, will be variable. The channel profile in this instance would be of triangular shape (Figure 8.7) with a fwhm equal to

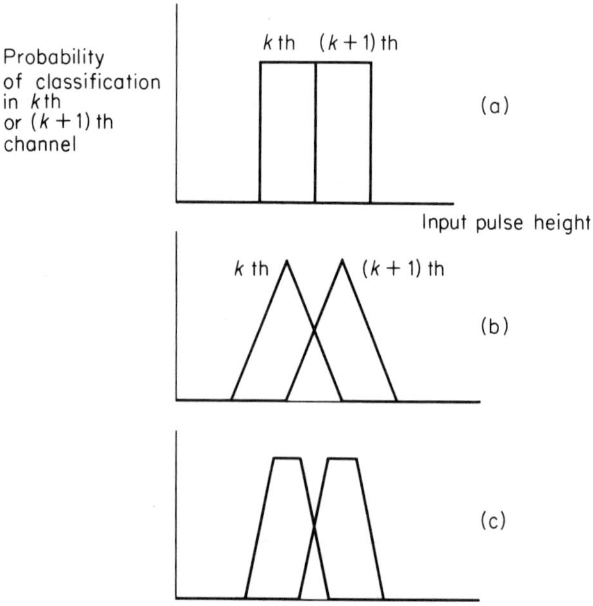

Figure 8.7 Examples of channel profiles. (a) Ideal; (b) Wilkinson ADC without phase correlated clock; (c) Typical profile observed in practice

the nominal channel width. With a phase correlated clock there is still the possibility that the last clock pulse may be imperfectly formed and cause uncertain triggering of the counting register. This can lead to effects that are cyclic with channel number, especially odd–even effects, which reflect the varying sensitivity of the register in the state prior to the arrival of the imperfectly formed last pulse. Finally there is always the problem of the introduction of noise, which will cause spreading of the channel profile. In most ADCs therefore, the channel profile has a trapezoidal shape, intermediate between the ideal rectangular shape and a triangle (Figure 8.7).

In the Wilkinson ADC the speed of conversion is limited by the maximum clock rate that can be processed by current digital circuit techniques. The fastest ADCs that have been fully described have used clock frequencies up to 100 MHz although it appears that this can be increased. The conversion time is of course proportional to k, the channel number.

In the binary successive approximation ADC of n channels the number of stages required is $\log_2 n$ and so the conversion time is proportional to this as a first approximation. At first sight it appears that a binary successive approximation ADC would have a conversion time much less than the Wilkinson, especially for large values of n. This is not necessarily true, as pointed out by Guillon,[14] since the time consuming operations in the Wilkinson ADC are digital operations whereas those in the successive approximation ADC are analogue operations. An analogue operation inevitably is slower than a digital operation, since time must be allowed for the transients to disappear. In a digital operation transients are of no importance, provided it is possible to distinguish between the '0' and '1' states. A further point is that the time allowed for each analogue operation must be increased for a successive approximation ADC with a large number of channels, since the precision of the analogue operations must be greater.

Of the Wilkinson type ADCs the staircase method would be expected to be slower than the linear discharge method since the former uses $2k$ analogue operations (switching on and off of the discharge) per conversion, although these would not be required to be high precision operations. Probably for this reason there have been no publications of designs of a staircase ADC having a speed approaching that of other methods.

The relative merits of the other types of ADC can best be seen by comparing the specifications of the designs described in the next section. In general, however, the Wilkinson ADCs give a good differential non-linearity and are almost universally used for $n \leqslant 1000$. The other types require special attention to the stability of the weights and often require some form of channel smoothing to give a differential non-linearity comparable to the Wilkinson ADC. Recent designs with a large number of channels, however, have given a faster speed than the Wilkinson types.

8.3.2 Sliding scale method of channel smoothing

The differential linearity of any ADC may be greatly improved by the use of the 'sliding scale' method of Cottini, Gatti and Svelto.[15,16] This technique is therefore particularly valuable for successive approximation types of ADC. The principle of the method is that if V is the height of the input pulse whose coding is ultimately required, then a subsidiary voltage E_s, generated by a subsidiary circuit, is added to V. The voltage $V + E_s$ is now fed into the ADC and the coding is found. If now the coding s for E_s alone is known and subtracted from the coding for $V + E_s$, the resultant coding is the one finally used for the input voltage V. In summary, an increment is added in analogue form prior to the ADC and subtracted off again in digital form after the ADC.

The point now is that the voltage E_s is made to take on various values. Considering input pulses of constant height V, if E_s takes on different values for each input pulse, then due to non-linearities in the ADC the final coding for V will not always be the same and may be spread over a channel or two. However, provided E_s takes on values such that s ranges over many values, the statistical distribution of the final channel numbers will be determined by a weighted average over part of the ADCs range and will therefore be insensitive to the characteristics (such as the width) of any one channel. The principle is therefore analogous to the measurement of an object with a rule of well defined length but with inaccurate divisions. If many measurements of the object are made with the rule but using different portions of the rule then an accurate measurement may be obtained by averaging the individual measurements.

The method is obviously most effective when s varies over a large range and Cottini *et al.* assumed that it would take on the values 0 to m with equal probability. This could be arranged in a practical set up by taking random numbers between 0 and m, or by a deterministic scheme whereby the value of s is incremented by one for each event analysed. Assuming the channel profiles to be rectangular, and that the width of the kth channel is ΔE_k, then it was shown[17] that the width of the kth channel after smoothing, $\Delta E'_k$, is given by

$$\Delta E'_k = \frac{1}{m+1} \sum_{s=0}^{m} \Delta E_{k+s}$$

Thus, roughly speaking, the variations in the widths of the channels will be reduced by a factor of $(m + 1)$.

Gatti *et al.*[18] have also considered cases where the values of s take on values given by distributions other than the rectangular one mentioned above. They derived the optimum form of the distribution for the case of the unsmoothed channel boundaries being independent random variables with a constant variance. They also considered the effect of the use of several distributions of s for other sources of error in the channel boundaries, and concluded that in some practical cases a slight improvement should be made if the distribution of s is given by a quadratic or a triangular shaped function.

In the original arrangement of Cottini *et al.* the code s was first generated, and was incremented by one for each input pulse analysed, and the voltage E_s was generated from s by a digital to analogue converter (DAC). In this instance the DAC will be an additional source of non-linearity but this will also be reduced by the smoothing effect. To avoid this source of non-linearity other workers have used a single ADC or DAC to consecutively process both E_s and $V + E_s$ or their codes. Thus, for instance, Bonsignori *et al.*[12] use the following scheme. Starting with the code s the voltage E_s was generated by a

DAC. The voltage $V + E_s$ was then formed and stored on a capacitor. The DAC was then used to generate an increasing voltage and when this equalled the voltage $V + E_s$ the input code was taken and the code s subtracted to give the final coding. Kandiah, Stirling and Trotman[19] started by generating E_s and used the same ADC to obtain codings of both V and $V + E_s$ consecutively. It follows that both these schemes using consecutive processing of V and $V + E_s$ (or their respective codes) take approximately twice the time as the original method.

8.4 Design examples of ADCs

8.4.1 Wilkinson type

A fast ADC of the Wilkinson type has been described by Goursky.[20] The ADC had up to 8192 channels and used a clock frequency of 100 MHz. Figure 8.8(a) shows the block diagram and it will be seen that means are available to set the range of pulse heights accepted for conversion. The pulse stretcher causes the peak value of the positive input pulse to be stored on the L.H. plate of the capacitor. 0·5 μs after the peak of the input pulse, a step of voltage is applied to the R.H. plate of the capacitor. The purpose of this is to provide analogue back biasing so that, by adjustment of the size of the step, the pulse height corresponding to channel zero may be adjusted. The linear discharge of the capacitor is then initiated at an instant synchronized to the phase of a free running quartz controlled clock oscillator. The Schmitt trigger senses the completion of the discharge and causes the discharge current to be stopped at the next clock pulse.

The pulse stretcher, Figure 8.8(b), basically used the principle of the amplified diode arrangement but transistors T2 and T3 could also act as a trigger circuit for sensing the instant when the capacitor was fully discharged by the constant current. T2, T3, T5 and T6 act as amplifier stages during the charging up of C. Immediately after the peak of the input pulse, diode D2 blocks and the capacitor plate is effectively isolated, except for the small contribution from the leakage currents of the diode and the FETs T7 and T10. At the same instant diode D1 becomes conducting and causes positive feedback between T2 and T3 and transforms the pair into a Schmitt trigger. This trigger only resets when the capacitor C has been discharged to near-zero and so provides the signal which defines the length of the train of clock pulses. The linear discharge of the capacitor is effected by T10 becoming conductive. The discharge current was stabilized to within 10^{-5}/°C.

The linear gate used the principle of Figure 8.9. Quiescently T_7 and T_8 are blocked and a current I_0 flows through T_9 into T_6. The input pulse is converted to a current signal I_s in Figure 8.9 so that a current $2I_s$ flows mainly through R_1, there being negligible current through R_2, and negligible output signal therefore. The presence of the negative gate signals on the bases of

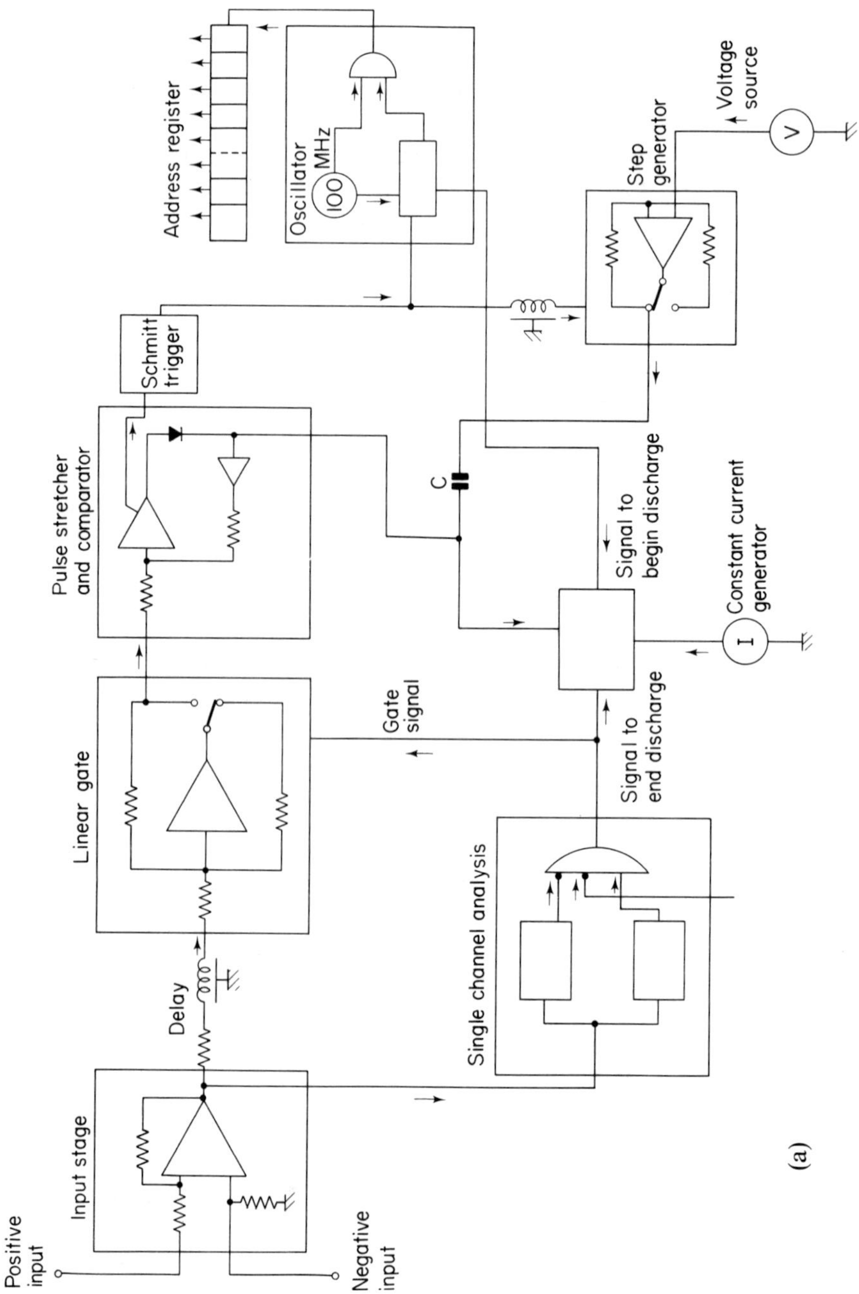
Positive input
Negative input
Input stage
Delay
Linear gate
Pulse stretcher and comparator
Schmitt trigger
Address register
Oscillator
100 MHz
C
Gate signal
Signal to end discharge
Signal to begin discharge
Single channel analysis
Constant current generator
I
Step generator
Voltage source
V

(a)

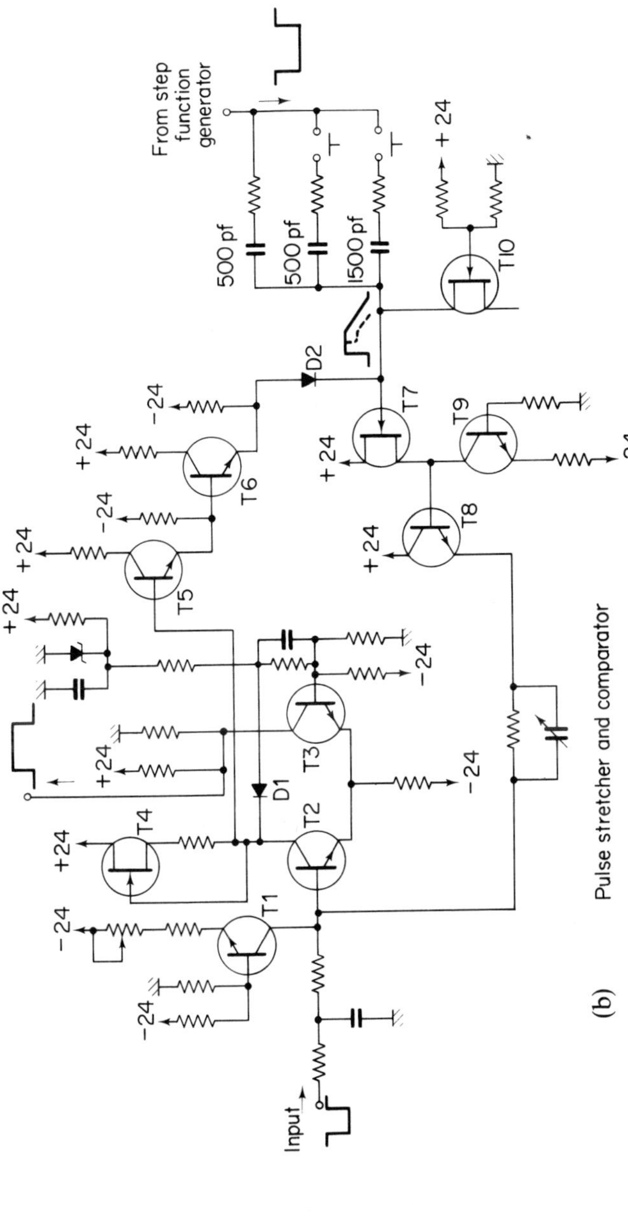

Figure 8.8 ADC of Goursky.[20] (a) Block diagram. (b) Circuit of the pulse stretcher

T_6 and T_7 causes T_6 and T_9 to block and T_7 and T_8 to conduct. Thus the signal current $2I_s$ now flows mainly through R_2 and an output is obtained. The circuit may be seen as essentially a symmetrical two way switch diverting the signal current into either R_1 or R_2. In either state negative feedback is assured by means of the resistors R_3 and R_4 which are connected to the amplifier input in a shunt feedback (operational amplifier) configuration.

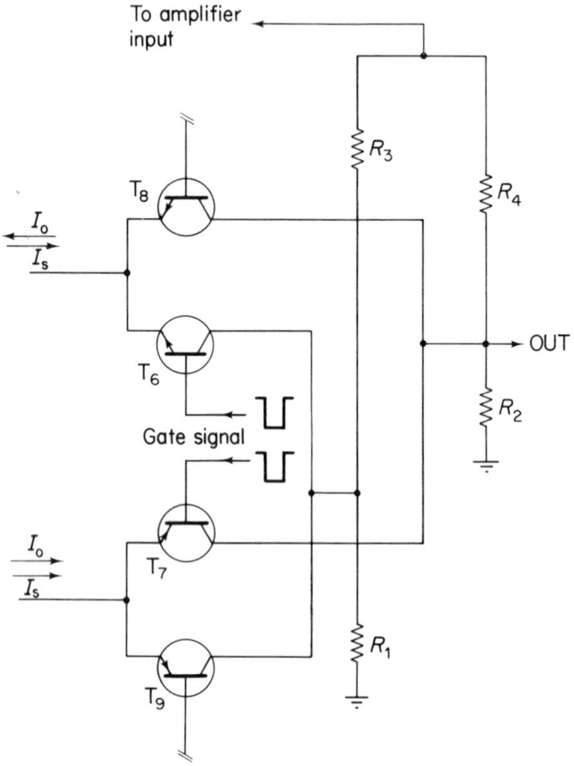

Figure 8.9 Principle of the linear gate used by Goursky[20]

For input pulses in the range 0 to ± 5 volts the stability of transmission when open was $5 \times 10^{-5}/°C$ and $0{\cdot}2\,mV/°C$ and the attenuation when closed was 10^{-3}.

The step generator was realized by a circuit similar to that used for the linear gate (Figure 8.9) but with T_7, T_9 and I_s omitted. In this case the current I_0, whose magnitude determines the size of the step, quiescently flows through R_2 and T_8, since T_6 is normally blocked. When a positive signal is applied to the base of T_6 it conducts and T_8 blocks, so diverting the current I_0 into R_1. The voltage across R_1 was used as the step and, after a pair of emitter follower stages, was applied to the lower plate of the capacitor. The stability of the step was 0·2 mV/°C.

The following summarizes the performance of the overall system: input DC coupled to ADC; clock frequency 100 MHz; available number of channels, N, 64 to 8192; mean conversion time $(5 + N/200)$ μs for $N \geqslant 1024$ and 10 μs for $N < 1024$; maximum input pulse height 2·5, 5 or 10 volts positive or negative with 0·1 to 10 μs rise time; integral and differential non-linearity $\pm < 10^{-4}$ and $\pm < 2 \times 10^{-3}$ respectively; stability of conversion $< 5 \times 10^{-5}$/°C; stability of baseline <0·2 mV/°C.

8.4.2 *Successive approximation type*

An example of the binary successive approximation type of ADC has been described by Kandiah, Stirling and Trotman.[19] The instrument used 13 stages of binary successive approximation to give 8192 channels, with a basic conversion time of 26 μs although, to achieve a differential non-linearity of <1 %, channel smoothing was used and this increased the conversion time to 60 μs.

In the conventional successive approximation method, a number of weights $E/2$, $E/4$, $E/8$, . . . , etc., E being the maximum pulse amplitude, are successively compared with, and if necessary subtracted from, the input pulse height or its residue. For n bit conversion n weights are required which must all be accurately defined. The system used by Kandiah *et al.* differs from the conventional scheme in that only one weight of magnitude $E/2$ is used; after the first comparison of V with this, and after subtraction if necessary, V or its residue is amplified by a factor of two and again compared with the same weight. This process is repeated a further $n - 1$ times to give n bit conversion. The advantage of this recycling of the signal residues through a single comparison circuit is that the number of high precision components is kept to a minimum. Figure 8.10 shows the principle of the method.

The sample and hold circuits are controlled by sampling pulses, not shown in Figure 8.10. Normally each sample and hold circuit produces an output that accurately follows the input signals but, when the sampling pulse is applied, the output holds at its value at the instant of arrival of the sampling pulse. For each of the n cycles of the successive approximation, the two sample and hold circuits in Figure 8.10 are provided alternatively with sampling pulses so that two sampling pulses per cycle are used.

The purpose of the gated OR circuit, essentially a two way switch, is first to allow the input signal to be admitted for the first comparison; it is then gated to the other position so as to provide a path for the recirculation of the residues in the other comparisons.

The input signal V from a pulse stretcher is therefore first presented to the $E/2$ level discriminator and if it is greater than $E/2$ a '1' is generated at the discriminator output, otherwise a '0' is generated, This is the coding process for the most significant bit. For the case $V > E/2$ the weight generator supplies a voltage $-E/2$ at the summing unit input, but if $V < E/2$ the weight generator supplies zero voltage. A sampling pulse is now applied to the sample and hold circuit No. 1, which therefore presents the voltage V at the other input to the summing unit. The voltage at the input to sample and hold circuit No. 2 is therefore $2(V - E/2)$ for $V > E/2$, or $2V$ for $V < E/2$. The second sampling pulse, applied to the sample and hold circuit No. 2, therefore causes this

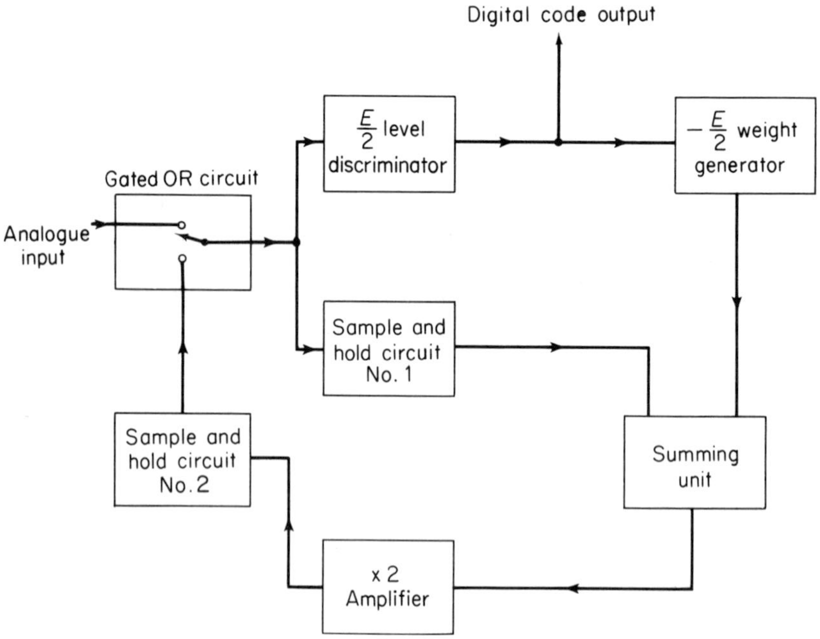

Figure 8.10 Principle of the ADC of Kandiah, Stirling and Trotman[19]

voltage to appear at its output; hence it is recycled through the gated OR switch, now in the lower position for the second comparison and it is then recycled for the second comparison and the generation of the next most significant bit. The process is repeated until all the bits have been generated.

Figure 8.11 shows the principle of the sample and hold circuits. The long tailed pair of matched transistors T_1 and T_2 compare the input and output levels and feed a current proportional to the difference through the common base transistor T_3 to the holding capacitor C. The $\times 1$ buffer amplifier has an FET input stage to preserve a high input impedance and, together with T_2 and T_3, completes a feedback loop. Since the open loop gain is high ($>10^5$) the closed loop gain is almost exactly unity. When the sampling pulse arrives, the currents through T_3 and T_4 are cut off and the output remains steady at the value immediately before the sampling pulse. This output voltage remains steady due to the nearly matched leakage currents of T_3 and T_4. The full version of the circuit incorporated complementary emitter followers inserted between T_3 and T_4 and the capacitor C to improve the response time during 'follow'. In the case of the sample and hold circuit No. 2 the summing unit and $\times 2$ amplification were also included by means of a slight modification to the feedback arrangement.

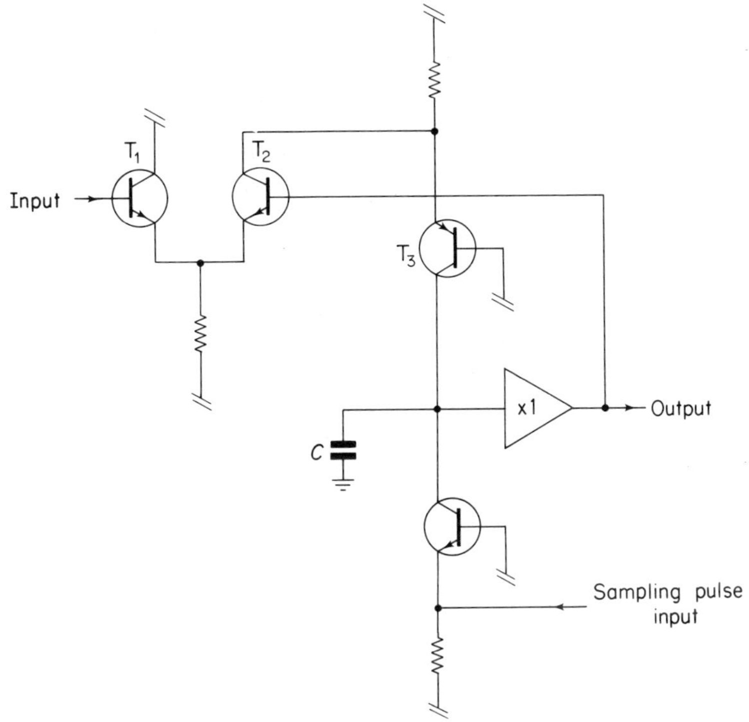

Figure 8.11 Principle of the sample and hold circuit used in the ADC of Kandiah, Stirling and Trotman[19]

The $E/2$ discriminator consisted of a difference amplifier with the input applied to the base of a long tailed pair and the voltage $E/2$ applied to the other base. The difference signal was amplified ~ 1000 times so that the output switched rapidly as the input signal exceeded the $E/2$ threshold. In order to reduce the differential heating in the two input transistors a special feedback circuit was used to maintain approximately equal power dissipation in them.

The overall converter required very few high precision components; three high stability high precision resistors were necessary to determine the $\times 2$ amplification and the weight generation, and two high stability resistors and two high stability potentiometers were required in the $E/2$ discriminator and summing circuit. The basic ADC had a differential non-linearity of $\pm <40\,\%$ which arose primarily from a dielectric absorption of charge by the insulating circuit board in the follow and hold circuits. Although there is some prospect of this charge absorption being reduced in future, the present design reduced the differential non-linearity to an acceptable value by the use of a channel smoothing method. The method used has already been described in principle (Subsection 8.3.2). In practice the incremental voltage E_s was generated first, and stored on an additional follow and hold circuit, not shown in the block diagram Figure 8.10. The code s was then determined while the value $V + E_s$ was stored on another sample and hold circuit, to be fed into the converter after the s coding was completed. The coding of E_s and of $V + E_s$ was taken to 15 bits but, after subtraction, the two least significant bits were discarded to give the final 13 bit coding for V. This sequential coding of E_s and $V + E_s$ increased the conversion time to 60 μs and gave a differential non-linearity of $\pm <1\,\%$.

Digital back biasing was employed to set the pulse height corresponding to channel zero. Accepting pulse heights in the range 0 to 6 volts the zero channel and full scale stability were $20\,\mu V/°C$ and $2 \times 10^{-5}/°C$ respectively. The channel profile was essentially rectangular and the ADC had a full scale stability of <40 p.p.m. when the pulse rate was varied from 500 Hz to 10 kHz.

Kurz[21] has outlined the design of a fast and precise 2048 channel ADC using three stages of successive approximations in four bit stages. The input voltage is compared simultaneously to fifteen levels $E/16$, $E/8 \ldots 15E/16$ and depending on which level is exceeded the corresponding amount is subtracted from the input signal; the residue is amplified by a factor of 12 and re-presented to the same level comparison circuit. After the second comparison and subtraction, the residue is again amplified by a factor of 12 before being compared to the same fifteen levels for the third and last time. It is apparent that that the total number of channels in principle is threfore $12 \times 12 \times 16 = 2303$; however, since channel smoothing was used with the variable s ranging over 0 to 128, the number of channels available for the actual classification of events after smoothing was less and was fixed to be 2048. The choice of the factor of amplification of 12 between stages of comparison was made so as to give a certain amount of redundancy in the conversion. This means that in the case of an input voltage corresponding to one of the primary stage levels $E/16$, $E/8, \ldots,$ etc., the ultimate channel coding might be obtained in different ways, depending on whether the practical realization of the relevant level in the comparison circuit is in error on the low side, or on the high side, or is precise. The effect of this redundancy is to reduce errors in the channel boundaries due to inaccurate realization of the comparison levels and also in the multiplication factor between stages. It does not, however, reduce errors arising in the formation of the residues after the first and second comparisons, which may arise from inaccuracy of the realization of the subtraction 'weights'.

A sliding scale method of channel smoothing was used and in this case the code *s* was sequentially advanced by one for each analysis event. *s* ranged from 0 to 128 and E_s was generated by an auxiliary DAC and so did not increase the conversion time beyond that of the main ADC. In all this led to a conversion time of 3 μs; the differential non-linearity was ~0·15 %. The channel profile was essentially rectangular and the instrument had a zero level stability of 6 p.p.m. per °C. It should be noted that the above figures exclude the contribution from any pulse stretcher and were obtained with a pulse generator.

8.4.3 Dual ramp Wilkinson

Patzelt[11] has described a 8192 channel ADC which uses two Wilkinson-type linear discharge processes. Figure 8.12 shows the principle of the method. The

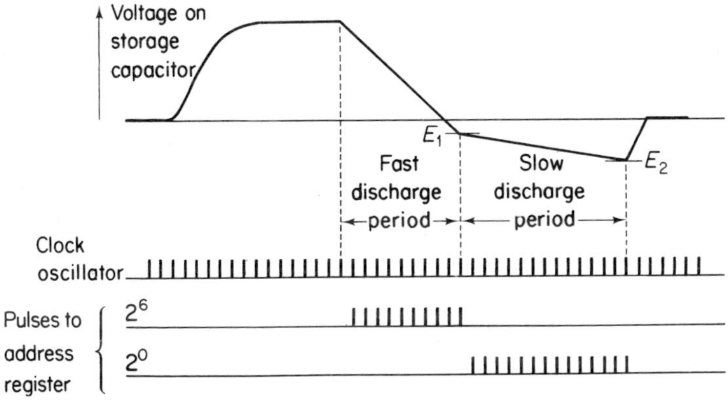

Figure 8.12 Principle of the ADC of Patzelt[11]

capacitor carrying the peak height of the input pulse is first discharged rapidly by a large current (6·4 mA), and during this fast discharge the memory address register is advanced by 64 units by each clock pulse. When the capacitor voltage reaches E_1 (about −200 mV) the next clock pulse causes the discharge current to be reduced to 0·1 mA and during this slow discharge period the memory address register is advanced by one for each clock pulse. When the capacitor voltage reaches E_2, the next clock pulse terminates the slow discharge. The method may be regarded as analogous to the making of first a coarse measurement with a rule with large divisions and then measuring the residue with a second rule with fine divisions. As discussed earlier, if the fast and slow discharge currents are not exactly in the ratio 64:1 this will cause overlap or underlap of the channels next to the coarse subdivision (i.e. at 64 channel intervals). To prevent the differential non-linearity from this source being greater than 1 %, the ratio of the fast to the slow discharge currents was realized with an accuracy of 1 part in 10^4.

It was important to stably synchronize the starting and stopping of the fast discharge period to the clock pulses. The clock frequency used was 20 MHz and so, during the fast discharge, one clock period of 50 ns corresponded to 64 channels. Thus it was necessary for the starting and stopping of the fast discharge to have a time jitter, with respect to the clock pulses, of less than

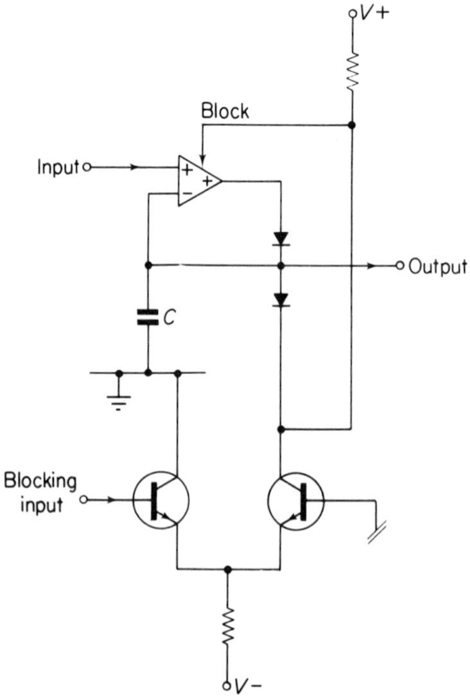

Figure 8.13 Principle of the pulse stretcher circuit of Patzelt[11]

0·1 ns in order to make the corresponding jitter in channel position less than 13 % of one channel width. Even when jitter is negligible there will be a systematic error since the first and last periods of the fast discharge will not correspond to exactly 64 channels, due to imperfect switching of the discharge current. Hence, in order to maintain linearity down to small channel numbers, the fast discharge is arranged to go down to a negative level E_1 (about −200 mV) so that in every conversion several periods of fast discharge are included.

The pulse stretching was realized by two stretcher stages placed in series The second of these is shown in Figure 8.13 and it will be seen to be of the amplified diode variety. In order that effects due to transients, and to stored charge on the diode, could be kept to a minimum a large storage capacitor C(4 nF) was used, and the forward diode current during charging was arranged to be small. This meant that the capacitor charging was slow and could not follow a rapid input pulse accurately. Another pulse stretcher was therefore placed prior to the main pulse stretcher which utilized a similar circuit except that the capacitor had a value of 470 pF, so the charging of this was sufficiently rapid to follow the input pulse.

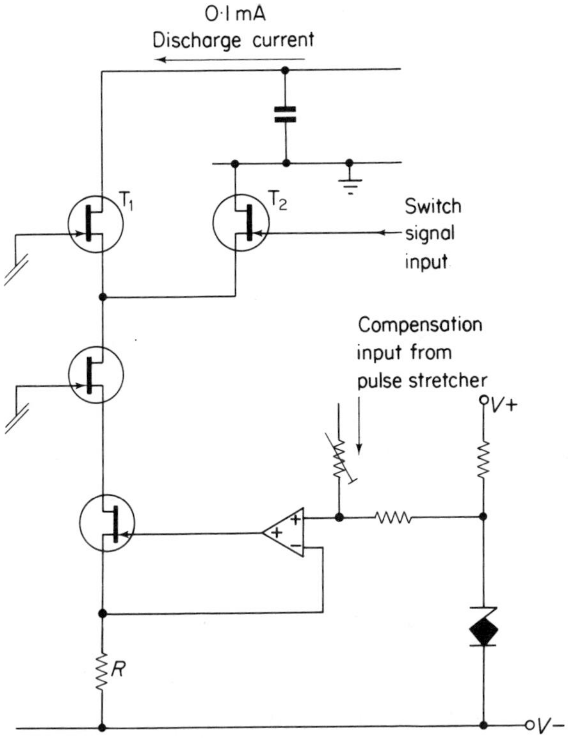

Figure 8.14 Principle of the discharge current generator and switch used in the ADC of Patzelt[11]

In the pulse stretcher circuit the amplifier used a pair of FETs for the differential input stage to give the necessary high input impedence together with high linearity and stability. The amplifier had a gain of ~500 and this was sufficient to reduce the forward resistance of the diode so that the capacitor voltage could take up a value within 1 mV of the input voltage. After sufficient time had elapsed (2.5 μs) for this to occur, the amplifier was blocked so as to isolate the capacitor. After a short time to allow the transients to decay, the fast discharge was initiated in coincidence with the next clock pulse.

During the discharge of the capacitor, the sensing of the equivalence of the capacitor voltage with E_1 or E_2 (Figure 8.12) was achieved by the use of integrated circuit differential amplifiers, which have the necessary properties of high differential gain and stability.

The fast (6·4 mA) and slow (0·1 mA) discharge currents were provided by current generators of 6·3 mA and 0·1 mA, the former of these being switched off at the termination of the fast discharge. Figure 8.14 shows the circuit of the 0·1 mA current generator. The amplifier, of gain $\sim$500, compares the voltage across R, due to the discharge current, with that across the temperature compensated Zener diode. In addition, an arrangement was included to compensate for stored charge effects in the dielectric of the main storage capacitor. The residual charge retained by the dielectric at the end of the fast discharge is approximately proportional to the input signal amplitude; to compensate for this the slow discharge current was increased by an amount proportional to the signal amplitude.

The switching of the discharge currents was accomplished by the differential pair of FETs T_1 and T_2 in Figure 8.14. The use of FETs meant that little of the discharge current was lost to the gates and that charge storage effects were negligible.

The clock oscillator was free running and quartz controlled. The logic part of the ADC was realized with fast integrated circuits. The overall ADC had a maximum conversion time of 12 μs, integral and differential non-linearity of $\pm < 2{\cdot}5 \times 10^{-4}$ and $\pm < 0{\cdot}5\%$ and the integral stability was 0·5 channels per °C. The channel profile was measured and found to be essentially rectangular.

8.5 Spectral stabilization

The stability of a spectrometry system may be improved by the use of a servo stabilizing system. Instabilities in the amplifier gain or ADC conversion gain, for instance, will cause a peak in the pulse height histogram to drift over a number of channels. Essentially the servo stabilizing technique is to continually sense the position of some suitable reference peak, compare its position with its correct position and generate an error signal corresponding to the difference in position. The error signal is then used to control the gain of the amplifier or ADC. In practice the reference peak selected may be a suitable peak already existent in the pulse height spectrum being investigated, or it may be superimposed on the existent spectrum by the use of either a pulse generator feeding into the amplifier input or an auxiliary radioactive source.

Figure 8.15 shows the principle of servo stabilization as most commonly implemented. The position of the reference peak in the pulse height spectrum is sensed by means of a pair of windows. If the system is initially set up with a gain G so that the reference peak is in its correct position without servo stabilization, then when the servo loop is connected, equal numbers of reference pulses will fall into both windows and so there will be no error signal generated in the servo loop and the gain G of the system will not be altered. Suppose, owing to instabilities in the system, the reference peak suddenly shifts to a lower position in the pulse height spectrum. In this case more reference pulses will fall into the lower window than the upper, and now the servo loop generates an error signal which causes the variable gain element to increase its gain. Eventually the gain is increased sufficiently to restore the reference peak to almost its original position, so that again there is little error signal and the system is in balance.

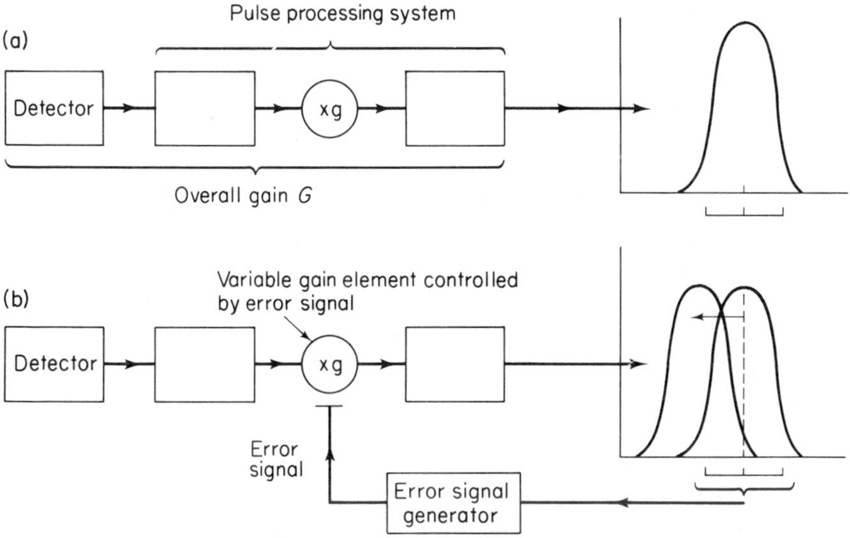

Figure 8.15 Spectrometry system used (a) without and (b) with servo stabilization of the gain of the system.

For a spectrometry system using a MCA, once a pulse has been coded in digital form by the ADC, no further sources of instability are possible. The use of an ADC in a pulse stabilization arrangement is therefore particularly attractive since the reference pulses are processed by the same ADC which is used for the processing of all pulses. Any drift in the ADC will therefore be encompassed by the servo stabilization. A further advantage is that the variable gain element can be realized by causing the conversion gain of the ADC to be varied, in a Wilkinson ADC for example, by varying the discharge current or the clock frequency. In addition, a second servo loop can be incorporated to stabilize the zero level (see later) and the whole of the servo circuitry can be buit into the MCA unit.

For systems using a single channel analyser, the circuit sensing the position of the reference peak is usually a separate unit. The servo stabilization therefore only encompasses the amplifier output; any drifts in the single channel analyser are not included. Such systems have been variously described and the gain of the detector–amplifier system may be controlled by, for instance, varying the voltage on the photo-multiplier tube[22, 23] or by the use of such devices as Raysistors.[24] In the present context however only systems using an ADC will be considered.

Since the servo stabilization should ideally encompass the detector as well as the amplifiers and the ADC, the best solution from this point of view is to use reference pulses derived from nuclear radiation entering the detector. The presence of a suitable peak in the existent pulse height spectrum cannot, however, be guaranteed and, if an auxiliary source of radiation is used, the size of the pulses produced cannot be easily changed to give a suitable value for differing pulse height ranges likely to be encountered in experiments. On the other hand, electronically generated reference pulses which are fed into the preamplifier input exclude the detector from the system encompassed in the servo stabilzation. Most published systems use this arrangement, however, since they allow the use of two different sizes of reference pulses which allow both the gain of the system and the zero level to be servo stabilized. For Ge(Li) detector systems, which probably represent the most critical application in view of the high resolution attainable, the gain of the detector appears to be fairly stable[25] so its exclusion from servo stabilization is probably not very important. A further advantage of electronically generated reference pulses is that their time position is known and so it is possible to gate the system so that the reference pulses are not written into the memory. In addition it is possible to test for the presence of a detector pulse at the instant of generation and so prevent any reference pulses which pile up on detector pulses from being processed by the servo control. Finally a solution which seems to combine the best of both the above systems is to use a light pulser. Unfortunately, according to Gere and Miller[26] a light pulser of high stability is not available so the problem then resolves into adequate stabilization of the light source.

Unlike the simple system of Figure 8.15 which stabilizes only the gain of the system, some published systems stabilize both the gain and the zero level. This is accomplished by using two different sized reference pulses which are generated alternatively, one of large value, which produces an error signal which controls the gain as in the example of Figure 8.15, and the other of small value which produces an error signal controlling the zero level of the ADC. Since the reference pulse controlling the zero level stabilization cannot be infinitely small there is inevitably some interaction between the gain control and zero control by the two servo systems. An incidental advantage of the zero stabilizing servo loop is that if the size of the small test pulse is deliberately increased from its normal value (i.e. that value in the correct proportion to the large test pulse to give the same ratio in channel position) then the resultant shift in the zero due to the servo action is equivalent to that obtainable with back biasing (Subsection 8.1.3).

Some consideration must now be given to the means by which the error signal in the servo loop of Figure 8.15 is generated. In the case of a system using an ADC we will suppose that the system is operating correctly when the reference peak is located on the boundary between channels m and $m + 1$. The reference pulses will have codings statistically distributed about their mean. It may be noted here that, even if electronically generated reference pulses of constant size are used, there will still be a statistical spread in size at the ADC input, due to noise in the system. The most common way of deriving the error signal is to cause the error signal to be incremented or decremented by a constant amount by each reference pulse; hence the gain is increased or decreased by a constant amount, Δg, by each reference pulse. If the coding of the reference pulse is m or less, the error signal, and hence the gain, is increased by the constant amount Δg. If the coding is greater than m the error signal causes the gain of the system to be decreased by Δg. Thus the servo loop will continually 'hunt' so as to tend to stabilize the reference peak on the boundary between channels m and $m + 1$. If the gain of the detector–amplifier–ADC chain, for example, decreases due to any cause then more reference pulses will be coded with m or less than will be coded with greater than m, and the error signal, and hence the gain, will be increased in steps until, on average, equal numbers of reference pulses fall in the two categories. In the scheme of Figure 8.15 this is equivalent to the use of two wide windows adjoining at the $m/m + 1$ channel boundary. With electronically generated reference pulses suitable gating circuits can be provided, so that only the reference pulse codings are accepted by the error signal generation circuitry; at the same time the codings are not written into the memory of the MCA. If the system is intended to operate on reference pulses from a nuclear source then it is necessary to restrict the window widths to the region of the reference peak.

A simple method of implementing the above scheme in a Wilkinson ADC is to choose m to be a number of value 2^x where x is an integer. When the ADC feeds the train of clock pulses, constituting the coding of the reference pulse, into the address register then, by inspection of the output of the xth stage of this register, the existence of the carry pulse to the next stage will indicate if the coding is greater than 2^x.

The choice of the constant amount Δg, by which the gain is incremented or decremented by each reference pulse, is critical. Since the reference pulse codings have a statistical size distribution the error signal, and hence the constant gain increments, Δg, will be statistically distributed between positive and negative increments. This resultant statistical variation of the gain of the system will cause some loss of resolution and the peaks in the pulse height spectrum will be broadened. The broadening from this source will be directly dependent on the magnitude of Δg chosen. If, on the other hand Δg is chosen to be small so as to make the loss of resolution small, then it is easy to see that the servo system will be slow to correct drifts in the gain of the system, i.e. the stabilization time constant will be large. If the drifts in the gain of the system can be represented by a RMS value, then owing to the slowness of the response of the servo loop the gain drifts will not be perfectly compensated by the servo action and this also will produce a component of line broadening in the pulse height spectrum. Thus the choice of too large a value of Δg will cause excessive loss of resolution from the high gain of the servo loop and its statistical fluctuations; the choice of a small value will cause excessive loss of resolution from drifts in the gain of the detector–amplifier–ADC system. The optimum value in a given system will depend on the magnitude and speed of the gain drifts in the system and on the width of the reference peak. The choice of the optimum value of Δg has been examined theoretically by several workers.[27, 28, 29] Hewitt[30] extended the theory to take into account the use of a second servo loop to stabilize the zero level.

The above discussion has all centred on one method, the step method of deriving the error signal from the digital codings of the reference pulses. Other methods[31, 32, 33] have been used from time to time. Dudley[28] showed that some methods of statistical processing of the reference pulse codings prior to error signal generation yielded no benefit although, for a reference peak of Gaussian shape, he was able to specify optimum window widths. Konrad[33] derived a figure of merit for any servo stabilization system and compared various methods of generating the error signal, including a method based on the use of two ratemeters to register reference pulses with codings $\leqslant m$ and $>m$.

Using the step method Klopf *et al.*[25] examined experimentally the effect of the choice of Δg where the gain drift was simulated. The servo stabilization was applied to an ADC having 1024, 2048 or 4096 channels. Electronically generated reference pulses were fed into the input of the charge sensitive preamplifier, which was part of a Ge(Li) spectrometer system. Large and small reference pulses of magnitudes equivalent to $\frac{1}{8}$ and $\frac{7}{8}$ of full scale were generated alternatively at a rate of 40 per second, and the error signals produced were used to control the zero bias level and the gain of the ADC respectively. When used to detect a 3·25 MeV gamma ray with temperature variations of $\pm 3\,°C$ the peak position shifted in 70 hours by 0·27 keV (0·01 %).

8.6 Additional facilities on MCAs

Modern commercial versions of MCAs may be obtained with a wide variety of facilities. Most have some form of oscilloscope or television type display of the accumulated pulse height spectrum. Essentially this requires sequential read-out of the contents of the memory locations into a digital to analogue converter; in some designs this takes place concomitantly with the accumulation of data, the latter having priority if it should happen that at any address both functions coincide. The contents of the memory may also be sequentially read out, either in digital form to a paper tape punch, magnetic tape recorder or typewriter, or in analogue form to a chart or XY recorder.

It is usually possible to split the available number of memory locations into different sections. In this way, in a 4000 channel instrument for instance, the instrument can be used to store four separate pulse height histograms, each of 1000 channels. For example, one section may be used for the storage of data from a source and another for the storage of the background. Alternatively, data from four detectors may be accumulated concomitantly by the use of a routing circuit. The incoming pulses from all detectors are passed into the same ADC but the resultant codings are routed to one of the four memory sections according to which detector furnished the signal to the routing circuit.

Simple arithmetic operations may be included in the facilities available. Examples of these are when it is possible to transfer, add or subtract the contents of one section of the memory from the corresponding contents of another section, and this is obviously of use in subtracting backgrounds. Other facilities allow integration of, for example, the number of counts in a peak, by adding up the total number of counts contained in channels between any pair of specified channel numbers. It is possible to arrange for the multiplication of the contents of all the channels of any section of the memory by a constant. This, together with a means for reading standard pulse height spectra into the memory from paper or magnetic tape, provides a means for spectrum stripping. The constant by which the standard spectrum is multiplied can be varied and, by trial and error, the value of it giving the best stripping can be found.

Most MCAs include additional circuitry to enable them to be used as multiscalers. In this mode of operation a n channel MCA may be used as n scalers, into any one of which incoming counts may be stored. The address, i.e. channel number, at which the count is to be stored can be advanced by a separate input pulse. The incoming pulses are first counted in a register and on receipt of the address advance pulse the contents of the register are stored in the next memory address. If the address advance pulse occurs at regular intervals then the contents of the memory give the time course of the counting rate. In this form it also finds application for slow time of flight studies.

In most MCAs, particularly those using Wilkinson type ADCs, the conversion time for an input pulse, and hence the dead time, is dependent on the magnitude of the pulse. The absence of a fixed dead time means that the normal method of correction for lost counts (Appendix B.5) cannot be applied. One method of avoiding the necessity of correcting for counting losses is by allowing the analysis to proceed for a known 'live' time, and this may be implemented as follows. The timer which controls the time the analysis is

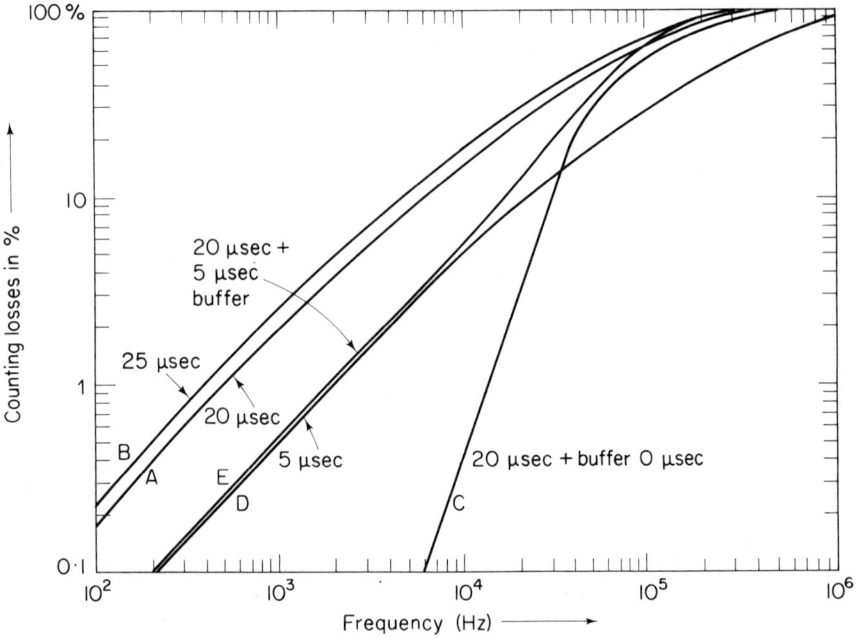

Figure 8.16 Counting loss plotted against mean input pulse rate for various cases, according to Koeman.[5] A. Converter with 20 μs conversion time without analogue store. B. As A, but with 25 μs conversion time (represents also the average conversion time of a system with 20 μs conversion time of a converter plus 5 μs processing time of an analogue store at high repetition rates). C. Converter with 20 μs conversion time + analogue store without processing time (imaginary). D. Converter with 5 μs conversion time (represents the average processing time of converter plus analogue store at low repetition rates). E. Converter with 20 μs conversion time and analogue store with 5 μs processing time

to run is driven by an oscillator via a gate. The MCA is 'dead' whenever the linear gate on its input is closed during the processing of an event; the oscillator gate is arranged to be closed by the same signal which operates the linear gate. Thus the oscillator only advances the timer during periods when the linear gate is open and so the analysis runs for a defined amount of live time, and no counting loss correction is necessary. The same gated oscillator is often used to operate a meter indicating the mean dead time expressed as a percentage of the actual (i.e. real) time.

There are two disadvantages in the use of live time. First, the actual or 'real' time that an analysis runs is a variable quantity since it depends on the counting rate. This means that the system cannot be used for automatic data taking where, for example, a source and background are alternatively analysed at preset times. Secondly, if the intensity or shape of the pulse height spectrum is changing during the analysis the accumulated spectrum cannot be properly interpreted.

One method of avoiding some of these problems is to fix the dead time at a large enough value and run the analysis for a known real time. A more satisfactory solution has been proposed by Harms[34] in which the analysis is run for a fixed real time and, for every event which is lost because of dead time, the memory contents of the MCA are increased in proportion to the pulse height distribution at that instant. This almost instantaneous correction of the accummulated spectrum for events lost can therefore be applied in the case of intensities or spectral shapes which change during the course of the analysis. The basis of the method was that if an input pulse is lost due to the MCA being dead, then when the next event is processed by the MCA into, say, channel k, the count content of channel k is increased, not by one, but by two. The extra counts accumulated in this way will be distributed according to the pulse height distribution at that instant and so should, on average, compensate for the lost events. One method of detecting lost events is to use an add–subtract register, the add input of which is operated by pulses arriving at the MCA input and the subtract input of which is operated each time a count is accumulated in the memory. If the content of the register is one or more, the double increment mechanism is activated after the conversion of the next event. Masters and East[35] have examined the method by computer simulation of a wide variety of pulse height spectra, including those with changing intensity and shape, and found that the technique accurately corrected for dead time effects.

Finally a useful facility for reducing dead time losses is the provision of an analogue buffer store. The function of this is to store the peak height of an incoming pulse until the ADC is free to accept and process it. In effect therefore this acts as a de-randomizing buffer store and the consequent reduction of counting loss is considered in Appendix B.5. The design of such a store has many problems in common with the design of the pulse stretcher and a recent example has been given by Koeman.[5] The store had a dead time of the order of 5 μs and Figure 8.16 shows the reduction in counting loss for the case of an ADC with a 20 μs conversion time. The magnitude of the pulse height was stored on a capacitor, the effects of dielectric charge absorption being offset by injecting a compensating current, and it was possible to store a 5 V signal for 100 μs with an accuracy of 0·6 mV.

References

1. T. L. Emmer, *IEEE Trans. Nucl. Sci.*, NS**12**, No. 1, 329 (1965).
2. K. Kandiah, *Nuclear Electronics*, Proc. Int. Conf., Belgrade, Vol. 2, p. 11, I.A.E.A., 1962.
3. P. Casoli and P. F. Manfredi, *IEEE Trans. Nucl. Sci.*, NS**16**, No. 1, 392 (1969)

4. I. Alleva, I. De Lotto, P. F. Manfredi and P. Maranesi, *ISPRA Nuclear Electronics Symposium*, Euratom Report EUR 4289, p. 193 (1969).
5. H. Koeman, *Nucl. Instr. and Meth.*, **86**, 301 (1970).
6. E. Gatti and F. Piva, *Nuovo Cimento*, **10**, 984 (1953).
7. E. Gatti, *Nuovo Cimento*, **11**, 153 (1954).
8. S. Colombo, C. Cottini and E. Gatti, *Nuovo Cimento*, **5**, 748 and 1739 (1957).
9. L. L. Lewyn, *IEEE Trans. Nucl. Sci.*, NS**14**, No. 1, 126 (1967).
10. S. Gorodetzky, P. L. Wendel, Ch. Ring and R. Armbruster, *Nucl. Instr. and Meth.*, **45**, 72 (1966).
11. R. Patzelt, *Nucl. Instr. and Meth.*, **70**, 61 (1969).
12. C. Bonsignori, S. Colombo, G. Granucci and V. Pellegrini, *Int. symp. nuclear electronics, Versailles*, Vol. 2, p. 88, S.F.E.R., 1968.
13. P. Lebail, *Int. symp. nuclear electronics, Versailles*, Vol. 2, p. 93, S.F.E.R., 1968.
14. H. Guillon, *Nucl. Instr. and Meth.*, **43**, 240 (1966).
15. C. Cottini, E. Gatti and V. Svelto, *Nucl. Instr. and Meth.*, **24**, 241 (1963).
16. C. Cottini, E. Gatti and V. Svelto, *Electronique Nucléaire*, Proc. Int. Symp. on Nuclear Electronics, Paris, S.F.E.R., 1963.
17. E. Gatti, P. F. Manfredi and V. Svelto, *Int. symp. nuclear electronics, Versailles*, Vol. 2, p. 89, S.F.E.R., 1968.
18. E. Gatti, P. F. Manfredi, V. Svelto and P. Thieberger, *IEEE Trans. Nucl. Sci.*, NS**16**, No. 4, 10 (1969).
19. K. Kandiah, A. Stirling and D. L. Trotman, *IEEE Trans. Nucl. Sci.*, NS**16**, No. 5, 3 (1969).
20. V. Goursky, *Int. symp. nuclear electronics, Versailles*, Vol. 2, p. 90, S.F.E.R., 1968.
21. R. Kurz, *ISPRA Nuclear electronics symposium*, Euratom Report EUR 4289, p. 179 (1969).
22. D. Williams, G. F. Snelling and J. Pickup, *Nucl. Instr. and Meth.*, **39**, 141 (1966).
23. C. Bacci, V. Bidoli and R. Baldini-Celio, *Nucl. Instr. and Meth.*, **57**, 100 (1967).
24. A. Pakkanen and F. Stenman, *Nucl. Instr. and Meth.*, **44**, 321 (1966).
25. P. Klopf, H. Meyer and H. Verelst, *Int. symp. nuclear electronics, Versailles*, p. 92, S.F.E.R., 1968.
26. E. A. Gere and G. L. Miller, *IEEE Trans. Nucl. Sci.* NS**13**, No. 3, 508 (1966).
27. R. A. Dudley and R. Scarpatetti, *Nucl. Instr. and Meth.*, **25**, 297 (1964).
28. R. A. Dudley, *Nucl. Instr. and Meth.*, **46**, 116 (1967).
29. J. M. Ferguson, *Nucl. Instr. and Meth.*, **58**, 318 (1968).
30. J. S. Hewitt, *IEEE Trans. Nucl. Sci.*, NS**17**, No. 1, 242 (1970).
31. R. L. Chase, *IEEE Trans. Nucl. Sci.*, NS**9**, No. 1, 119 (1962).
32. M. Matoba and I. Kumabe, *Nucl. Instr. and Meth.*, **74**, 70 (1969).
33. M. Konrad, *Semiconductor Nuclear Particle Detectors and their Circuits*, p. 731, U.S. Nat. Acad. Sci., Publication 1593, Nuclear Science Series Report No. 44 (1969).
34. J. Harms, *Nucl. Instr. and Meth.*, **53**, 192 (1967).
35. C. F. Masters and L. V. East, *IEEE Trans. Nucl. Sci.*, NS**17**, No. 3, 383 (1970).

9

Multiparameter and Computer Analysis

9.1 Multiparameter analysis

9.1.1 Introduction

The multichannel analyser is an example of an instrument classifying events by a single parameter, pulse height, into various channels or codings. Multiparameter analysis is the classification of events in terms of a combination of two or more of their parameters. An example of this is when two detectors are used to detect coincidences and it is necessary to have pulse height information about both detections. Such a scheme can be realized by means of the arrangement shown in Figure 9.1. Figure 9.2(a) shows representative results for the ^{60}Co coincidence spectrum, where the display is arranged so that the pulse height from detector 1 is plotted along the X axis; that from detector 2 is plotted along the Y axis and the intensity of any point represents the number of counts corresponding to the particular combination of X and Y coordinates. Figure 9.2(b) shows the same results displayed in a

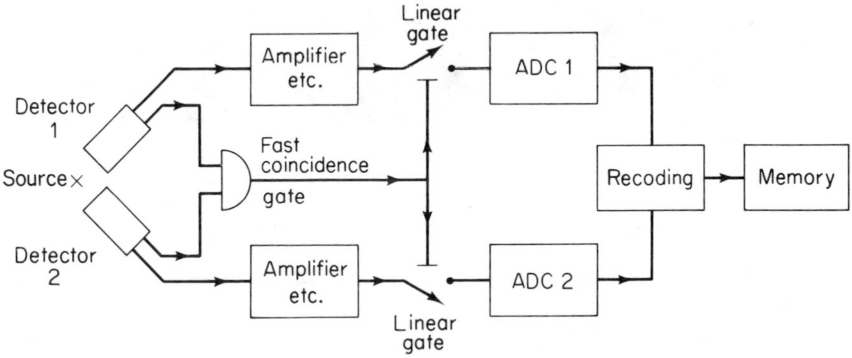

Figure 9.1 Scheme for classifying and recording coincident events in terms of the pair of pulse heights (i.e. giving the two dimensional coincidence pulse height spectrum)

different form as an isometric plot, with the vertical ordinate representing the number of counts.

In other types of experiment, parameters other than pulse height may be of interest. Examples of such parameters are those specifying particle identification information, time information, and in arrays of many detectors which yield information about the spatial or angular position of the incident particle.

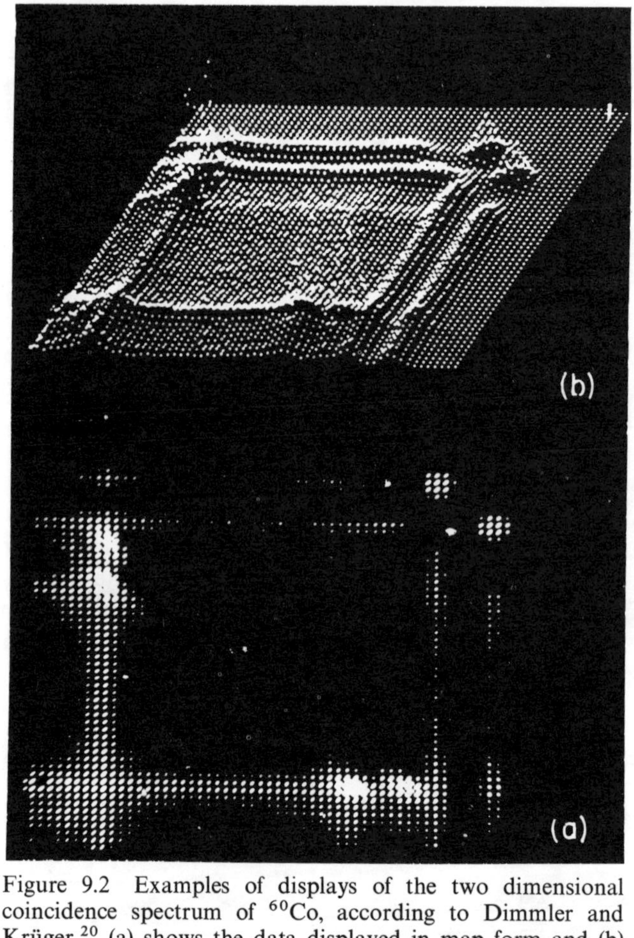

Figure 9.2 Examples of displays of the two dimensional coincidence spectrum of ^{60}Co, according to Dimmler and Krüger.[20] (a) shows the data displayed in map form and (b) shows the same data displayed isometrically

Generally, before digitizing the data, a certain amount of processing of the signals is necessary. This includes testing for coincidence, testing whether the parameter falls in the range of interest etc. Assuming such criteria are satisfied, parameters such as pulse height are then digitized by an ADC and parameters such as time are digitized via a time to digital converter or TAC plus ADC. Certain other types of information may not need the use of an ADC. Thus the information that a particle activates detector number k out of an array may be made available as some form of digital coding of k.

For the sake of simplicity we will, for the most part of this chapter, confine our attention to two parameter analysis systems. In this case, the information of interest is contained in the pair of digital codings corresponding to the two parameters of interest. This pair of digital codings will be termed the 'descriptor' of the event. Ideally, the object of the analysis is to keep tallies of the number of times each of the possible descriptors occurs during the experiment.

Probably the main difficulty in multiparameter analysis is to achieve a sufficient memory capacity to tally all the possible descriptors. If both parameters are classified into 100 channels, the number of possible combinations, i.e. of descriptors, is 10^4. If a memory is provided with 10^4 word capacity then each possible descriptor can be associated with one memory location; the number of events accumulated for any descriptor can be stored as the word of the corresponding memory location of that descriptor. This is essentially the principle used in most MCAs and, to preserve a reasonable working speed, this dictates the use of a fast memory such as ferrite core memory. This type of storage of the data will be termed descriptor–address storage, since each memory address is uniquely associated with each of the possible descriptors. To realize such a system it is only necessary to take the pair of parameter codings making up the descriptor for the event and, by comparatively simple re-coding circuitry, cause them to generate the corresponding memory address code. The memory content (i.e. stored word) at that address is then incremented by one by the same type of read–increment–write cycle used in MCAs.

If, however, both parameters are coded in 1000 channels then there are a possible 10^6 descriptors and the cost of a fast memory of 10^6 word capacity becomes prohibitive. Several solutions to this dilemma have been proposed. One possibility is to make a tape recording of the descriptors as they are produced and to perform the sorting of these later using a general purpose digital computer. The tape may be played back several times, each time a different subset of descriptors being sorted, and by this means it is possible to sort a large number of possible descriptors using a computer of limited memory capacity. There is obviously a practical limit to the number of times a tape may be re-run into the computer however.

Before recording the descriptors on the tape it is preferable to 'de-randomize' the data to some extent. In effect, this means providing several buffer registers before the tape recorder, so that, if two or more descriptors are produced in rapid succession, they can be temporarily stored in these registers while the first is being recorded on the tape. This therefore decreases the number of events lost through the dead time of the tape writing process and, for a continuously running tape, makes more efficient use of the tape. Alexander *et al.*[1] have considered the effect of various numbers of buffers in reducing counting losses, and some of their results are given in Appendix B.5. Alexander and Leng[2] have given details of a complete system for realizing a buffered tape system. Egelstaff and Rae[3] have given a general survey of magnetic tape data recording systems. The magnetic tape may be run continuously or stepwise and the tape replay for the subsequent analysis may be either speeded up or slowed down, according to the requirements of the experiment and the analysis, by a factor of up to 100. Using 1 inch tape it was possible to record a 25 bit word laterally across the tape by means of a multi-track recording head and their system allowed a maximum of 200 words per inch to be recorded.

More recently McClure *et al.*[4] have described a system in which the data is recorded in a form compatible with the current I.B.M. format for $\frac{1}{2}$ inch tape. This uses a step advance transport and a 12 bit data word could be recorded in two six bit segments, each recorded laterally across the tape. With this system it was possible to record a data rate of 500 words (12 bit) per second at a density of 556 words per inch of tape.

The disadvantage of the use of tape recording is that the results are not immediately available so that adjustments to the experiment or rectification of faults cannot be made. Where the number of possible descriptors is large and immediate processing of data is required, there are two alternatives. The first alternative is to use a processor with sufficient memory capacity; in practice this means the on-line use of a large general purpose computer. It may be noted that, even in this case, the fast memory capacity is usually not sufficient to allow descriptor–address allocation of the fast memory. Instead the main storage is located in a drum or disc store, probably allocated on a descriptor–address basis, and this is periodically updated from the fast memory store. Further discussion of such large computer systems is given in Section 9.3.

The second alternative avoids the use of a large store by storing the tallies of certain 'selected' descriptors which only constitute a small fraction of the total possible descriptors. The resultant partial storage of the experimental information may be adequate itself (in the average experiment a large fraction of the possible descriptors will not occur anyway), or it may at least be adequate for the purposes of monitoring an experiment in which the data is also tape recorded for later full analysis.

The selection of the descriptors for which tallying is to be performed may be by direct control of the experimenter or, quite commonly, it may be automatically performed by the analyser, on the basis of the most commonly occurring descriptors. In an experiment having a field of 10^6 possible descriptors if, for example, the memory capacity of the analyser is only sufficient to tally the counts of a selected 4000 descriptors, then the simplest system of selecting the 4000 most frequent descriptors is on the basis of the first 4000 different descriptors to arrive at the beginning of the experiment. This 'first come first served' principle will not truly select all the most frequent descriptors since there remains the possibility of the early chance occurrence of comparatively infrequent descriptors. This is particularly troublesome with methods using zone selection (see the next section) since the chance occurrence of a single statistically infrequent descriptor will cause a whole zone of unwanted descriptors to be selected for further tallying. The situation can be improved by the following procedure. After an initial period of the experiment the tallies of all descriptors are each reduced by one. If the tally of any descriptor is thereby reduced to zero then that descriptor is discarded and the memory location made available for new descriptors. The process can be repeated so that the final selection of 4000 descriptors approximates more nearly to those which are the most frequent.

The association of the selected 4000 descriptors with specific memory locations cannot now be done on the same basis of the simple descriptor–address storage discussed earlier, since the selected descriptors are different for each experiment and so their codes cannot be permanently associated with corresponding memory address codes. Souček[5] has comprehensively reviewed the various techniques for assigning memory locations to the selected descriptors and the following sections describe some of the methods used for the purpose.

The methods of partial classification of certain selected descriptors may use a unit specially built for the purpose or may make use of a small general purpose computer, possibly controlled by a stored program. Most of the systems described have used small computers having memory capacities up to 10,000 words. The comparatively large scale production of such computers usually makes this a cheaper alternative than a specially built unit of similar memory capacity, particularly since its cost can be borne partially by other users. A further advantage is the increased flexibility if part of the memory is used for a stored program.

9.1.2 Optical display

For the display of data corresponding to any pair of parameters, some form of optical display is usually desirable. This may be of the contour map form shown in Figure 9.2(a) where the channel numbers of the individual parameters are displayed as the X and Y coordinates on a cathode ray tube and the number of counts for each descriptor is displayed by a spot of proportional intensity. An alternative is to use an isometric type of display, where the channel number of one parameter corresponds to a deflection on the CRT along the X axis, the channel number of the second parameter to a deflection along the diagonal and the number of counts of any descriptor to the Y deflection (Figure 9.2(b)). In either system of display, the voltages used for the deflection of the beam and control of the intensity are derived by periodically causing the contents of the memory to be sequentially read out. From the descriptor code the two component channel numbers are derived; the appropriate voltages for the control of the CRT are generated by three digital-to-analogue converters.

A useful facility for use with optical display systems of this type is the so-called 'light pen'. The purpose of this is to allow any descriptors of interest in the visual display to be so designated that the descriptor and its tally can be read out in digital form. The light pen contains a photocell optically coupled to the tip and it can be placed on the point of interest on the CRT screen. Since the CRT display is generated sequentially, by means of a coincidence circuit connected to the light pen output it is possible to cause the register contents corresponding to a particular point of the CRT screen to be made available so the details of channel numbers and count content can be read out through a suitable output device.

9.2 Partial storage of selected descriptors

9.2.1 Associative storage

The principle of associative storage is as follows. Assuming descriptors are selected on the basis of first come first served, the first descriptor to occur is ascribed to the first memory location where a record is kept of the code of the descriptor and of the number of times it has occurred (i.e. one initially). The second descriptor to occur, assuming it is different from the first, has its code and tally (again one initially) stored in the second memory location. As long as different descriptors arrive, their codes and tallies are similarly stored in successive addresses. Every time a descriptor arrives it is necessary to examine the contents of the memory to see if this descriptor has occurred before and, if so, the count content in the relevant memory location is increased by one, otherwise the descriptor is stored in the next empty location.

It will be obvious that a large part of the memory in associative storage is used up by the storage of the codes of the selected descriptors (Figure 9.3(b)). This is in distinction to the descriptor–address type of storage, where the whole of the memory capacity can be used for the storage of the tallies (i.e. counts) of the descriptors (Figure 9.3(a)).

In the associative system, the arrival of a descriptor necessitates a search through the contents of the memory to see if that descriptor has already been selected. It will be apparent that this searching procedure is much more time consuming than the direct recording used in the descriptor–address system. To keep the number of locations searched to a minimum Souček[6] used a so-called tree algorithm. The first event arrives (Figure 9.4) and its descriptor code A is recorded in memory location 1. An event with descriptor code B arrives next and B is compared with A. We will assume that B is greater than A, in which case the memory address at which B is stored (i.e. 2) is stored as

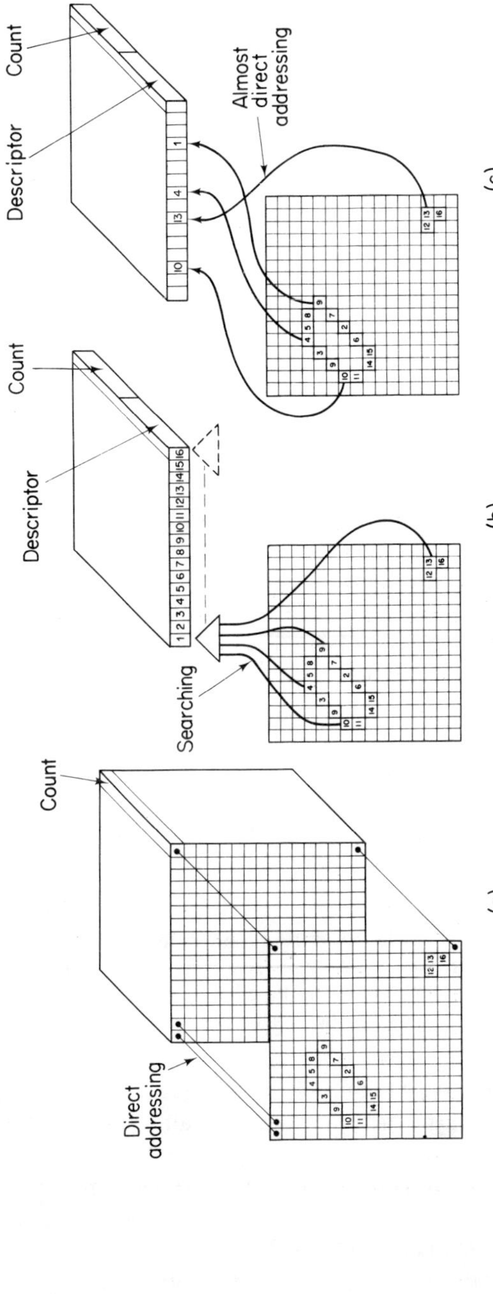

Figure 9.3 Allocation of the memory in (a) descriptor-address storage, (b) associative storage and (c) by the transformation method. Figure taken from Souček, Bonačić and Čuljat[9]

the 'higher node address' in location 1. If a third event C arrives and C < A, then C is placed in address 3, and in address 1 is recorded the fact that the 'lower node address' is 3. Thus each memory location stores the descriptor code, the lower node address, the higher node address and the tally of counts. A fourth descriptor D is compared to A; assuming it is greater, it is directed to the higher node address, 2, and then it is compared with the descriptor stored there, i.e. with B. Assuming D > B the storage address of D (i.e. 4) is stored in location 2 as the higher node address. Figure 9.4 shows

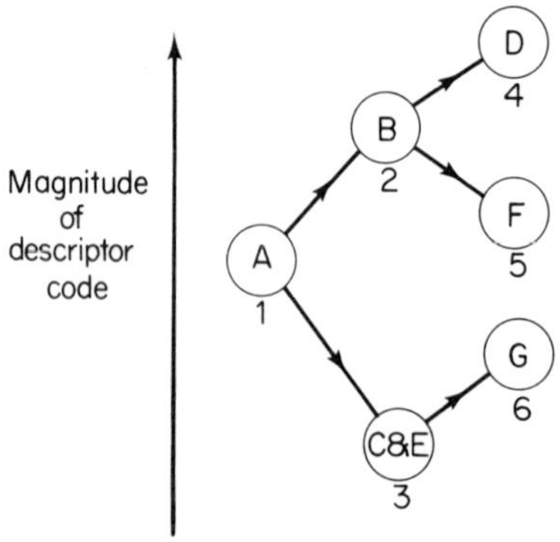

Memory location	Information stored			
	Descriptor assigned	Higher node address	Lower node address	Count
1	A	2	3	1
2	B	4	5	1
3	C (& E)	6	–	2
4	D	–	–	1
5	F	–	–	1
6	G	–	–	1

Figure 9.4 Schematic representation of the tree algorithm. The table gives the information stored after the first seven events, labelled A, B, . . , G in order of arrival, have been processed

schematically the relative magnitude of the descriptor codes and gives a table of the contents of the memory locations after arrival of the first seven descriptors, A, B, C, . . ., G.

With m memory locations the average number of comparisons necessary, using the tree method, would be approximately $\log_2 m$, against $m/2$ for a sequential memory search. For a 4096 channel memory this means an average number of comparison of approximately 12, as against 2048 for a sequential search method. In the design by Souček[6] a SDS 910 small general purpose computer having a 8192 word (i.e. 8 K) memory was used, and the processing of the data was controlled by a stored program. The system could accept a descriptor field of 1024 × 1024 (i.e. 1,048,576 or 2^{20}), and selected for storage the first 4000 descriptors to arrive. Up to 4096 events of any selected descriptor could be tallied in the memory and the system could handle an input rate of up to 1000 events per second.

The essential disadvantage of the above method is the poor utilization of the total memory capacity for the actual storage of the counts. This, of course, arises from the necessity for recording in each memory location the descriptor and upper and lower node addresses, as well as the tally of counts of that descriptor. An improvement in the utilization of the memory in this respect is provided by the 'most active zone' method.

9.2.2 *Associative zone storage*

The zone method, proposed by Hooton,[7] divides the field of possible descriptors into blocks or zones of, for example, $4 \times 4 = 16$ descriptors. Selection is now based on complete zones. At the beginning of the experiment suppose a descriptor occurs. The zone into which this descriptor falls is then selected, so that this and any other of the 16 descriptors falling in the zone are subsequently tallied and recorded. The process of zone selection continues until no more memory space is available to record classifications in any more new zones. The method was implemented by Best[8] using a PDP 8 computer having a 4096 word (of 12 bits) memory. Out of a descriptor field of 1024 × 1024, 128 zones of 4×4 descriptors could be selected (i.e. 2048 descriptors). The system was implemented by using part of the memory as an 'index'. Another part of the memory was used solely to store the counts of all the descriptors falling in the selected zones. The index section of the memory recorded the selected zone descriptors and also, for each zone descriptor, the address in the counts section of the memory where the counts of the first descriptor in the zone were to be stored. It was not necessary to record the addresses for the storage of counts of the other descriptors falling into the zone since these addresses could be generated using the least significant bits of the descriptor code.

An incoming descriptor had its zone descriptor compared with those stored in the index. If a match was found then the tally of that descriptor, maintained in the 'counts' section of the memory, was incremented by one, otherwise the zone descriptor was stored in the index (provided space is available). The system could tally up to 2^{18} counts of each selected descriptor and most of the memory capacity was available for the accumulation of counts. The system could handle mean input rates up to 10,000 per second and Figure 9.5 shows an optical display, where the selected zones of 4×4 descriptors can be clearly seen. The disadvantage of the zone selection principle

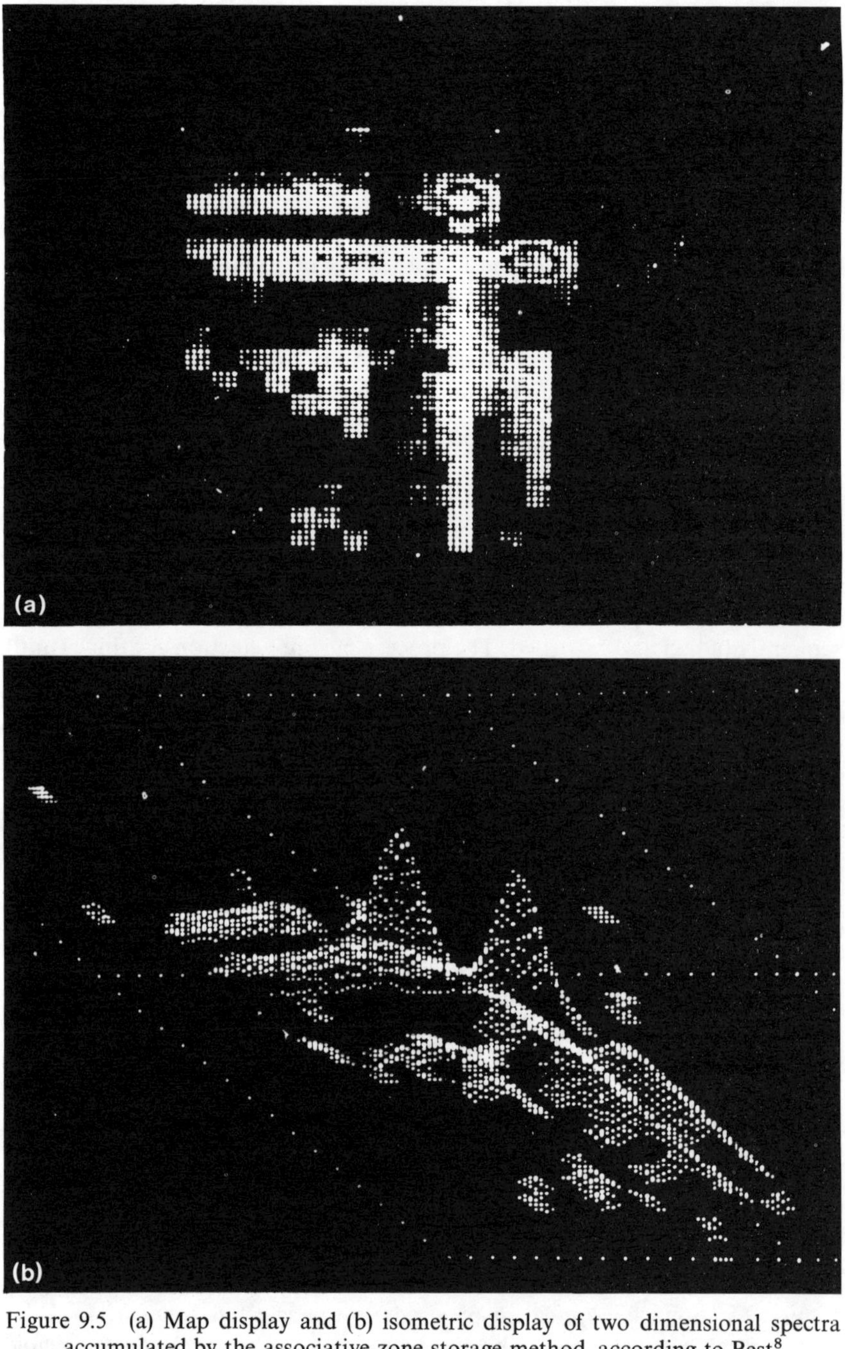

Figure 9.5 (a) Map display and (b) isometric display of two dimensional spectra accumulated by the associative zone storage method, according to Best[8]

is that it is necessarily a coarse selection and there is a greater risk of missing whole sections of descriptors which may be of low intensity but nevertheless of interest.

9.2.3 Transformation method

A rather different means of storing the selected descriptors has been utilized by Souček, Bonačić and Čuljat.[9, 10] As with associative storage of individual descriptors, the first descriptors to arrive in the experiment are the ones which are selected, are recorded in the memory and are subsequently tallied during the rest of the experiment. The assignment of memory locations to the selected descriptors however is done randomly, but reproducibly, by means of a 'transformation box' (Figure 9.3(c)). An incoming descriptor of a certain code is therefore initially assigned by the box to a specific memory address. Since the number of possible descriptors is much greater than the number of memory locations there is a possibility of two or more descriptors being transformed to the same address. The transformation is therefore only used to indicate the memory address at which to start searching. A limit is set to the number of memory locations searched. If the incoming descriptor is not found in any of these it is stored in an empty location, if available, within the search region, otherwise it is discarded. This system can be made to give a better memory utilization, and a small average number of searches per incoming descriptor, than using the tree algorithm, although this is attained at the expense of some descriptors which should be selected being lost due to the unavailability of an empty location within the search region. The system was realized using a SDS 930 small general purpose computer. Out of 128×128 possible descriptors 1024 selected descriptors could be stored with a much smaller number of searches than in other methods.

9.2.4. Use of contents-addressable memory

The systems described so far have all used a conventional ferrite core memory to store the selected descriptors and the counts of each. In each case the arrival of a descriptor necessitated a search of the contents of the memory, or part of it, to find where the counts were to be stored, if at all. Due to the destructive read-out nature of ferrite core memories, this search necessarily had to be a serial search. Some systems have been described, however, which use a different type of memory element which can be non-destructively read out and, by special arrangements, an incoming descriptor can be compared in parallel (i.e. simultaneously) with the contents of all the memory locations. Such a memory will be termed 'contents-addressable'. The resultant speeding up of the comparison process is obtained at the expense of the use of a special purpose memory unit, since this type of memory is not currently available in general purpose digital computers.

In the system described by Best *et al.*[11] the contents-addressable part of the memory used transfluxor memory elements. This element differs from the conventional ferrite toroid in that it has a figure of 8 shape, with the two loops of the '8' of differing size, and in essence the direction of magnetization in the major and minor loops can be set independently. The direction of magnetization in the major loop determines whether the device is in the '1' state or the '0' state and the 'write' current is passed through a wire passing through the major loop. Interrogation of the state of the device is accomplished by passing a biphasic current 'read' pulse through a wire in the minor aperture, when a 'sense' wire through the minor aperture gives differing signals according to the state of magnetization in the major loop. A pair of such transfluxors were used for each bit, one having the complementary state of the other. When it is necessary to compare some input with the state of the element, according to whether the input is '0' or '1' the read current is directed to one or the other of the transfluxor pair so that the sense wire always gives an output only if there is a mismatch between the input and the state of the transfluxor pair.

The system operated by selecting 256 zones, of 4×4 descriptors each, from a possible field of 1024×1024 descriptors. The selected zone descriptors were stored in a 256 word 16 bit transfluxor memory. A conventional ferrite core memory, of 4096 word 20 bits, formed the store for the counts of the 4096 descriptors in the selected zones. An incoming descriptor had its zone descriptor compared simultaneously with all the tranfluxor memory words and if a match was found in any location the address of this location, together with the least significant bits of the descriptor, were used to generate the address in the ferrite core memory where the counts of that descriptor were stored. Using this system the time necessary to process any descriptor was only 10 μs.

Ofengenden *et al.*[12] described a system using a contents-addressable memory made up of elements similar to the transfluxor. In this instance the main store for both selected descriptors and counts was a magnetic drum memory. All events arriving had their descriptors recorded in a 32 word 26 bit contents-addressable memory. Each cycle of the drum caused the selected descriptors to be read out serially and each of these was compared simultaneously with all the contents of the contents-addressable memory. If a match was found, the count corresponding to that descriptor maintained in the drum store was incremented by one. At the end of each drum cycle the contents of the contents-addressable memory were cleared. This system could accept up to 2^{16} counts of the selected 4096 descriptors out of a field of 1024×1024 and could accept up to 6000 events per second.

The use of a contents-addressable memory, in principle, offers an ideal solution for the tallying of selected descriptors. The problem to date has been the high cost of this type of memory. However, there are many other problems in information retrieval systems which would benefit from the availability of a contents-addressable memory; this has recently resulted in the commercial availability of several types based on semiconductor elements. There is some hope therefore that this type of device might offer an economical solution to the problem of rapid tallying of selected descriptors.

9.3 Large computer systems

The use of a large computer system allows the tallying of a large number of possible descriptors without the necessity for the partial processing schemes discussed in Section 9.2. Even so, in the case of a large number of possible descriptors, the fast memory of a large computer is not large enough to store all the data in a convenient form and it is necessary to make use of backing stores, such as magnetic drum or disc. The fast memory acts as an intermediary buffer, rapidly accepting incoming data and temporarily storing it, while the backing store is continually updated at a necessarily slower speed. It is essentially in the organization of these operations that systems differ, and Souček[13] has reviewed some current techniques.

To a considerable extent the scheme adopted will depend on the characteristics of the hardware, particularly of the backing store. For example, Adam[14] described a system using an IBM 360–50 computer together with a disc store. Each disc could store the tallies (each a word of up to 16 bits) of 3×10^6 descriptors. Each disc used a 10 track recording head which could be positioned over any one of 200 'bands'. Each track of each band could store 3625 words, of 8 bits, around the circumference and each address was uniquely associated with a descriptor's tally. The access time to a randomly chosen address on the disc was made up of an average of 75 ms positioning time for the head and an average disc rotation time of 12·5 ms. Thus the essential organizational problem was to avoid lengthy disc accessing times by using the fast memory to sort the incoming descriptors, so that a whole block of descriptors, corresponding to a band on the disc, could be updated together in the same revolution of the disc. By using the fast memory as a system of buffers which, when empty, return to a pool ready for new incoming data, it was possible to accept an input rate of up to 2000 descriptors per second.

The essential limitation on data rate in most schemes is set by the capacity of the fast buffer memory. Souček[13] has discussed this factor and describes a method of disc buffering which can give increased data acceptance rates. The method limits the occupation of the fast memory in the disc read–update–write operation, by causing the data of part of a band to be written back much earlier, but into a different disc address. The data therefore does not have a fixed location in the disc store but is continually being 'shunted' along.

Quite apart from the functions of sorting and storage of descriptors, large computers are often used to do some arithmetical and monitoring operations on the incoming data. The programming of a large computer can often be done in a high level language such as FORTRAN and this is valuable wherever the program is required to be altered frequently, particularly when on-line programming is used during an experiment to amend the data taking or analysis in some way. A further advantage of a large computer system is the availability of associated peripheral equipment such as magnetic drum or disc stores, magnetic tape units, printers, tape punches, CRT display devices etc. Program subroutines may be incorporated, and called up during the experiment, to enable certain aspects of the data to be analysed, examined and displayed, so that adjustments to the experiment may be made.

An extremely large number of publications have outlined different computer systems at many establishments and it is impossible to review them here. Many make use of a central processing unit (CPU) which serves several on-line experiments and off-line programs on a time shared basis. This involves some sort of interrupt routine and a priority list, so that an event occurring in any experiment, whose data is temporarily stored in buffer stores, causes, according to its priority, the CPU to interrupt its program and to read in the data for the event and take any other necessary action such as clearing the register, etc. Often a small 'satellite' computer is used for each experimental terminal to provide some preliminary processing of the data, to take some of the load off the CPU.

As an example of a large computer installation, that utilized on the Stanford two mile linear accelerator as described by Boyarski[15] will give some idea of the possibilities. This system used a SDS 9300 computer with 32,000 words of core memory, together with a magnetic disc store, 3 magnetic tape machines, card reader and punch, line printer, teletype, CRT display and plotter. The computer, as well as analysing data from nuclear events, also provided extensive monitoring of the accelerator and other equipment and was used to monitor and control the current through the magnets of a charged particle spectrometer.

When used on-line the computer was dedicated entirely to a single experiment. In addition, all nuclear event and monitoring data was logged by recording on magnetic tape, to allow further analysis off-line where necessary. Most of the programming was done in FORTRAN IV language.

In experiments for studying electron scattering off nuclei and photo-production of unstable particles in gamma–nucleon collisions, a magnetic spectrometer was used. Figure 9.6 shows the arrangement of detectors for a spectrometer which incorporated eleven magnets. The electron or photon beam strikes a target and scattered or produced particles enter the spectrometer where they are focused on to the P (momentum) focal plane and the T (angle) focal plane, in the vertical and horizontal directions respectively. By placing scintillation counters (41 in the P plane and 32 in the T plane) the momentum and production angle of the particle are measured. Other hodoscopes X and F determine the angles at the P and T foci. Particle identification is accom-

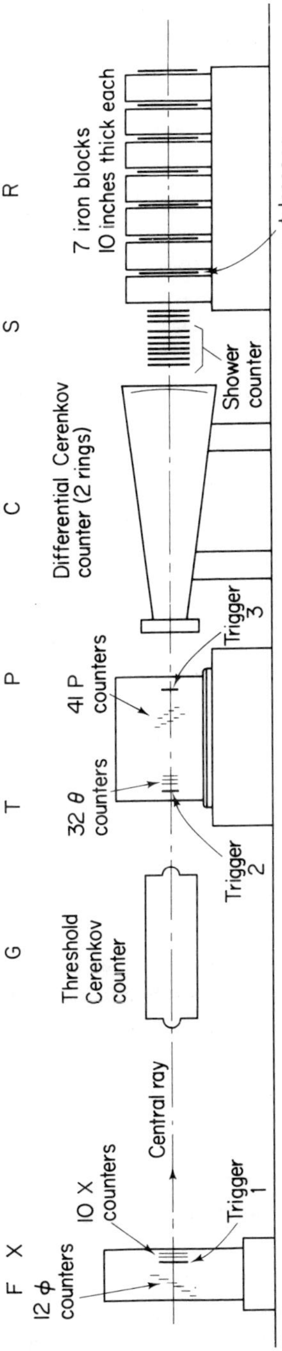

Figure 9.6 The particle detection arrangement used with the Stanford two mile linear accelerator, according to Boyarski[15]

plished by three gas Cerenkov counters G and C (for π, K, protons), a sandwiched lead/scintillation shower counter S (for electrons) and an iron/scintillator stack range telescope R (for muons).

The SDS 9300 can accept a 24 bit word parallel input. In order to receive or send data from or to many different devices, a multiplexer was used which served 300 device terminals. Whenever the program called for reading in of data from a device terminal the multiplexer, on receipt of the appropriate device terminal code, would cause a data word of up to 24 bits from that terminal to be read into the computer. Event data from the hodoscope counters was stored in buffer stores in readiness for transmission by the multiplexer when commanded by the computer. The particle identification detectors had their pulse heights coded by ADCs whose registers could likewise be interrogated by the computer through the multiplexer. In all, 8 ADCs were available for digitizing pulse heights or for use with time to amplitude converters. 40 high speed scalers were provided whose contents could likewise be interrogated. Finally, as well as the nuclear event data, device terminals were allocated to enable the computer to monitor such parameters as beam characteristics, Cerenkov counter gas pressures, beam slit positions and magnet currents.

Between 50% and 100% of the events were analysed on-line depending on the incoming data rate. The analysis was completed, where necessary, by later off-line computing. The on-line program consisted of several major sub-programs such as data logging, display program, equipment status checks and data analysis. Extensive use of program interrupts was made so that a priority system for various operations could be established. Thus the top priority was the reading in and storing of data of a new event and so this eventuality caused that part of the program which happened to be running to be interrupted, in order to change to the data reading sub-program. The next priority was the eventuality of a scaler overflowing, which again caused reading in of the scaler contents. The logging of the data on to magnetic tape was done by means of two 360 word sections of the memory acting as buffer stores. When one section was full, the incoming data was diverted to the second and the contents of the first were recorded on the tape. These changeover operations also have a high priority in the list of program interrupts. Up to 10,000 words per second could be logged on to the tape. Further down the priority list is the display sub-program and further down still is the data analysis sub-program.

For any experiment several analysis sub-programs could be used. In this way it was possible to monitor different aspects of the event data and make adjustments to the program on-line. Table 9.1 gives a typical list of analysis programs that were used in a single experiment, any of which could be displayed on the CRT. Figures 9.7 and 9.8 show some examples of displays generated on the CRT. In Figure 9.7 the Z axis denotes the number of

Table 9.1 List of typical analysis programs as described by Boyarski[15]

1.	P (momentum) hodoscope distribution
2.	T (angle) hodoscope distribution
3.	X hodoscope distribution
4.	F hodoscope distribution
5.	P versus T distribution
6.	SR (shower pulse height versus range hodoscope)
7.	Eight correlated pulse height distributions from three ADCs connected to the Cerenkov counters
8.	2 dimensional Cerenkov pulse height spectrum
9.	Missing mass display for one spectrometer setting
10.	Missing mass display for a combination of many settings
11.	List of accumulated counts for various particle types
12.	List of constants with names for the run
13.	Status of magnets on the spectrometer

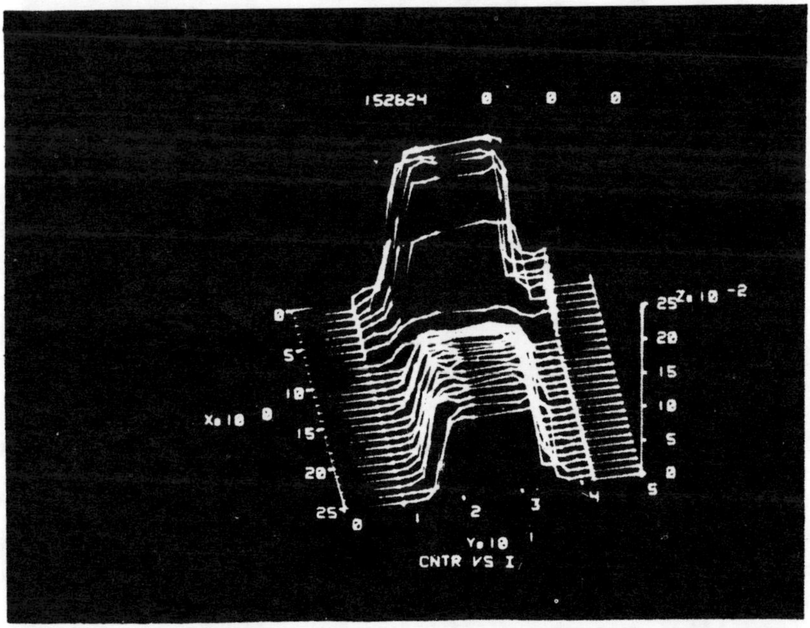

Figure 9.7 Computer display, according to Boyarski[15] showing timing or delay curves for 20 counters. The Z axis gives the number of counts for a certain delay (Y axis) and counter number (X axis)

counts, the X axis denotes counter number and the Y axis the delay in nanoseconds of the master trigger signal. In Figure 9.8 the Z axis denotes counts, the Y axis the pulse height from the shower counter and the X axis the range in the iron/scintillator stack, allowing identification of electrons and mesons (or protons).

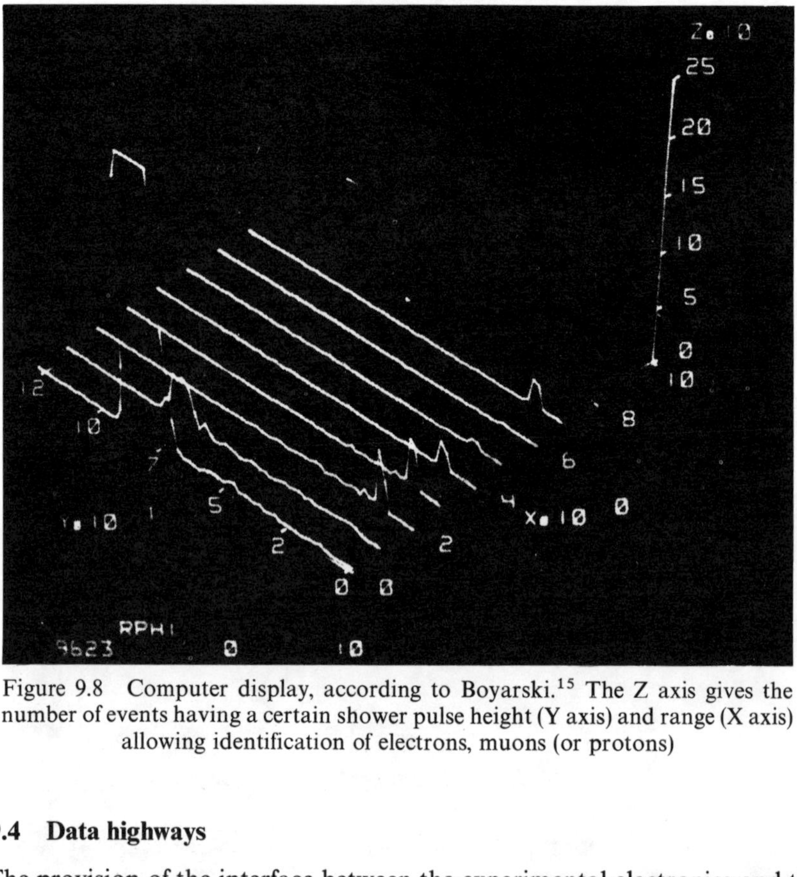

Figure 9.8 Computer display, according to Boyarski.[15] The Z axis gives the number of events having a certain shower pulse height (Y axis) and range (X axis) allowing identification of electrons, muons (or protons)

9.4 Data highways

The provision of the interface between the experimental electronics and the computer is usually a major problem, since it must often be specially designed for the particular combination of experiment and computer system. The problem has been simplified to some extent with the advent of 'data highways'[16] to take commercially available modules. Thus the CAMAC system which has been adopted as a European standard[17, 18, 19] specifies a certain mechanical and electrical format for data handling systems.

The CAMAC system comprises of modules which fit into a standard 'crate' in which digital data operating at standard logic levels can be transferred by a data controller. Computer control of the data transfers is accomplished through a 'master controller' which contains the necessary interfacing circuits. Thus only this interface is specific to the computer and all other units are independent of the type of computer being used.

The types of module fall into several categories. Input modules accept analogue or other signals and process or convert the information to dataway standards. Examples of these are ADCs, time to digital converters and counters. Output modules provide an interface from the dataway to devices such as printers, paper tape punches, tape recorders and CRT displays. Some types of peripheral equipment such as teleprinters, magnetic disc units and core stores require the use of modules handling both input and output functions. Some of these modules may themselves contain limited storage facilities for processing input and output data. Other types of module include clock oscillators and devices for multiplexing and controlling analogue signals (e.g. linear gates). Finally, modules may be used to perform data processing tasks such as binary to BCD conversion, multiply or divide, and floating point calculations.

The standard CAMAC module is equipped with an 86 way plug connector at the rear. Up to 23 such modules may be fitted into the crate which is provided with mating 86 way socket connectors which are wired up in a defined manner to form a 'dataway'. A crate also contains space for a 'controller' unit which is also provided with rear plug connectors to connect to the dataway. The main function of the controller unit, which itself is under control of a computer, is to control operations on the modules in the crate.

The wiring of the dataway is arranged so that 24 bussed 'read' lines and 24 bussed 'write' lines are available at the 86 way connector at each module position and these same lines are also available to the controller. All data transfers take place from a module to the controller or vice versa. To transfer data from a module to the controller, after the appropriate command signals from the controller the module outputs a 24 bit data word on to the common read lines, from which it is accepted by the controller. Similarly, after the appropriate command, data transfers from the controller to a module take place on the common write lines.

The remaining pins of the 86 way module connectors are concerned with the command, control and information signals and also provide the power supplies to the modules. As with the read and write lines, most of these pins are bussed on lines common to all modules, however each module connector has two pins assigned to it, designated the N pin and the L pin, which are unique to that module in that they are wired only to the controller unit. The purpose of the N pin is to allow the controller to designate that module for the purpose of any operation. This facility for addressing the module must be coupled with the provision for sub-addressing a unit within the module (since a module may contain several separate units such as registers) and also for sending a signal to specify the function (e.g. read, write, test status, etc.)

to be performed. Thus a command signal from the controller will comprise of a '1' signal appearing on the N line appropriate to the module of interest, together with signals on 4 bussed sub-address lines and on 5 bussed function lines. The actual execution of the command is controlled by a sequence of two strobe pulses appearing on two bussed strobe lines. The first of these is used for any operation which does not require changing the conditions on the dataway (e.g. read or write operations) and the second strobe pulse can initiate such operations as the clearing of a register, status testing etc.

A bussed line is also necessary to carry a 'busy' signal to staticize dataway conditions to reduce crosstalk during read or write operations. Another bussed line is provided to allow an addressed module to signal a response (indicating, for example, data ready or register overflow) and three other bussed lines allow non-addressed operations of initialize, clear and inhibit to be performed.

So far all operations mentioned above have been initiated from the controller to the module. In some instances it may be necessary for the module to signal to the controller. An example of this is where a module has processed some experimental information and is ready for some further operation. This is the purpose of the L (for 'look at me') lines, one assigned to each module, by which any module may signal to the controller that it is awaiting attention.

In a single crate system the controller may include the interface to the computer or the interface may be a separate unit. If more than one crate is used it is necessary to couple the individual crate controllers via a 'branch highway' to a branch driver, which is in turn interfaced to the computer. The branch driver then coordinates the data flow and control functions of the individual crate controllers. Since the crates may be located some distance apart, and signal transit times become important, operations on the branch highway involve 'handshake' routines, in which the branch driver requires a positive response from the addressed crate before a transfer operation is commenced.

References

1. T. K. Alexander, H. G. Reddering and J. M. Kennedy, *Chalk River Report CREL* 779, Atomic Energy of Canada, Ltd. (1959)
2. T. K. Alexander and J. Leng, Chalk River Report CREL 1036, Atomic Energy of Canada, Ltd. (1961).
3. P. A. Egelstaff and E. R. Rae, *Proceedings, Conference on Automatic Acquisition and Reduction of Nuclear Data*, p. 83, EANDC, Karlsruhe, 1964.
4. D. A. McClure, N. S. Kendrick and J. W. Lewis, *Nucl. Instr. and Meth.*, **103**, 453 (1972).
5. B. Souček, *IEEE Trans. Nucl. Sci.*, NS **16**, No. 5, 36 (1969).
6. B. Souček, *Rev. Sci. Instr.*, **36**, 750 (1965).

7. I. N. Hooton, *IEEE Trans. Nucl. Sci.*, NS **13**, No. 3, 553 (1966).
8. G. C. Best, *IEEE Trans. Nucl. Sci.*, NS **13**, No. 3, 566 (1966).
9. B. Souček, V. Bonačić and K. Čuljat, *Nucl. Instr. and Meth.*, **66**, 202 (1968).
10. V. Bonačić, B. Souček and K. Čuljat, *Nucl. Instr. and Meth.*, **66**, 213 (1968).
11. G. C. Best, S. A. Hickman, I. N. Hooton and G. M. Prior, *IEEE Trans. Nucl. Sci.*, NS**13**, No. 3, 559 (1966).
12. R. G. Ofengenden, F. N. Berezin, P. V. Lapshuk and A. M. Shchur, *Pribory i Teknika Eksperimenta*, May–June 1968, p. 68. English translation, *Instruments and experimental techniques*, p. 576 (1968).
13. B. Souček, *IEEE Trans. Nucl. Sci.*, NS**17**, No. 4, 20 (1970).
14. J. Adam, *Nucl. Instr. and Meth.*, **73**, 89 (1969).
15. A. M. Boyarski, *Proceedings for the Skytop conference on computer systems in experimental nuclear physics*, p. 139 (1969).
16. A. Senator, I. N. Hooton, G. L. Miller, H. P. Lie and E. A. Gere, *Proceedings for the Skytop conference on computer systems in experimental nuclear physics*, p. 394 (1969).
17. *Euratom Report EUR* 4100e (1969).
18. *Euratom Report EUR* 4600e (1970).
19. CAMAC Tutorial issue, *IEEE Trans. Nucl. Sci.*, NS**18**, No. 2 (1971).
20. G. Dimmler and G. Krüger, *Proceedings, Conference on Automatic Acquisition and Reduction of Nuclear Data*, p. 393, EANDC, Karlsruhe, 1964.

Appendix A

Transform Methods in Transient Analysis

This Appendix contains a simplified treatment of the use of Fourier and Laplace methods. Several monographs are available that treat the subject more rigorously.[1,2,3] The approach that is adopted here is modelled on that of Lathi[3] to whom reference should be made for amplification of the various points discussed.

A.1 The general differential equation

Consider the circuit of Figure A.1 with a transient signal $v_1(t)$ applied. We wish to find the output signal $v_2(t)$. Using Kirchoff's law for the algebraic sum of the currents at node A in Figure A.1 we have

$$C\frac{dv_2}{dt} - C\frac{dv_1}{dt} + \frac{v_2 - v_1}{R_1} + \frac{v_2}{R_2} = 0 \tag{A.1a}$$

i.e.

$$\frac{dv_2}{dt} + \frac{1}{C}\left(\frac{1}{R_1} + \frac{1}{R_2}\right)v_2 = \frac{dv_1}{dt} + \frac{v_1}{CR_1} \tag{A.1b}$$

This is a linear differential equation with constant coefficients which may be solved by standard methods. For a more complex linear circuit it may be shown that the unknown current or voltage may always be found by solving a similar equation. In general, if $f_1(t)$ is the input current or voltage, the unknown current or voltage $f_2(t)$ may be found by solving an equation of the type

$$a_m\frac{d^m f_2}{dt^m} + a_{m-1}\frac{d^{m-1} f_2}{dt^{m-1}} \ldots + a_0 f_2 = b_n\frac{d^n f_1}{dt^n} + b_{n-1}\frac{d^{n-1} f_1}{dt^{n-1}} \ldots + b_0 f_1 \tag{A.2}$$

where the coefficients $a_m, a_{m-1}, \ldots, a_0, b_n, b_{n-1}, \ldots, b_0$ are constants. The solution to this type of equation consists of the sum of two parts, a 'particular integral' which depends on the form of the input $f_1(t)$, and a 'complementary function' or transient part. The transient part is the solution of Equation (A.2), with the R.H.S. put equal to zero, and is of the form $k_1 e^{\nu_1 t} + k_2 e^{\nu_2 t} + \ldots + k_m e^{\nu_m t}$ where $\nu_1, \nu_2, \ldots, \nu_m$ are the roots of the equation

$$a_m \nu^m + a_{m-1} \nu^{m-1} + \ldots + a_1 \nu + a_0 = 0$$

and $k_1, k_2, \ldots, k_m$ are arbitrary constants. This part of the solution is independent of the input $f_1(t)$ except through the setting of the arbitrary constants. It will be seen later that the m values of $\nu_1, \nu_2, \ldots, \nu_m$ may be identified with the poles of the circuit.

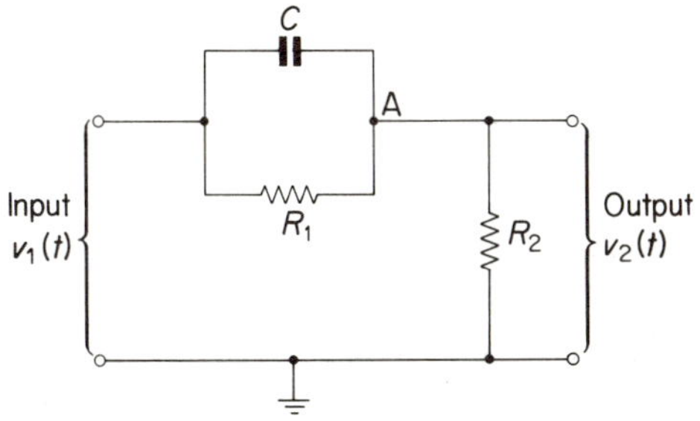

Figure A.1 Example of a simple network

In the case of Equation (A.1) the transient part of the solution has only one term, of the form

$$v_2(t) = k_1 e^{\nu_1 t} \tag{A.3}$$

where

$$\nu_1 = -\frac{1}{C}\left(\frac{1}{R} + \frac{1}{R_2}\right).$$

Consider an input of the form shown in Figure A.2, that is, $f_1(t) = 0$ for $t \leqslant 0$ and $f_1(t) = v_1(t) = E_1 e^{-gt}$ for $t > 0$. For the region $t > 0$ a solution of Equation (A.1b) may be found for the input dependent part (or particular integral) having the form

$$v_2(t) = E_2 e^{-gt} \tag{A.4}$$

It will be seen by substitution in Equation (A.1b) that E_2 has the value

$$E_2 = E_1 \left[\frac{g + \phi_1}{g + \nu_1}\right] \text{where } \phi_1 = -\frac{1}{CR_1} \text{ and } \nu_1 = -\frac{1}{C}\left(\frac{1}{R_1} + \frac{1}{R_2}\right)$$

The complete solution is then the sum of Equation (A.3) and Equation (A.4). In order to set the arbitrary constant k_1, we note that at $t = 0$ there is no voltage across the capacitor, so $v_1 = v_2$. The complete solution for $t > 0$ is therefore

$$v_2(t) = E_1 \left[\frac{g + \phi_1}{g + \nu_1}\right] e^{-gt} + E_1 \left[\frac{\nu_1 - \phi_1}{\nu_1 + g}\right] e^{\nu_1 t} \tag{A.5}$$

If the input voltage had been of the form

$$v_1(t) = 0 \text{ for } t \leqslant 0$$

$$v_1(t) = E_{11}\,e^{-g_1 t} + E_{12}\,e^{-g_2 t} \text{ for } t > 0$$

the expression for the output voltage would have terms containing $e^{-g_1 t}$ and $e^{-g_2 t}$. In general, for an input consisting of a sum of such exponential terms, the output will be the sum of the transient part and of a part containing a sum of exponentials having the same exponents as the input.

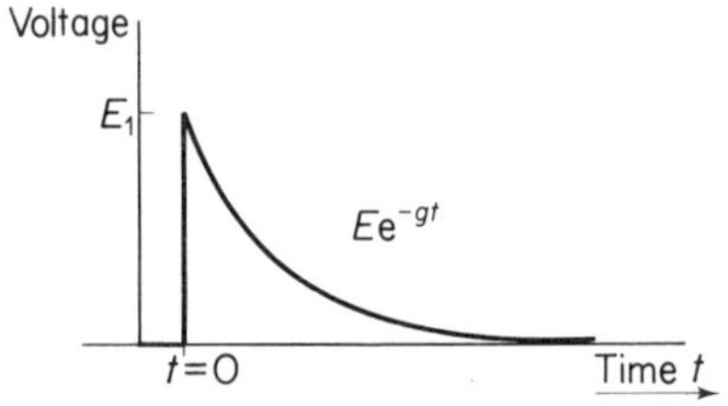

Figure A.2 Input voltage waveform

The ease with which the solution was found when the input was a sum of exponential terms suggests the possibility of representing an arbitrary input as a sum of exponential terms. This is the technique of the Fourier and Laplace transform methods. The types of input considered so far have been of the type e^{-gt} for $t > 0$, i.e. having the exponential shape only in the region $t > 0$. We will now consider the response of a circuit to an input having the form of a 'pure' or 'eternal' exponential, that is, one having the form e^{st} over all time $-\infty \leqslant t \leqslant \infty$. It is as a sum of these eternal exponentials that the Fourier and Laplace methods express an arbitrary input function.

A.2 Excitation by an eternal exponential

When the input has the form $A_1\,e^{st}$ (for $-\infty \leqslant t \leqslant \infty$) under certain conditions the transient part of the solution may be neglected and the complete solution is just the input dependent part. The transient part of the solution consists of a sum of terms $k_1\,e^{\nu_1 t} + k_1\,e^{\nu_2 t} + \ldots + k_m\,e^{\nu_m t}$ and the input dependent part will be a single term containing e^{st}. If we now assume that s (or the real part of s if s is complex) is greater than $\nu_1, \nu_2, \ldots, \nu_m$ (or the real parts of $\nu_1, \nu_2, \ldots, \nu_m$ if any are complex), the transient part of the solution will decay away more rapidly, or increase less rapidly, than the input dependent part. Since the input started at $t = -\infty$, at the time of interest the transient components would have vanished; the complete solution for this special case would be the input dependent term only. The essential condition on s was $\mathrm{Re}(s) > \mathrm{Re}(\text{any } \nu_r)$; this is termed the dominance condition and will always be assumed to be satisfied in the following treatment. The solution is therefore found by simply substituting $f_2(t) = A_2\,e^{st}$ in the differential equation Equation (A.2) and finding A_2. A_2 is seen to be given by

$$\frac{A_2}{A_1} = \frac{b_n s^n + b_{n-1}s^{n-1} + \ldots + b_0}{a_m s^m + a_{m-1}s^{m-1} + \ldots + a_0} = \frac{b_n(s-\phi_1)(s-\phi_2)\ldots(s-\phi_n)}{a_m(s-\nu_1)(s-\nu_2)\ldots(s-\nu_m)} \tag{A.6}$$

this quantity being termed the transfer function. The numerator and denominator of the transfer function have been factored to show the poles $\nu_1, \nu_2, \ldots, \nu_m$ and the so-called zeros $\phi_1, \phi_2, \ldots, \phi_n$ of the circuit.

For the example of Figure A.1 with an input voltage $v_1(t) = E_1\,e^{st}$ the output voltage is $v_2(t) = E_2\,e^{st}$ and substitution in Equation (A.1a) gives

$$sC(E_2 - E_1) + (E_2 - E_1)/R_1 + E_2/R_2 = 0 \tag{A.7}$$

and hence the transfer function, $H(s)$, is

$$H(s) = \frac{E_2}{E_1} = \frac{s-\phi_1}{s-\nu_1} \text{ where } \phi_1 = -\frac{1}{CR_1} \text{ and } \nu_1 = -\frac{1}{C}\left(\frac{1}{R_1} + \frac{1}{R_2}\right)$$

Equation (A.7) could have been written down directly if instead of $C\left[\frac{dv_2}{dt} - \frac{dv_1}{dt}\right]$, which represents the current through the capacitor, $sC(E_2 - E_1)$ was written. In other words, it could have been obtained if the capacitor had been imagined as possessing an impedance $1/sC$ in the circuit and applying Kirchoff's law as before. A further example will make this clear.

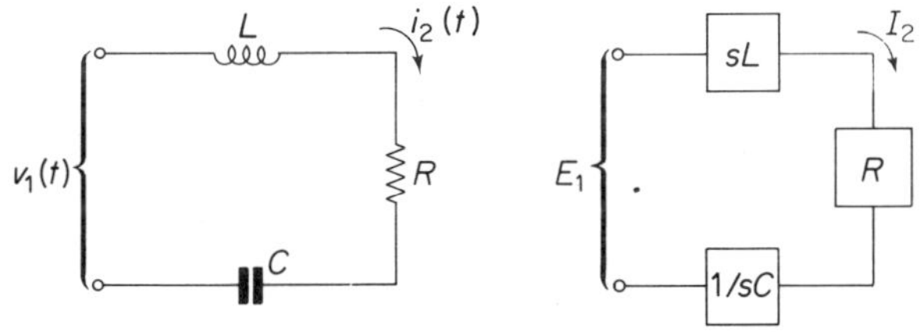

Figure A.3 Simple circuit containing resistance, capacitance and inductance

The circuit of Figure A.3 has a differential equation relating the voltage $v_1(t)$ applied and the resulting current $i_2(t)$

$$\frac{dv_1}{dt} = L\frac{d^2 i_2}{dt^2} + R\frac{di_2}{dt} + \frac{i_2}{C} \tag{A.8}$$

If $v_1(t) = E_1\,e^{st}$ then $i_2(t) = I_2\,e^{st}$ and substitution in Equation (A.8) gives

$$E_1 = sLI_2 + RI_2 + I_2/sC \tag{A.9}$$

and hence the transfer function is

$$H(s) = \frac{I_2}{E_1} = \frac{1}{sL + R + 1/sC} \qquad \text{(A.10)}$$

Equation (A.9) and (A.10) could have been obtained by assuming impedances of sL, R and $1/sC$ for the inductance, resistance and capacitance as shown in Figure A.3, together with an applied voltage of E_1 and a resultant current of I_2. Further investigation reveals that such impedances behave in a similar way to resistors in DC circuits and, for example, may be combined in the usual way when connected in series or parallel.

The transfer function for any circuit containing only lumped parameter components can always be expressed as a ratio of two polynomials in s, as in Equation (A.6), which can then be factored to give the poles and zeros. It should be noted however that the poles and zeros may be real or complex, but if they are complex they must occur in complex conjugate pairs in order to give real coefficients $a_1, a_2, \ldots, a_m, b_1, b_2, \ldots, b_n$ in the differential equation, Equation (A.2). The poles and zeros of a circuit are often shown on a plot in the complex plane (the s plane). For the circuit of Figure A.3 the transfer function, Equation (A.10), may be factored to give

$$H(s) = \frac{s}{L(s - v_1)(s - v_2)} \text{ where } v_1 = -\frac{R}{2L} + \sqrt{\frac{R^2}{4L^2} - \frac{1}{LC}}$$

$$\text{and } v_2 = -\frac{R}{2L} - \sqrt{\frac{R^2}{4L^2} - \frac{1}{LC}}$$

and the poles and zeros are plotted in the s plane in Figure A.4.

It is possible to express other types of signal as a sum of terms of the type e^{st}. Thus a steady sinusoidal signal $v_1(t) = V_1 \cos \omega_1 t$ may be re-expressed

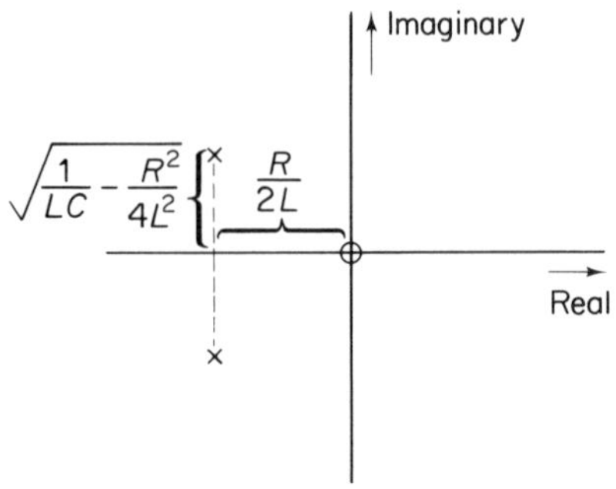

Figure A.4 Position of the poles (X) and zeros (O) in the complex plane for the circuit of Figure A.3, on the assumption that $1/LC > R^2/4L^2$

$v_1(t) = \frac{1}{2}V_1(e^{j\omega_1 t} + e^{-j\omega_1 t})$. Putting $s = \pm j\omega_1$ we have as transfer function $H(s) = H(\pm j\omega_1) = |H| e^{\pm j\theta}$ say. We have therefore

$$v_2(t) = \tfrac{1}{2}V_1 |H| (e^{j\omega_1 t + \theta} + e^{-j\omega_1 t - \theta}) = V_1 |H| \cos(\omega_1 t + \theta) \qquad \text{(A.11)}$$

$H(j\omega)$ is the same as the complex transfer function in ordinary complex AC theory (because it is obtained by replacing resistances R, capacitances C and inductances L by impedances R, $1/j\omega C$ and $j\omega L$ respectively) and Equation A.11 is clearly the expected result.

A.3 The Fourier transform

It was shown above that the representation of a signal by a sum of eternal exponential terms enables the output signal to be found. The Fourier and Laplace transforms provide two systematic methods for expressing signals as a continuous sum of exponentials. The Laplace method expresses the signal as a sum of exponentials of the type e^{st} where s is complex ($s = \sigma + j\omega$), while the Fourier method expresses the signal as a sum of exponentials of the type $e^{j\omega t}$.

The Fourier transform $\mathscr{F}(j\omega)$ of $f(t)$ is given by

$$f(t) = \frac{1}{2\pi}\int_{-\infty}^{\infty} \mathscr{F}(j\omega)\, e^{j\omega t}\, d\omega \qquad \text{(A.12)}$$

Thus $\mathscr{F}(j\omega)$ may be regarded as a spectral density function, that is, as determining the relative intensities of the component sinusoidals in the spectrum of ω. $\mathscr{F}(j\omega)$ may be found from $f(t)$ by

$$\mathscr{F}(j\omega) = \int_{-\infty}^{\infty} f(t)\, e^{-j\omega t}\, dt \qquad \text{(A.13)}$$

Equations (A.13) and (A.12) define the Fourier transform of $f(t)$ and the inverse transform of $\mathscr{F}(j\omega)$ respectively, and may be indicated by the notation

$$f(t) \leftrightarrow \mathscr{F}(j\omega)$$

Suppose the input signal $f_1(t)$, which has the Fourier transform $\mathscr{F}_1(j\omega)$, is applied to the input of a circuit having a transfer function $H(s)$ to the exponential signal e^{st}, as defined in the previous section. An input signal of the form $\mathscr{F}_1(j\omega_1)\, e^{j\omega_1 t}$ would therefore have an output $H(j\omega_1)\mathscr{F}_1(j\omega_1)\, e^{j\omega_1 t}$. The output $f_2(t)$ for the input $f_1(t)$ would therefore be the sum of all the exponential components, that is

$$f_2(t) = \frac{1}{2\pi}\int_{-\infty}^{\infty} H(j\omega)\mathscr{F}_1(j\omega)\, e^{j\omega t}\, d\omega \qquad \text{(A.14)}$$

Equation (A.14) is equivalent to taking the inverse transform of $H(j\omega)\mathscr{F}_1(j\omega)$. To summarize

if $$f_1(t) \leftrightarrow \mathscr{F}_1(j\omega)$$

then $$f_2(t) \leftrightarrow H(j\omega)\mathscr{F}_1(j\omega)$$

Thus the process of taking the Fourier transform of $f_1(t)$ may be seen as the breaking of $f_1(t)$ into its spectrum of component frequencies as shown in Figure A.5, which are then selectively filtered by the variation of $H(j\omega)$ with ω; then the components are combined at the output to give $f_2(t)$.

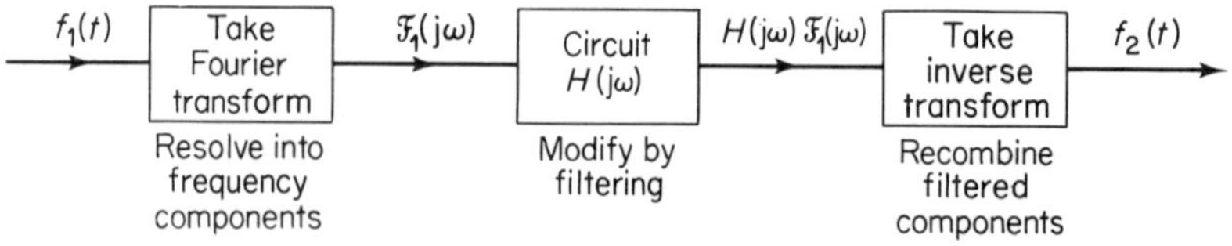

Figure A.5 Physical analogy for the use of the Fourier transform method

It will be noticed that ω ranges over negative values as well as positive values. However, as $f_1(t)$ must be real, and since $\mathscr{F}_1(j\omega) = \int_{-\infty}^{\infty} f_1(t)\,e^{-j\omega t}\,dt$ and the only imaginary quantity in $\mathscr{F}_1(j\omega)$ is $j\omega$, it follows

$$\mathscr{F}_1^*(j\omega) = \mathscr{F}_1(-j\omega)$$

Therefore we have

$$f_1(t) = \frac{1}{2\pi}\int_{-\infty}^{\infty} \mathscr{F}_1(j\omega)\,e^{j\omega t}\,d\omega = \frac{1}{2\pi}\int_{0}^{\infty} \mathscr{F}_1(j\omega)\,e^{j\omega t}\,d\omega + \frac{1}{2\pi}\int_{0}^{\infty} \mathscr{F}_1(-j\omega)\,e^{-j\omega t}\,d\omega$$

$$= \frac{1}{2\pi}\int_{0}^{\infty} \mathscr{F}_1(j\omega)\,e^{j\omega t}\,d\omega + \frac{1}{2\pi}\int_{0}^{\infty} \mathscr{F}_1^*(j\omega)\,e^{-j\omega t}\,d\omega = \begin{matrix}\text{Real}\\ \text{part of}\end{matrix}\ \frac{1}{\pi}\int_{0}^{\infty} \mathscr{F}_1(j\omega)\,e^{j\omega t}\,d\omega$$

and the latter formulation is sometimes used as a definition of the inverse Fourier transform.

A similar result may be obtained for $f_2(t)$, the signal at the output of any network. Since $H(j\omega)$ for any network contains no imaginary quantities other than $j\omega$, it follows that $H^*(j\omega) = H(-j\omega)$ and, by a similar reasoning as that used above, it follows that

$$f_2(t) = \frac{1}{2\pi}\int_{0}^{\infty} H(j\omega)\,\mathscr{F}_1(j\omega)\,e^{j\omega t}\,d\omega + \frac{1}{2\pi}\int_{0}^{\infty} H^*(j\omega)\,\mathscr{F}_1^*(j\omega)\,e^{-j\omega t}\,d\omega$$

and so is real, as expected, and may be written

$$f_2(t) = \underset{\text{part of}}{\text{Real}}\ \frac{1}{\pi}\int_0^{\infty} H(\mathrm{j}\omega)\,\mathscr{F}_1(\mathrm{j}\omega)\,\mathrm{e}^{\mathrm{j}\omega t}\,\mathrm{d}\omega$$

We have so far assumed that the Fourier transform of $f_1(t)$ does exist and that the dominance condition is satisfied. For the Fourier transform to exist the integral in Equation (A.13) must be finite; since $|\mathrm{e}^{-\mathrm{j}\omega t}| = 1$ a sufficient but not necessary condition for this can be seen to be

$$\int_{-\infty}^{\infty} |f(t)|\,\mathrm{d}t \text{ must be finite.} \tag{A.15}$$

The dominance condition is satisfied if the poles of the transfer function $v_1, v_2, \ldots v_m$ have real parts less than the real part of $\mathrm{j}\omega$, that is, less than zero. This is always the case for any stable circuit. Thus, sufficient but not necessary conditions to include both the existence of the Fourier transform and the dominance condition may be stated

$$\left.\begin{aligned} &\int_{-\infty}^{\infty} |f_1(t)|\,\mathrm{d}t \text{ must be finite} \\ \text{and}\quad &\int_{-\infty}^{\infty} |f_2(t)|\,\mathrm{d}t \text{ must be finite} \end{aligned}\right\} \tag{A.16}$$

These conditions limit the class of problems to which the Fourier method may be applied and for this reason the Laplace method is sometimes to be preferred.

The Fourier transforms, inverse transforms and the transfer functions of circuits may often be found by recourse to standard tabulations and by the use of various theorems. Some examples of these were given in Figure 2.13 and Figure 2.14. It may be noted that the Fourier transform of a step function, in view of the condition in Equation (A.15), only exists in the limit. It may be obtained from the transform of the function $f(t) = 0$ for $t < 0$ and $f(t) = \mathrm{e}^{-gt}$ for $t > 0$, and letting $g \to 0$. Thus

$$\mathscr{F}(\mathrm{j}\omega) = \lim_{g\to 0}\frac{1}{g + \mathrm{j}\omega} = \frac{1}{\mathrm{j}\omega} \text{ provided } \omega \neq 0$$

The value of $\mathscr{F}(\mathrm{j}\omega)$ at $\omega = 0$ may be shown to be $\pi\delta(\omega)$ where $\delta(\omega)$ is the delta function of Dirac.

A selection of useful theorems for the Laplace transform are given in Figure A.6 and these can generally be applied to Fourier transforms by the substitution of $\mathrm{j}\omega$ for s, $\mathscr{F}(\mathrm{j}\omega)$ for $\mathscr{L}(s)$, and $H(\mathrm{j}\omega)$ for $H(s)$ respectively, provided of course that the Fourier existence condition is preserved.

A.4 The Laplace transform

The Laplace transform $\mathscr{L}(s)$ of a function $f(t)$ is given by

$$\mathscr{L}(s) = \int_{-\infty}^{\infty} f(t)\, e^{-st}\, dt \tag{A.17}$$

where $s = \sigma + j\omega$ and the inverse transform is

$$f(t) = \frac{1}{2\pi j} \int_{\sigma - j\infty}^{\sigma + j\infty} \mathscr{L}(s)\, e^{st}\, ds \tag{A.18}$$

or, in the short notation $\mathscr{L}(s) \leftrightarrow f(t)$. Thus the signal is expressed as a continuous sum of exponentials e^{st} where s may range in the complex plane; this has no simple physical analogy, unlike the case of the Fourier method. Equation (A.18) is a line integral in the complex plane and the path would normally be chosen to make the integral converge. This extra degree of freedom means that Laplace transforms can be found for a wider variety of functions than in the case of the Fourier transform and the dominance condition can be satisfied if s is large enough. The inverse transform is in general non-unique unless the region of convergence for s is specified. If however $f(t) = 0$ for $t < 0$, the inverse transform is unique and, since most types of $f(t)$ of practical interest can be arranged to be in this form, the problem of specifying the region of convergence does not arise. Although it is not difficult to take into account signals where $f(t) \neq 0$ for $t < 0$, this case will not be discussed further and in the following discussion it will always be assumed that $f(t) = 0$ for $t < 0$.

The procedure for using the Laplace method is essentially the same as for the Fourier method. If $\mathscr{L}_1(s)$ is the Laplace transform of the input $f_1(t)$, and the transfer function of the circuit is $H(s)$, then the output signal $f_2(t)$ is obtained by taking the inverse transform of $H(s)\mathscr{L}_1(s)$. Thus

$$f_1(t) \leftrightarrow \mathscr{L}_1(s)$$

$$f_2(t) \leftrightarrow H(s)\mathscr{L}_1(s)$$

As in the case of the Fourier transform, the taking of the transform, and of the inverse transform, may often be done by using standard tables and theorems. The transform exists if the integral in equation (A.17) is finite. A sufficient but not necessary condition for this is that there exists a real positive finite number M and a real finite number α such that

$$f(t) = 0 \quad \text{for } t < 0$$

$$|f(t)| < Me^{\alpha t} \quad \text{for} \quad t > 0$$

and this clearly is satisfied for many signals of practical interest.

A.5 Rules for finding transforms and inverse transforms

Several examples of transform pairs have already been given in Figure 2.13. A particularly useful rule exists for finding the inverse transform of functions of the type

$$\frac{N(s)}{(s - v_1)(s - v_2)\ldots(s - v_m)}$$

This was discussed in Subsection 2.2.1 for the case where $N(s)$ is a polynomial in s of degree less than the denominator of the function.

As an example, the circuit of Figure A.1 was shown to have a transfer function $H(s) = (s - \phi_1)/(s - v_1)$ where $\phi_1 = -1/C_1R_1$ and $v_1 = -(1/C) \times (1/R_1 + 1/R_2)$ and if the input, as before, is $f_1(t) = 0$ for $t < 0$ and $f_1(t) = E_1 e^{-gt}$ for $t > 0$, Figure 2.13 gives the Laplace transform $\mathscr{L}_1(s)$ of the input as $E_1/(s + g)$. The output is therefore obtained by taking the inverse transform $E_1(s - \phi_1)/(s + g)(s - v_1)$ which, using the rule mentioned above gives

$$f_2(t) = \frac{E_1(g + \phi_1)}{(g + v_1)} e^{-gt} + \frac{E_1(v_1 - \phi_1)}{(v_1 + g)} e^{v_1 t}$$

which is the same as the solution obtained earlier in Equation (A.5).

Finally there are several theorems which are often of value when using the transform methods and some of these are summarized in Figure A.6.

	Function $f(t)$	Laplace transform $\mathscr{L}(s)$
Addition	$a_1 f_1(t) + a_2 f_2(t)$	$a_1 \mathscr{L}_1(s) + a_2 \mathscr{L}_1(s)$
Time differentiation	$\frac{df(t)}{dt}$	$s\mathscr{L}(s)$
Time integration	$\int_0^t f(t)\,dt$	$\frac{\mathscr{L}(s)}{s}$
Frequency integration	$\frac{f(t)}{t}$	$\int_s^\infty \mathscr{L}(s)\,ds$
Time shift	$f(t - t_0)$	$\mathscr{L}(s)\,e^{-st_0}$
Frequency shift	$f(t)\,e^{s_0 t}$	$\mathscr{L}(s - s_0)$
Scaling, $a > 0$	$f(at)$	$\frac{1}{a}\mathscr{L}\left(\frac{s}{a}\right)$
Time convolution (see Subsection 2.2.2)	$f_1(t) * f_2(t)$	$\mathscr{L}_1(s)\,\mathscr{L}_2(s)$

Figure A.6 Summary of some useful theorems relating to transform methods

So far the transform method has been presented as a method of characterizing the input signal as a continuous sum of complex frequencies which are filtered by the circuit and then recombined to give the output time signal. However, it is possible to view the transform method as a purely mathematical device to solve the circuit differential equations. To illustrate this, the example of Figure A.1 will be taken. The differential equation for this circuit was given in Equation (A.1.b) and if we take the Laplace transform of both sides of this equation we obtain

$$s\mathscr{L}_2(s) + \frac{1}{C_1}\left(\frac{1}{R_1} + \frac{1}{R_2}\right)\mathscr{L}_2(s) = s\mathscr{L}_1(s) + \frac{1}{CR_1}\mathscr{L}_1(s)$$

where $\mathscr{L}_1(s)$ is the Laplace transform of the input time function and $\mathscr{L}_2(s)$ that of the output time function. We have therefore

$$\mathscr{L}_2(s) = \frac{s + 1/CR_2}{s + (1/C)(1/R_1 + 1/R_2)}\mathscr{L}_1(s) = \frac{s - \phi_1}{s - \nu_1}\mathscr{L}_1(s)$$

and if we were to take the inverse transform of both sides of this equation we would obtain an expression for the output time function which is the same as that obtained by the method given earlier. Thus the transform can be regarded as an equation solving routine and it is not necessary to attach any physical significance to s. This approach has the advantage that the properties and conditions of the transform method can be formally elucidated without recourse to the physical problem.

References

1. M. Schwartz, *Information Transmission, Modulation and Noise*, 2nd ed., McGraw-Hill, 1970.
2. B. M. Brown, *Mathematical Theory of Linear Systems*, Chapman and Hall, 1965.
3. B. P. Lathi, *Signals, Systems and Communication*, Wiley, 1965.

Appendix B

Statistics of Counting

B.1 General considerations

We will assume that in an experiment lasting a time t the number of nuclear events is measured. We will further assume that this number is not subject to any losses due to imperfect functioning of the apparatus as might arise from, for example, the finite dead time of the apparatus. The number of counts measured is, of course, a statistical variate and if the experiment were identically repeated a different value of the number would probably be measured.

We define $P(n)$ as the probability of n counts occurring during the experiment. We will define m as the true mean count (or expectation value) for the experiment and σ^2 as the variance

where
$$m = \sum_{n=0}^{n=\infty} P(n)n$$

$$\sigma^2 = \sum_{n=0}^{n=\infty} P(n)(m - n)^2$$

The true mean m is a quantity that would be approximated to if the experiment were identically repeated many times and the average number of counts obtained. The true mean can never be obtained exactly from a finite series of experiments and, like the probability $P(n)$ and variance σ^2, is an abstraction in the present context. It is worth noting however that it is possible to construct a mechanical device in which the quantities $P(n)$, m and σ^2 are known exactly (within the limits of the available precision of construction) but this is only possible with the additional knowledge about the internal workings of the mechanism, and this is not available to us in the case of nuclear phenomena.

Before considering the form of $P(n)$ two useful theorems will be presented which do not require any assumptions on the form of $P(n)$. These theorems relate to two statistically independent variates. Two variates n_1 and n_2 are said to be independent if the probability that either one of them will assume a prescribed value does not depend on the value assumed by the other. The following theorems are simply proved and are dealt with in any elementary textbook on statistics:

If the distribution of n_1 has a true mean m_1 and variance σ_1^2
and the distribution of n_2 has a true mean m_2 and variance σ_2^2
then $(n_1 + n_2)$ has a mean $(m_1 + m_2)$ and a variance $(\sigma_1^2 + \sigma_2^2)$ (B.1)
and $(n_1 - n_2)$ has a mean $(m_1 - m_2)$ and a variance $(\sigma_1^2 + \sigma_2^2)$ (B.2)

The theorems are useful in several contexts. If the number of counts from a single source are measured over two time periods, then, provided the source has negligible decay, the two measurements are statistically independent and the above results may be used. In the determination of the counts from a source where a background is present, it is usual to measure the main counts (source + background) as n_1, and the background counts alone as n_2, and take the difference. Where statistical independence between n_1 and n_2 may be assumed the above theorems may be used. Practically, the rare occasions when independence may not be assumed arise, for instance, when the background count is determined after the main count and some decay of the background has occurred. A further instance is when the main count and background count are measured simultaneously by an arrangement of two detectors, such that a single background event activates both detectors, and a non-background event only activates the main detector.

B.2 Radioactive decay—the binomial distribution

Radioactive decay appears to be a chance phenomenon, that is, if the probability of an undecayed atom decaying in a period of time dt is $\lambda\,dt$, then λ is a constant and is independent of any other processes occurring in the sample. It follows that, if an initially undecayed atom is selected and observed for a period of time t, the probability of it decaying in this period is p, where $p = (1 - e^{-\lambda t})$. If we have a sample initially containing N_0 undecayed atoms then we wish to know the probability $P(n)$ of observing n disintegrations or counts in the time t. This is given[1] by the binomial distribution and is

$$P(n) = \frac{N_0!}{(N_0 - n)!n!} p^n(1 - p)^{N_0 - n} \tag{B.3}$$

This distribution has a mean of $m = pN_0 = N_0(1 - e^{-\lambda t})$ (B.4)

and a standard deviation

$$\sigma = (N_0 p(1 - p))^{\frac{1}{2}} = (me^{-\lambda t})^{\frac{1}{2}} \tag{B.5}$$

The half life $T_{\frac{1}{2}}$ is, of course, that value of t for which $m = N_0/2$, and from Equation (B.4) we have $\frac{1}{2} = 1 - e^{-\lambda T_{\frac{1}{2}}}$ so $\lambda T_{\frac{1}{2}} = 0{\cdot}693$. All the above results are exact. For the case where the half life is much longer than the experiment we have $\lambda t \ll 1$ and the true mean m becomes

$$m \simeq N_0 \lambda t$$

and

$$\sigma \simeq m^{\frac{1}{2}}.$$

B.3 The Poisson distribution

For the case $\lambda t \ll 1$ (half life large compared with the duration of the experiment) we have already seen that the mean count $m = N_0 \lambda t$ and the standard deviation $\sigma = m^{\frac{1}{2}}$. If we make the additional assumption that $N_0 \gg n$ (this will usually be assured in any case if $\lambda t \ll 1$) then the probability distribution Equation (B.3) simplifies to the Poisson distribution:

$$P(n) = \frac{m^n e^{-m}}{n!} \tag{B.6}$$

In the case of two independent variates which are summed, we have already seen how the mean value of the sum and the standard deviation may be derived from the means and standard deviations of the individual variates (Equation (B.1)). If in addition the individual variates are described by Poisson distributions, then the sum of the variates is itself also described by a Poisson distribution (see Weatherburn[2] for example)

$$P(n_1 + n_2) = \frac{(m_1 + m_2)^{(n_1 + n_2)} e^{-(m_1 + m_2)}}{(n_1 + n_2)!}$$

It may be noted that the distribution of the difference of two independent variates each described by a Poisson distribution is not given by a Poisson distribution but, as discussed in the next section, it can usually be approximated by a normal (i.e. Gaussian) distribution.

For reasonably large values of n, the Poisson distribution, Equation (B.6), may be approximated[1] by a simpler expression which is sufficiently accurate for most purposes. Using Stirling's approximation

$$n! = \sqrt{2\pi n}\, n^n e^{-n}$$

we have

$$P(n) = (2\pi n)^{-\frac{1}{2}} (m/n)^n e^{-(m-n)}$$

For $n = 10$ the error in $P(n)$ is only 0·8 % and for $n = 100$ the error is 0·08 %.[1]

B.4 Optimum apportioning of time with source and background

We will assume that the background has a mean rate of R_b counts per unit time and is measured for a time t_b so that the mean number of counts is

$R_b t_b$ and the standard deviation of the counts measured is $\sqrt{R_b t_b}$, It follows that the standard deviation σ_b of the measured rate is therefore $\sqrt{R_b/t_b}$. If a measurement of the source plus background is additionally made, which has a mean rate of R_{s+b}, for a time t_{s+b}, then the mean number of counts measured is $R_{s+b}t_{s+b}$. The number of these measured counts due to the source plus background has a Poisson distribution as discussed in the previous section and therefore has a standard deviation of $\sqrt{R_{s+b}t_{s+b}}$, so that the standard deviation in the measured rate is $\sigma_{s+b} = \sqrt{R_{s+b}/t_{s+b}}$.

To find the rate R_s due to the source alone, we subtract the measured background rate from the measured source + background rate. The corresponding standard deviation σ_s in the calculated rate of the source alone, according to Equation (B.2), is

$$\sigma_s = (\sigma_b^2 + \sigma_{s+b}^2)^{\frac{1}{2}} = (R_b/t_b + R_{s+b}/t_{s+b})^{\frac{1}{2}} \qquad \text{(B.7)}$$

The problem is to apportion a fixed amount of time T between t_b and t_{s+b} so as to make σ_s a minimum.

We have $t_b + t_{s+b}$ = constant = T so $\mathrm{d}t_b/\mathrm{d}t_{s+b} = -1$, and for σ_s to be a minimum we have $\mathrm{d}\sigma_s/\mathrm{d}t_b = 0$. Hence, from Equation (B.7) we have, at minimum σ_s,

$$\frac{t_b}{t_{s+b}} = \left(\frac{R_b}{R_{s+b}}\right)^{\frac{1}{2}}$$

and this represents the optimum apportioning of the available time. The standard deviation with the optimum time apportioning is then seen to be

$$\sigma_s = \frac{\sqrt{R_b} + \sqrt{R_{s+b}}}{\sqrt{T}} \qquad \text{(B.8)}$$

The statistical theory of errors is usually based on the assumption of variates described by normal distributions and most analyses of nuclear data tacitly make this assumption. The normal distribution of the variate x is a continuous distribution and gives the probability, $\mathrm{d}P(x)$, that x lies in the interval $\mathrm{d}x$, by

$$\mathrm{d}P(x) = \frac{1}{\sigma\sqrt{2\pi}}\,\mathrm{e}^{-(x-\bar{x})^2/2\sigma^2}\,\mathrm{d}x$$

The normal distribution is symmetrical about the mean and, unlike the Poisson distribution, is specified by two parameters, the mean $\bar{x}$ and the standard deviation σ.

Provided the mean, m, is large enough, the binomial and the Poisson distributions may be approximated by the normal distribution with $\sigma = \sqrt{m}$. Figure B.1 shows the Poisson distribution for $m = 10$ and the normal distribution for $\bar{x} = 10$, $\sigma = \sqrt{10}$. It will be observed that the Poisson distribution is slightly asymmetric.

If $x_1, x_2, \ldots, x_q$ are statistically independent variates each distributed normally with standard deviations $\sigma_1, \sigma_2, \ldots, \sigma_q$, then it can be shown[2] that the variate $\sum_{i=1}^{q} c_1 x_1$, where the c_i are all constants, is also normally distributed with a variance of $\sigma^2 = \sum_{i=1}^{q} c_i^2 \sigma_i^2$. In the present context, this result shows that the calculated rate for the source alone is normally distributed, provided the measured rates are derived from sufficient numbers of counts to justify the approximations of their distributions as normal.

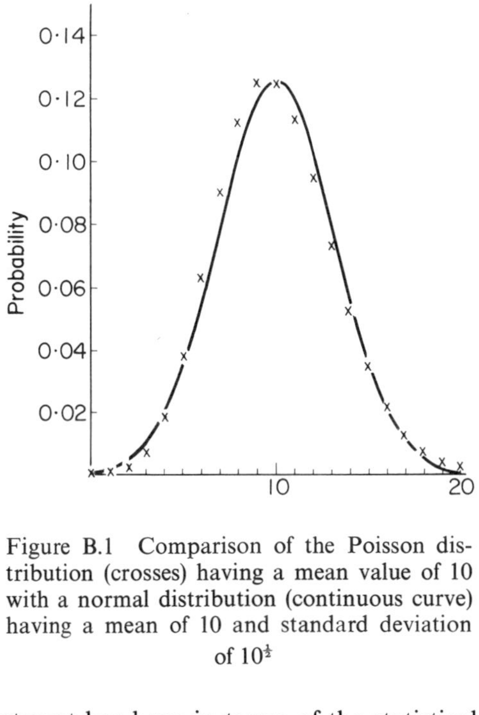

Figure B.1 Comparison of the Poisson distribution (crosses) having a mean value of 10 with a normal distribution (continuous curve) having a mean of 10 and standard deviation of $10^{\frac{1}{2}}$

So far the treatment has been in terms of the statistical variation of the observed counting rate or of some derived quantity. Although Equation (B.6), for example, gives the probability of a certain number of counts occurring, the experimenter is usually interested in the obverse: given a certain observed count, what can be said about the probability of the source intensity lying in a certain range? A similar problem arises when a source intensity is to be determined in the presence of background, or where the ratio of two source intensities is required. A rigorous treatment of these cases has been given by Goldanskii *et al.*[3] As expected, under suitable conditions, the required distributions approximate to the normal and have standard deviations the same as those given by the simple reasoning used above.

B.5 Counting losses due to dead time

If we assume that a circuit element has a dead time of τ then, if a pulse is processed by that circuit element, it is incapable of processing another pulse for a period of time τ after the first pulse. If R is the mean rate of pulses at the input, then let R' be the mean rate at the output where, of course, $R > R'$. Since R' is the rate at which pulses are processed by the circuit, it follows that in unit time the circuit is 'dead' for a total time of $R'\tau$. The number of incident pulses occurring in that time is, on average, $RR'\tau$ and all these are lost due to the dead time. We therefore have

$$R - R' = RR'\tau$$

So

$$R = \frac{R'}{1 - R'\tau} \qquad \text{or} \qquad R' = \frac{R}{1 + R\tau}$$

It will be apparent that some counts are always lost due to the dead time. If the dead time is, say, 10 μs then the loss from a true counting rate of 10^4 per second is 10%. This loss arises essentially from the random nature of the arrival times of the pulses; it will be seen that the same circuit could handle a counting rate of up to 10^5 per second of regularly occurring pulses with no loss at all. Hence a useful technique in reducing counting losses is to place a de-randomizing circuit before any circuit having a large dead time. This takes the form of one or more buffer stores which temporarily store the pulses until they can be processed.

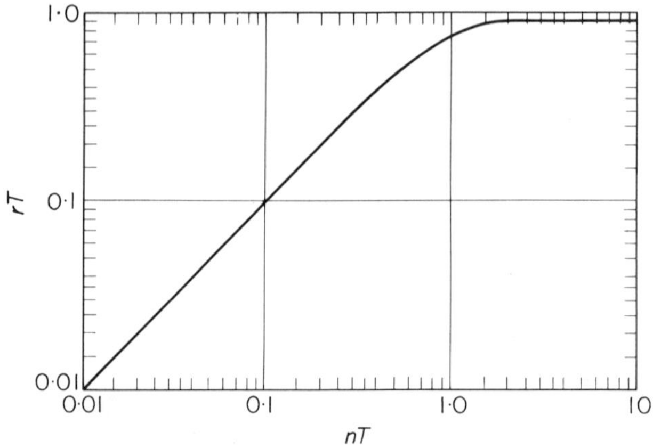

Figure B.2 Counting loss with a single buffer store, from *Nuclear Pulse Spectrometry* by R. L. Chase.[4] Copyright 1961 by McGraw–Hill Book Company Inc. Reproduced by permission of the McGraw–Hill Book Company. T is the circuit dead time, n the mean input pulse rate and r the accepted pulse rate

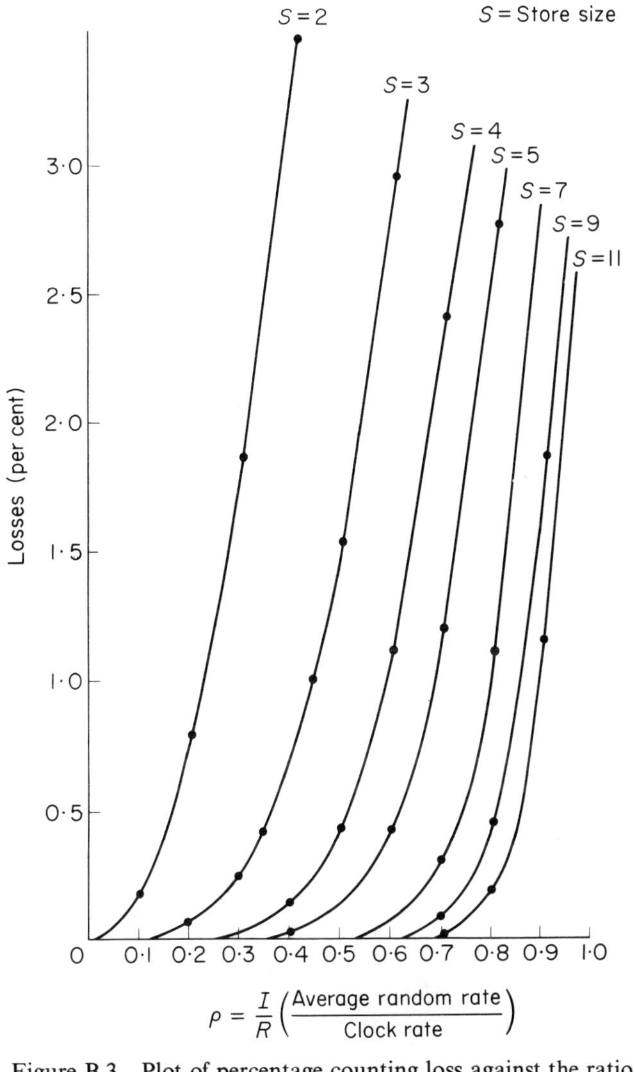

Figure B.3 Plot of percentage counting loss against the ratio of the mean input rate to output rate, according to Alexander, Reddering and Kennedy.[5] Different curves relate to different numbers of buffer stores

An analogue buffer store is sometimes used at the input of a multichannel pulse height analyser (MCA). Normally, the value of the height of an incoming pulse is stored in the buffer. If the MCA is free to process the event, the pulse height information is immediately passed to it so that the buffer is vacated for the next event. If the MCA is already processing a previous event, the newly arrived event is stored in the buffer until the MCA is free to process it. Thus events are only lost when both buffer and MCA are occupied. Chase[4] has analysed the case where the buffer store has negligible access time and the MCA has a dead time τ; Figure B.2 shows some of his results for the mean pulse rate accepted by the MCA plotted against the mean input rate. For the example considered earlier, that of a dead time of 10 μs, the counting loss with a mean input rate of 2×10^4 per second is less than 0·5 %.

When analogue storage (i.e. preserving information about pulse height) is not required and only digital storage is necessary, it is fairly simple to provide more than one buffer store. Examples where digital buffer stores are useful are in the storage of information in ferrite core memory or on magnetic tape, so that losses due to the writing time of the storage media may be reduced. Alexander *et al.*[5] have calculated the losses to be expected when various numbers of buffer registers are provided. They assumed that randomly arriving data words were fed sequentially into the buffer stores and were read out at regular time intervals compatible with the following processing. The read-out must of course be higher than the mean input rate and an event would only be lost if it arrives to find all the buffers full, a contingency whose likelihood reduces with the total number of buffers. Figure B.3 shows a selection of their results for various numbers of buffers and gives the percentage of incoming events which are lost plotted against the ratio of the mean input rate to the buffer read-out rate.

B.6 Accidental coincidence rates

If two independent sources of pulses of mean rates R_1 and R_2 are fed into the two inputs of a coincidence circuit having a coincidence resolving time of $\pm\tau$, then we wish to know what the resulting mean rate R_a at the coincidence circuit output will be due to accidental coincidences. Consider a single pulse present at input 1. If a pulse occurs at input 2 either preceding or following that at input 1 by periods of τ, then a coincidence will be registered by the circuit. The number of pulses at input 2 in any interval of duration 2τ is, on average, $2R_2\tau$. It follows therefore that the total accidental coincidence rate R_a is given by $2R_1R_2\tau$.

In an experiment designed to detect coincident pairs of particles (true coincidences) from a single source, there will be an additional component from accidental coincidences; the observed coincidence rate must be corrected for this accidental rate to give the true rate. The true rate R_t is obtained by first measuring the true + accidental rate R_{t+a}, and subtracting from it the accidental rate R_a. We now wish to examine the influence of the source strength on the standard deviation of the resultant true rate. We will assume that a fixed amount of experimental time is available and the situation is analogous to that already discussed for the case of a source and background. Assuming the time is apportioned in the optimum way, the fractional standard deviation in the true rate is given by

$$\frac{\sigma_t}{R_t} = \frac{\sqrt{R_a} + \sqrt{R_{t+a}}}{R_t\sqrt{T}}$$

If the activity of the source is M disintegrations per unit time, then the true rate R_t will equal bM, where b is a constant depending on the relative intensity of the coincidences in the source, the efficiencies of the detectors and

Ma/b	Relative fractional standard deviation σ_t/R_t
∞	1·00
2	1·11
1	1·21
0·5	1·37
0·1	2·16
0·01	3·50

Figure B.4 The fractional standard deviation of the calculated true coincidence rate, on a relative scale, for various values of the source strength M

the solid angles subtended by the counters at the source. The accidental rate R_a can analogously be expressed as aM^2 where a is another constant. Substituting these values in, we have

$$\frac{\sigma_t}{R_t} = \frac{\sqrt{aM^2} + \sqrt{aM^2 + bM}}{bM\sqrt{T}}$$

On examining this expression we see that the fractional standard deviation is a minimum for $M = \infty$. However, Figure B.4 shows the values of the fractional standard deviation for other values of M relative to its value at $M = \infty$. It may be seen that little is gained in using a value of M greater than b/a. The value of $M = b/a$ corresponds to that where the accidental rate $R_a(=aM^2)$ is equal to the true rate $R_t(=bM)$ and for practical purposes represents the maximum useful source strength.

The use of the constants a and b conceals the dependence of the fractional standard deviation σ_t/R_t on quantities such as detector solid angle and efficiency. It may be easily seen, however, that large solid angles and efficiencies cause the fractional standard deviation to be reduced.

References

1. R. D. Evans, *The Atomic Nucleus*, McGraw-Hill, 1955.
2. C. E. Weatherburn, *Mathematical Statistics*, Cambridge University Press, 1952.
3. V. I. Goldanskii, A. V. Kutsenko and M. I. Podgoretskii, *Counting Statistics of Nuclear Particles*, Hindustan Publishing Corporation, Delhi, 1962. (Translated from the Russian version.)
4. R. L. Chase, *Nuclear Pulse Spectrometry*, McGraw-Hill, 1961.
5. T. K. Alexander, H. G. Reddering and J. M. Kennedy, Atomic Energy of Canada Ltd., Chalk River Report CREL 779 (1959).

Index